Stimuli-Responsive Nanomedicine

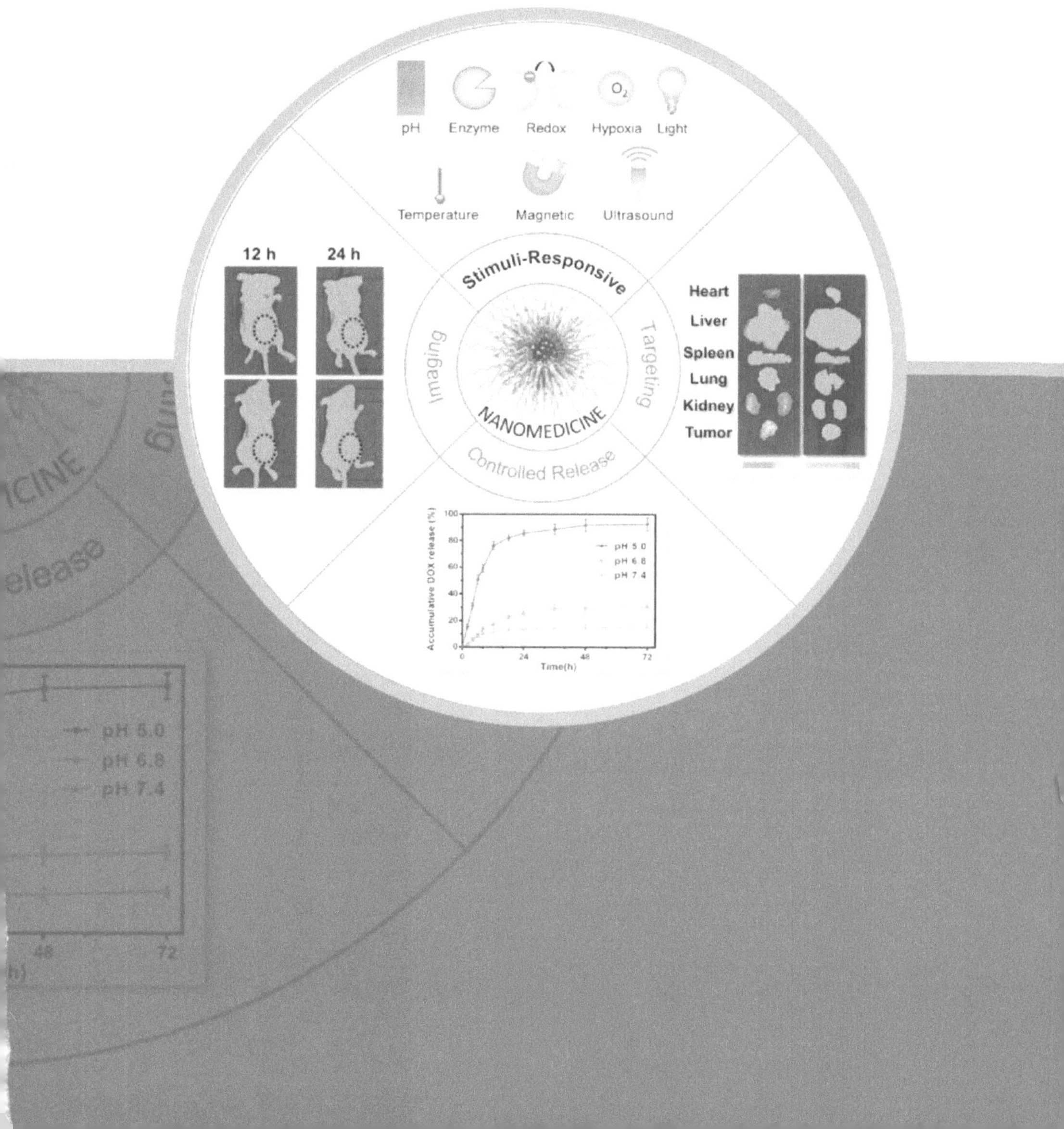

Jenny Stanford Series on Biomedical Nanotechnology

Series Editors

Vladimir Torchilin and Mansoor Amiji

Titles in the Series

Published

Vol. 1
Handbook of Materials for Nanomedicine
Vladimir Torchilin and Mansoor Amiji, eds.
2010
978-981-4267-55-7 (Hardcover)
978-981-4267-58-8 (eBook)

Vol. 2
Nanoimaging
Beth A. Goins and William T. Phillips, eds.
2011
978-981-4267-09-0 (Hardcover)
978-981-4267-91-5 (eBook)

Vol. 3
Biomedical Nanosensors
Joseph Irudayaraj, ed.
2013
978-981-4303-03-3 (Hardcover)
978-981-4303-04-0 (eBook)

Vol. 4
Nanotechnology for Delivery of Therapeutic Nucleic Acids
Dan Peer, ed.
2013
978-981-4411-04-2 (Hardcover)
978-981-4411-05-9 (eBook)

Vol. 5
Handbook of Safety Assessment of Nanomaterials: From Toxicological Testing to Personalized Medicine
Bengt Fadeel, ed.
2014
978-981-4463-36-2 (Hardcover)
978-981-4463-37-9 (eBook)

Vol. 6
Handbook of Materials for Nanomedicine: Lipid-Based and Inorganic Nanomaterials Vladimir Torchilin, ed.
2020
978-981-4800-91-4 (Hardcover)
978-1-003-04507-6 (eBook)

Vol. 7
Handbook of Materials for Nanomedicine: Polymeric Nanomaterials
Vladimir Torchilin, ed.
2020
978-981-4800-92-1 (Hardcover)
978-1-003-04511-3 (eBook)

Vol. 8
Handbook of Materials for Nanomedicine: Metal-Based and Other Nanomaterials
Vladimir Torchilin, ed.
2020
978-981-4800-93-8 (Hardcover)
978-1-003-04515-1 (eBook)

Vol. 9
Stimuli-Responsive Nanomedicine
Lin Zhu, ed.
2021
978-981-4800-70-9 (Hardcover)
978-0-429-29529-4 (eBook)

Forthcoming

Vol. 10
Inorganic Nanomedicine
Bhupinder Singh Sekhon, ed.

Vol. 11
Nanotechnology for Personalized Cancer Treatment
Julia Ljubimova, ed.

Vol. 12
Translation Industrial Nanotechnology
Thomas Redelmeier, ed.

Vol. 13
Microfluidics for Biomedicine
Tania Konry, ed.

Vol. 14
Nanopreparations for Intracellular Targeting
Swati Biswas, ed.

Vol. 15
Clinical Nanomedicine: Lessons Learnt from Doxil
Yechezkel Barenholz, ed.

Vol. 16
Electrical Interactions in Drug Delivery
Ambika Bajpayee, ed.

Jenny Stanford Series on Biomedical Nanotechnology **Volume 9**

Stimuli-Responsive Nanomedicine

edited by
Lin Zhu

JENNY STANFORD
PUBLISHING

Published by

Jenny Stanford Publishing Pte. Ltd.
Level 34, Centennial Tower
3 Temasek Avenue
Singapore 039190

Email: editorial@jennystanford.com
Web: www.jennystanford.com

British Library Cataloguing-in-Publication Data
A catalogue record for this book is available from the British Library.

Stimuli-Responsive Nanomedicine

ISBN 978-981-4800-70-9 (Hardcover)
ISBN 978-0-429-29529-4 (eBook)

Contents

Preface

Nanotechnology has revolutionized many aspects of science. Thanks to the advances achieved in materials science, physics and chemical engineering, and better understanding in physiology and pathology, a wide variety of nanoparticle-based systems have been developed for various purposes of biomedical applications. Nanomedicine, as one of the emerging therapeutics, demonstrates its superior physicochemical and biological properties to conventional medicines.

Nanomedicines or nano-sized drug delivery systems have evolved from passive drug targeting and drug release mediated mainly by the physicochemical properties of the nanocarriers, to active drug targeting and intracellular drug delivery benefited from discovery of tissue, cell or intracellular compartment-specific ligands.

Nowadays, we are entering the era of "on-demand" drug delivery which can exert diagnostic and/or therapeutic effects in response to specific biological, chemical, or physical stimuli. Stimuli-responsive nanomedicines that respond to the abnormalities of the diseased tissues and cells (e.g., pH, redox potential, enzyme), and externally applied stimuli (e.g., heat, magnetic field, light), have become one of the "smart" strategies for drug delivery.

This book contains 11 chapters. Chapter 1 briefly introduces the types, recent progress, and important applications of stimuli-responsive nanomedicines. Chapters 2–9 individually discuss various types of stimuli-responsive nanomedicines in terms of their formulation and preparation, physical and chemical properties, biomedical applications, and challenges. Chapters 10 and 11 focus on the major applications of stimuli-responsive nanomedicines on cancer and non-cancer diseases. With contributions from active researchers and experts in the field, this book provides a fundamental and comprehensive overview of stimuli-responsive nanomedicines. It also covers major cutting-edge findings and most impressive achievements in stimuli-responsive drug delivery, imaging and therapy. It will be a valuable book for students, researchers, and other professionals in biomedical sciences.

We would like to express our sincere gratitude to all the authors for their contributions in this book. We also acknowledge the publisher for their support during the book preparation.

Lin Zhu
Kingsville, Texas, USA

Chapter 1

Overview of Stimuli-Responsive Nanomedicine

Qing Zhou,* Qing Yao,* Jiao Wang,* and Lin Zhu

Department of Pharmaceutical Sciences,
Irma Lerma Rangel College of Pharmacy,
Texas A&M University Health Science Center,
Kingsville, Texas 78363, USA

lzhu@tamu.edu

1.1 Introduction

Nanoscale systems have recently received tremendous attention, particularly in the field of biomedical research [1, 2]. Although drug molecules can be manipulated to improve their properties at drug discovery stage, the compatible drug delivery strategies have to be applied to facilitate the clinical applications and achieve the desirable therapeutic outcomes. In this regard, the drugs are formulated into various dosage forms or drug delivery systems.

*These authors contributed equally.

Stimuli-Responsive Nanomedicine
Edited by Lin Zhu

ISBN 978-981-4800-70-9 (Hardcover), 978-0-429-29529-4 (eBook)
www.jennystanford.com

Among them, the nanoparticle-based medicine or nanomedicine is the most important one.

Therapeutic or imaging agents can be encapsulated into the inner layer or core of nanoparticles or absorbed on the surface of nanoparticles via various mechanisms. Once loaded, these molecules can be masked by the nanomaterials and drugs' physicochemical properties as well as some other properties, such as pharmacokinetic and biodistribution profiles, are expected to be improved [3]. The most commonly used nanoparticles/nanocarriers include liposomes [4], micelles [5], dendrimers [6], lipid or polymeric nanoparticles [7, 8], emerging inorganic nanoparticles, like quantum dots, gold nanoparticles, carbon nanotubes, etc. [9–11], and macromolecule-based conjugates, such as antibody–drug conjugates (ADC) and drug–polymer conjugates (e.g., polyethylene glycol (PEG)-protein conjugates and PEG–siRNA conjugates) [12–14]. These particles typically range from 10 to 100 nm, which is considered safe and effective for most administration routes, including oral, systemic (e.g., intravenous), inhalation, ocular, and transdermal/transmucosal routes.

Thanks to the advances achieved in materials science and engineering, a broad range of nanoparticles with various sizes, morphologies, architectures, and surface properties have been developed. Moreover, the chemical or physical engineering on the surface or backbone of the nanomaterials provides additional opportunities for controlling the performance of nanomedicine. One of the most successful modification technologies is PEGylation. Due to the "stealth" property, the PEG–modified nanoparticles are able to escape the capture by the mononuclear phagocyte system (MPS) [15], leading to the prolonged blood circulation and the enhanced permeability and retention (EPR) effect-mediated "passive" tumor targeting [15, 16]. This is extremely useful for the tumor-targeted delivery of drugs and imaging agents. The nanoparticles, such as polymeric micelles, are able to extravasate into the tumors through the gaps between endothelial cells and accumulate there due to poor lymphatic drainage.

To achieve site- or cell-specific drug delivery, the nanoparticles need to be further engineered by the targeting ligands (such as monoclonal antibodies), "on-demand" drug release/delivery moieties (such as stimuli-sensitive moieties), and the intracellular delivery moieties (such as tissue or cell-penetrating proteins or peptides).

The idea of the stimuli-responsive drug delivery comes from the fact of abnormalities in the diseased tissues or cells. For example, in the tumor microenvironment, the abnormalities include acidic pH, altered redox potential, and upregulated proteins. As shown in Fig. 1.1, these internal conditions as well as external stimuli such as hyperthermia, magnetic field, light, and ultrasound, can be employed to design a stimuli-responsive drug delivery system for on-demand and/or targeted drug delivery [17].

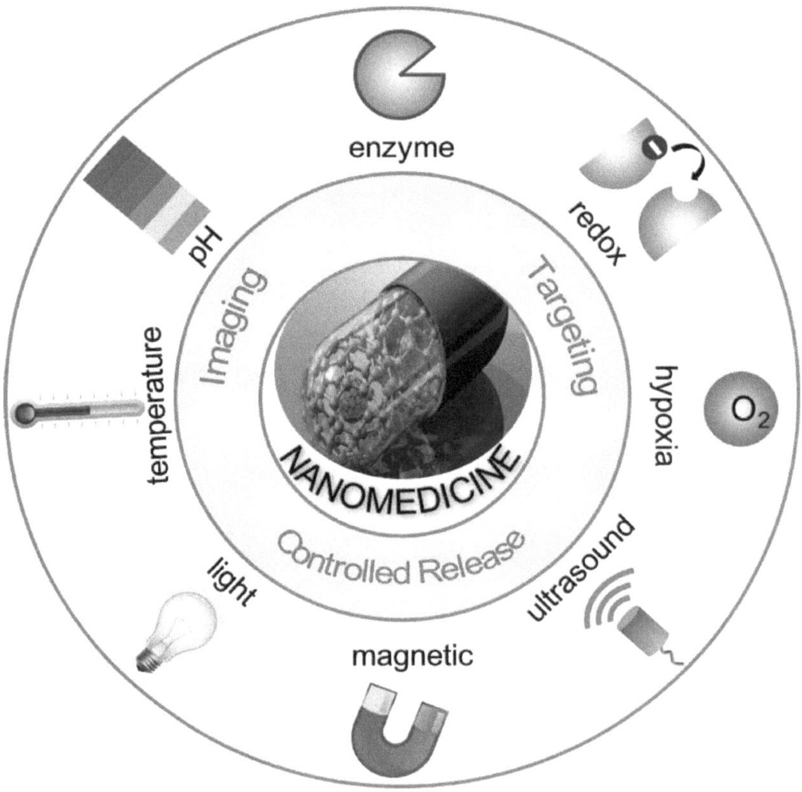

Figure 1.1 Schematic illustration of stimuli-responsive nanomedicine.

Generally, to construct a stimuli-responsive nanomedicine, first, we have to understand the pathological abnormalities/local stimuli or external stimuli and desirable clinical outcomes; and second, the biocompatible nanomaterials should be available for use, which may undergo conformational change, hydrolytic/enzymatic cleavage/degradation, or changes in their physico-chemical properties in response to corresponding stimuli. These physical and chemical alterations allow for the destabilization or rearrangement of the nanoparticles or "activation" of the blocked functions, resulting in the drug accumulation, on-demand drug release, and/or drug uptake at the disease site and/or diseased cells. A great deal of work has been dedicated to devising stimuli-sensitive polymers, biomacromolecules, and other nanomaterials. Using these "intelligent" materials, a wide variety of stimuli-responsive nanomedicines have been developed. Here, we will overview the recent progress, highlight the important applications, and predict the future trends of stimuli-responsive nanomedicine.

1.2 Typical Stimuli and Stimuli-Responsive Nanomedicines

The stimuli-responsive nanomedicine can employ the internal abnormalities in the specific tissues, cells, or even intracellular compartments, such as the abnormal pH [18], specific proteins/enzymes [19], redox potential [20], or hypoxia [21, 22]; and the physical stimuli that can be artificially applied from outside of the body, such as temperature [23], light [24], magnetic field [25], or ultrasound [26] (Fig. 1.1).

1.2.1 Internal Stimuli and Stimuli-Responsive Nanomedicines

Significant alterations in the physiological condition are often the vital hallmarks for certain diseases, such as cancer and inflammatory diseases, rendering them attractive targets for designing stimuli-responsive systems. Here, we discuss the stimuli-responsive nanosystems that take advantages of the internal stimuli, including pH, enzymes, redox potential, and oxygen. The typical internal stimuli-responsive nanomedicines are summarized in Table 1.1.

Table 1.1 Examples of typical internal stimuli and stimuli-responsive nanomedicines

Acting stimuli	Characteristics of stimuli	Responsive nanomaterials	Cargoes	Applications	Refs
pH	Gastric pH: 1.0–3.5; Small-intestinal pH: 7.5–8.0; Large-intestinal pH: 5.5–7.0.	Polysaccharide nanoparticles	Insulin	Oral delivery of insulin	[27]
	Inflamed tissues (infarcted myocardium): pH6-7	Hydrogel: p(NIPAAm-co-PAA-co-BA)	bFGF	Injection to the infarct zone	[28]
	Tumor extracellular pH: 6.5–7.2	iNPG-pDOX Liposomes: DSPE-KLA-DMA	DOX PTX	Cancer chemotherapy	[29] [30]
Enzyme	Upregulated tumoral MMPs	Liposomes: PEG-peptide-DOPE Silica nanoparticles: MSN-SS-CD-peptide-PASP	Rh-PE DOX	Tumor-targeted drug delivery	[31] [32]
	Elevated proteolytic enzymes in inflamed tissues	Hydrogel: TGMS-TAC	Tacrolimus	Local injection of immunosuppressants	[33]
Redox	In the tumor: Intracellular redox: 10 mM; Extracellular redox: 2–10 μM High ROS	HA-ss-DOCA PAMAM–S–S–NAC PATK polyplexes	PTX NAC Plasmid DNA	Tumor targeting	[34] [35] [36]
Hypoxia	Low blood oxygen level in ischemic tissues	Water-soluble lipopolymer	pEpo-SV-VEGF plasmid	Ischemic myocardium targeting	[37]
	Hypoxia in the solid tumor	Azobenzene nanoparticles 2-nitroimidazole derivative	GFP siRNA DOX	Tumor-targeted gene and drug delivery	[38] [22]

Abbreviations: bFGF, basic fibroblast growth factor; DOX, doxorubicin; PTX, Paclitaxel; NAC, N-acetylcysteine; HA-ss-DOCA, hyaluronic acid-deoxycholic acid; PAMAM–S–S–NAC, poly(amidoamine) (PAMAM) dendrimer-NAC; PMNT, poly[4-(2,2,6,6-tetramethylpiperidine-N-oxyl)aminomethylstyrene].

1.2.1.1 pH-responsive nanomedicines

Under the normal physiological conditions, significant pH gradients exist in the organs, such as the gastrointestinal (GI) tract, and are also found in the intracellular compartments, such as lysosomes (4.5–5), endosomes (5.5–6), Golgi apparatus (6.4), and cytosol (7.4) [39]. Under the pathological conditions, such as cancer, the local acidification is common in the cancerous tissues where the rapid proliferation of cancer cells outruns the blood supply, causing a deficit of nutrients and oxygen and the accumulation of lactic acids by glycolysis rather than oxidative phosphorylation [40]. These remarkable pH alterations work as the basis of designing the pH-responsive systems.

The pH-sensitive materials usually undergo the physical or chemical changes in response to pH changes, including swelling, shrinking, dissociation, degradation, or membrane fusion and disruption, which are mainly the results of the protonation of ionizable groups or the degradation of acid-cleavable bonds [41–43]. Thus, one common strategy to fabricate the pH-responsive nanocarriers is to incorporate the ionizable groups, like amines, carboxylic acids, and phosphoric acids, into the materials including polymers, lipids, and other nanomaterials, to facilitate the nanocarriers to ionize at certain pH values [44]. Usually, the ionization of these groups/moieties can change materials' solubility or hydrophilicity/hydrophobicity. For amines, they are uncharged and hydrophobic under neutral and alkaline conditions, while the protonated ones are hydrophilic in mildly acidic conditions. The protonation of amines may affect the nanoparticles' hydrophilic-hydrophobic equilibrium and induces their dramatic structural transformation, leading to the nanoparticles' swelling and increased solubility, which may induce drug release. For example, pH-responsive doxorubicin (DOX)-loaded micelles were prepared from the poly-L-histidine (pHis)-b-PEG and poly(L-lactic acid) (PLLA)-b-PEG [45]. The pHis has a pK_a value near 7.0 and shows the revisable hydrophilic to hydrophobic transition in consistent with its protonated and deprotonated states. At a pH above 7.4, the assembled micelles were stable, while at pH below 7.0, they were gradually destabilized due to the protonation of the pHis block in the micelle core [46, 47]. Because of this property, the micelles could deliver

DOX to the tumor sites with an acidic pH. Another popularly used strategy to prepare a pH-responsive nanocarrier is to use the pH-labile bonds/functional groups. The pH-labile linkers, such as benzoic imine, hydrazine, silyl ether, ortho ester, and Schiff base, have been a part of the polymeric components in the pH-responsive systems [44]. They are stable in the normal tissues and bloodstream, but undergo dissociation at lower pH. These bonds/linkers may present in the polymer backbone or at the site between the hydrophilic and hydrophobic blocks in amphiphilic polymers. For example, paclitaxel-loaded amphiphilic micelles composed of the block copolymer (PEG-b-PMME) were prepared [48]. Because the PEG-b-PMME contained acid-labile six-membered orthoester rings in its side chains, the acidic pH could induce the hydrolysis of orthoester rings resulting in disruption of the micelles and consequent drug release from the micelle cores. In another example, poly(L-lysine) (PLL) was conjugated with PEG by a benzoic imine linker to form a pH-sensitive self-assembling polymer for delivery of DOX [49]. At a pH of 7.4 (in the normal physiological condition), the assembled micelles were stable and had a long circulation time in the blood, while at a pH of 5.0~6.5 (in the tumor microenvironment), the PEG was deshielded and the positively charged PLL was exposed via the hydrolysis of the benzoic–imine bond. The positive charge of the micelles facilitated the cellular uptake. In addition, therapeutic agents can be directly conjugated to the polymers through the aforementioned acid-labile linkers to render the pH sensitivity to the prodrugs/conjugates for the tumor-targeted drug delivery. In a study, DOX was conjugated to the side chains of poly(lactic-co-glycolic acid) (PLGA) via an acid-labile hydrazine linker, to avoid the efflux pump-mediated drug excretion [50].

In addition to the organic materials, the inorganic materials, like calcium phosphate (CaP), zinc oxide (ZnO), mesoporous silica nanoparticles (MSNs), and liquid metal, have emerged as promising pH-responsive carriers for drug delivery and bioimaging. Most of these materials are insoluble at physiological pH and are dissolved as nontoxic ions in the acidic microenvironment, such as endo/lysosomes and solid tumors [51–54]. For example, He et al. fabricated a pH-sensitive MSN that showed an acid-triggered drug release and the improved cytotoxicity against cancer cells [55].

1.2.1.2 Enzyme-responsive nanomedicines

Enzymes play significant roles in biological processes due to their exceptional specificity and outstanding catalytic properties. Some diseases, like cancer or inflammatory diseases, frequently occur with the abnormal expression of specific types of enzymes, like matrix metalloproteinases (MMPs), phospholipases, glycosidases, and esterases, in the restricted diseased tissues. Thus, owing to the roles that enzymes play in different biological processes, the disease-associated enzyme dysregulations have recently become an emerging therapeutic target. Using enzymes as a trigger has a number of benefits because most enzymes catalyze chemical/biological reactions under moderate conditions (normal body temperature, neutral pH, and aqueous solutions), whereas many non-enzymatic conventional chemical reactions may not happen under the same conditions. Besides, enzymes possess exquisite selectivity for their substrates, allowing for specific, sophisticated, biologically inspired chemical reactions [19].

On the basis of these principles, a wide variety of nanomaterials, like polymeric materials, phospholipids or inorganic materials, have been developed as the enzyme-responsive systems for drug delivery [19, 41]. These nanomaterials usually contain enzyme-cleavable moieties in their backbone and/or side groups, including (i) the ester bonds sensitive to certain phosphatases, intracellular acid hydrolases and several other esterases; (ii) the azo linkers sensitive to colonic bacterial enzymes; and (iii) the amides sensitive to hydrolytic proteases [56].

Here, we only use MMPs which are highly associated with tumor growth, invasion, and metastasis [57] as an example to illustrate the enzyme-responsive nanomedicine. Due to the remarkable MMP upregulation in the tumor, some of the well-defined MMPs have been used as biomarkers for cancer diagnosis and prognosis and as the therapeutic targets for development of MMP inhibitors. Unfortunately, the MMP inhibitors are not successful in fighting cancer due to their severe toxicity and drug resistance [58, 59]. In contrast, MMPs have showed promise as site-specific biological stimuli for bioresponsive drug delivery

[31, 60–64] and imaging [65–69]. In a recent study, a short octapeptide which was cleavable by MMP2, was inserted between the PEG corona and TAT peptide-functionalized liposomes to improve the tumor targetability (Fig. 1.2A) [31]. In this design, the PEG was used to shield and protect other nanoparticles' surface functionalities from the nonspecific interaction with nontumor tissues or biomolecules. Only in the tumor, the upregulated MMP2 could cleave the MMP2-sensitive linker and deshielded the PEG, so as to expose the functional moieties on the liposomes for cancer cell endocytosis. In another study, the MMP2-sensitive linker and cell penetrating peptide were covalently linked together as a novel MMP2-activatable cell penetrating moiety to simplify the components of the nanoparticles. Via this dual-functional peptide linker, the anticancer drug, DOX, was conjugated with the PEG to form a polymer-drug conjugate which was able to specifically and intracellularly deliver drugs to the MMP2-overexpressed cancer cells [61]. In a follow-up study, an "all-in-one" micellar nanoparticle was developed as a universal platform to deliver the physically encapsulated anticancer drugs. The "all-in-one" system showed high stability even after the MMP2-induced cleavage, resulting in the improved cellular uptake and tissue penetration and prolonged in vivo drug retention in the tumor (Fig. 1.2B) [62]. To further improve drugs' tumor specificity, an MMP2 and folate receptor (FR) dual targeted micelles were developed, which showed the MMP2-dependent cellular uptake, penetration, and cytotoxicity. In vivo, the micelles showed much higher tumor targeting effect compared to the nonsensitive micelles and FR-targeted micelles (Fig. 1.2C) [63]. Interestingly, in these studies, for the first time, the structure, PEG-peptide-PE, was found to inhibit the P-glycoprotein (P-gp)-mediated drug efflux [64]. In addition, the extracellular enzyme-responsive systems with the sheddable PEG corona have been investigated for gene and siRNA delivery [70, 71]. The use of the MMP activatable cell-penetrating peptide (ACPP) which was invented by Tsien's group is another promising strategy for MMP-responsive drug delivery and have showed successes in various platforms [65–69].

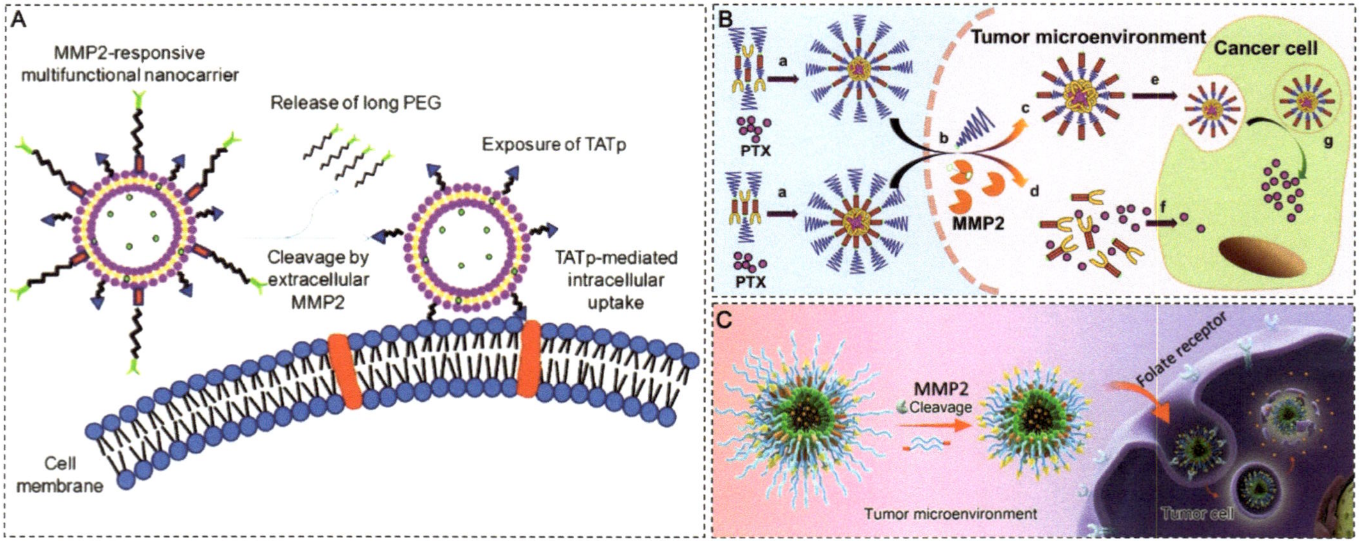

Figure 1.2 The MMP2-responsive drug delivery and tumor targeting. (A) MMP2-sensitive multifunctional liposomal nanocarrier; (B) Drug delivery mechanisms of the PEG2k-ppTAT-PEG1k-PE micelles ("all-in-one" MMP2-sensitive micelles, upper panel) and PEG2k-ppTAT-PE micelles (unstable MMP2-sensitive micelles, lower panel); (C) MMP2-sensitive FR-targeted delivery of dasatinib to the tumor [31, 62, 63].

1.2.1.3 Redox potential-responsive nanomedicines

Differences in redox potential exist at both the tissue and cellular levels. For example, the glutathione (GSH)/glutathione disulfide (GSSG) couple has been verified as the most abundant redox couple in animal cells, where glutathione is found at a level that is two to three orders of magnitude higher in the cytosol than that in the extracellular space [72, 73]. Furthermore, studies with a rodent model have revealed higher glutathione concentrations in tumor tissues compared with those in normal tissues [74, 75]. Low reducing GSH level is always associated with high reactive oxygen species (ROS) level, which is found in many pathological conditions [76–79]. In general, low ROS levels regulate cell-signaling pathway and promote cell proliferation, whereas high levels of ROS induce the non-specific damage to the protein, lipid, and DNA, the essential biomolecules for living organisms. The redox/ROS imbalance has been found in many diseases/disorders including autoimmune disease, cardiovascular disease, neurodegenerative disease, cancer, stroke, etc. The nanomaterials that are redox or oxidation-sensitive have been discovered/prepared for delivery of the drugs to the aforementioned diseases [80, 81].

Disulfide bond, prone to rapid cleavage by GSH, is commonly used for construction of the redox-responsive nanomaterials [20]. It is known that the concentrations of GSH in the tumor extracellular space (~2–20 μM) and intracellular compartments (~0.5–10 mM) are significantly higher compared with those in healthy tissues and cells. Therefore, the disulfide-containing nanomaterials have been prepared to deliver many drugs to the tumor [82]. The redox-responsive micellar nanoparticles may be prepared by (i) self-assembly of the amphiphilic copolymers containing disulfide bonds; or (ii) incorporation of the GSH-sensitive cross-linking agents in either the shell or core of the nanoparticles post-nanoparticle preparation [83–86]. These micelles usually undergo the reductive degradation and rapid disassembly under the reducing environment. For example, Fan and coworkers reported the redox-sensitive, disulfide-containing poly(amido amine)-graft-PEG (SS-PAmAm-g-PEG) as a carrier for DOX delivery [85]. The polymeric micelles were stable in the aqueous solution, while they were rapidly disassembled in the presence of the reductive dithiothreitol (DTT). In another study, the poly(ε-benzyloxycarbonyl-L-lysine) was PEGylated via a disulfide linkage to form the redox-sensitive self-assembling polymer, mPEG-SS-PzLL. The mPEG-SS-PzLL micelles underwent the PEG shedding and micelle disassembly via the cleavage of disulfide bonds, facilitating the intracellular release of the encapsulated payloads [86]. In addition, the redox sensitivity may be obtained by the modification of the nanoparticles with other redox-responsive motifs, including the diselenide linkage, cis,cis,trans-diamminedichlorodihydroxy-platinum (iv) (DHP) or cis,cis,trans-diamminedichlorodisuccinato-platinum (DSP), and trimethyl-locked benzoquinone (TMBQ) [87–89].

Oxidation-responsive nanomedicines are prepared to target ROS, such as hydrogen peroxide (H_2O_2) and hydroxyl radicals. A major class of oxidation-responsive materials is sulfur-based [56]. For example, the researchers conjugated oxidation-convertible poly(propylene sulfide) (PPS) with PEG to form an amphiphile for oxidation responsiveness [90]. In another study, the thioketal-based nanoparticles possessing the ROS sensitiveness were able to specifically and efficiently deliver the anti-TNFα siRNA to the sites of intestinal inflammation, resulting in the significant gene silencing after oral administration [91]. Besides, emerging responsive motifs such as boronic ester groups and phenylboronic acid (PBA) derivatives have attracted considerable attention [92–95]. For example, arylboronic esters were introduced on the dextran and RNase A via the hydroxyl groups and lysine residues, respectively, for oxidation-responsive protein release and activity recovery [96]. Quantification of the disease-associated ROS is a useful strategy for the diagnosis of a wide variety of diseases and also provides a direction of the treatment [97]. The use of ROS-responsive materials for ROS detection represents an appealing approach for developing the enzyme-free ROS sensors.

1.2.1.4 Hypoxia-responsive nanomedicines

Hypoxia, widely existing in various pathological conditions, is considered a hallmark of ischemic diseases, including ischemic heart disease, limb, and stroke [98–100]. The main cause of hypoxia is the decreased blood supply into the tissue, which leads to the shortage of oxygen and nutrient. Besides, hypoxia is also a common feature of the tumor microenvironment due to the uncontrolled angiogenesis. Solid tumors grow uncontrollably fast while their newly formed vasculatures fail to provide sufficient oxygen and nutrient. Thus, the central regions of the solid tumor are commonly under the hypoxic condition. The hypoxic cancer cells divide more slowly than their well-oxygenated counterparts, making them less susceptible to conventional chemotherapy that targets fast-dividing cells. Hypoxia can also aggravate the tumor with acquired resistance to other treatments, such as radiotherapy, causing the treatment failure and cancer relapse [101, 102]. Hypoxia has been extensively exploited as a target for developing various diagnostic and therapeutic agents.

Many nanotherapeutic strategies have been developed for targeting the hypoxic microenvironment of diseases, especially the solid tumor [103, 104]. Commonly, the hypoxia-sensitive motifs, such as nitroimidazole and azobenzene derivatives, can be incorporated in the nanomaterials as bioreductive linkers to achieve the tumor-targeted drug and gene delivery [105]. For example, Thambi et al. constructed the micelles based on nitroimidazole moieties [22]. In this system, the nitroimidazole derivative was covalently linked to the water-soluble carboxymethyl-dextran to produce an amphiphilic polymer, which could self-assemble into nanoparticles for encapsulation of hydrophobic drugs. The formed nanoparticles were stable in physiological conditions and capable of selectively releasing the loaded drugs under hypoxic conditions due to the hypoxia-triggered hydrophobic-to-hydrophilic transition and resultant nanoparticle destabilization. In another study, Perche et al. designed a siRNA nanocarrier by incorporating an azobenzene group between PEG and poly(ethyleneimine) (PEI) polymer segments, which could be "activated" in the hypoxic environments [38]. When the nanocarriers got into the hypoxic tumor, the azobenzene bond was cleaved which deshielded the PEG "corona." The remaining PEI/siRNA complex nanoparticles with the exposed positive charge enhanced the cellular uptake and gene silencing both in vitro and in vivo.

1.2.2 External Stimuli and Stimuli-Responsive Nanomedicines

Unlike traditional therapeutics which often accompanies with undesirable side effects, external stimuli-responsive nanomedicine is usually non-toxic or biologically inert in non-target sites and would exert the therapeutic action only under specific external stimuli, and thus can offer optimal clinical outcomes with greatly reduced adverse effects. In addition, these external stimuli-responsive therapies can also be easily combined with other existing therapeutics to achieve synergistic therapeutic effects. In this section, the nanomedicines that use the externally applied stimuli, including temperature, light, magnetic field, and ultrasound, to control drug delivery and release, are discussed. The typical external stimuli-responsive nanomedicines are summarized in Table 1.2.

1.2.2.1 Temperature-responsive nanomedicines

The local temperature in specific tissues can be elevated under certain diseased conditions, such as infection or inflammation, or through exposure to an external heat source generated by electromagnet [106], laser [107] or high intensity focused ultrasound (HIFU) [108]. Therefore, temperature may act as either an external or internal stimulus, and is among the most often used stimuli in stimuli-responsive nanomedicines [41]. Higher than the normal temperature in the inflamed or cancerous tissues has been noticed for decades, but the difference is not outstanding [109]. Hyperthermia in the range of 40–45°C can induce a greater local blood flow and increase the vasculature permeability as compared to those at the normal body temperature. So far, local hyperthermia in combination with chemotherapy has drawn a lot of attention. However, the temperature difference between normal and diseased tissues is still not sufficient for most thermosensitive nanopreparations. In contrast, the external heat source/device can precisely control the temperature and its location, which is popularly used in drug delivery and hyperthermia therapy.

The drug delivery mechanism of the temperature- or thermo-responsive nanomedicine is mainly based on the temperature-dependent alteration in nanomaterials' solubility and/or phase transition. When the temperature is around the upper critical solution temperature (UCST) or lower critical solution temperature (LCST), the shrinkage/water squeezing or expansion/swelling may occur and this swelling-shrinkage behavior is usually

Table 1.2 Examples of typical external stimuli and stimuli-responsive nanomedicines

Acting stimuli	Sites or diseases	Responsive nanomaterials	Cargoes	Applications	References
Temperature (Inflammation-induced hyperthermia or external heat sources)	Cancer	Hydrogels: OSM-PCLA-PEG-PCLA	PTX	Temperature-dependent sol-to-gel phase transition and drug release	[158]
	Spinal cord damage	Magnetite/PEO–PPO–PEO nanoparticles	Eosin Y and ibuprofen	Temperature-controlled "gate" for loading and release of cargoes	[159]
	Cancer	Poly(EOEOVE)-containing liposomes	DOX	Temperature-responsive drug release at > 40°C	[160, 161]
Light	Cancer	Liposomes (HAuNS-TSL)	HAuNS and DOX	Photothermal tumor ablation combined with chemotherapy	[162]
	Cancer	Azobenzene-tethered DNA	DOX	Photon-manipulated drug release	[163]
	Cancer	Gold nanorods with oligonucleotide-capped silica shell	DOX and siRNA	NIR light-responsive intracellular drug and siRNA release	[164]

(*Continued*)

Table 1.2 (*Continued*)

Acting stimuli	Sites or diseases	Responsive nanomaterials	Cargoes	Applications	References
Magnetic field	Cancer	GO–IONP	DOX	Tumor targeting, thermal therapy, and MRI	[165]
	Cancer	Super-paramagnetic nanoparticles	DOX	Magnetic field-responsive drug delivery/release	[166]
	Cancer	$MnFe_2O_4$ nanocrystals	DOX	Imaging-guided magnetic targeting	[167]
Ultrasound	Cancer	Polymer-grafted silica nanoparticles	DOX	Ultrasound-triggered drug release	[168, 169]
	Thrombotic diseases	PEG-gelatin/t-PA nano-complexes	t-PA	Ultrasound-triggered drug release	[170]
	Diabetes	Chitosan-coated and alginate-coated PLGA nanoparticles	Insulin	Ultrasound-triggered drug release	[171]

Abbreviations: PCLA, poly(α-caprolactone-co-lactide); OSM, sulfamethazine oligomer; PPO, poly(propylene oxide); EOEOVE, poly[2-(2-ethoxy) ethoxyethyl vinyl ether; pcCPP, photolabile-caged cell-penetrating peptide; HAuNS, hollow gold nanospheres; TSL, thermal sensitive liposome; GO–IONP, superparamagnetic graphene oxide–iron oxide hybrid nanocomposite; t-PA, tissue-type plasminogen activator.

reversible [110, 111]. The well-known examples of temperature-sensitive polymers include the poly(N-isopropylacrylamide) (PNIPAM), poly(ethylene oxide), poly(propylene oxide) (PPO), PLA (homo- and copolymers), proteins, and polysaccharides [112]. The polymer-solvent interactions in the case of PNIPAM decrease when the temperature is above the LCST due to the increased hydrophobicity of the polymers, while the polymers, like polypeptides and polysaccharides, usually undergo the coil-helix transition when the temperature is below the UCST. The PNIPAM-based thermo-responsive nanomedicine with a substantial swelling ratio and thermal reversibility, has been extensively studied. Because of its low LCST (about 32°C), the chemical modification of PNIPAM is usually needed to adjust its thermo sensitivity (i.e., LCST) through the covalent attachment of the hydrophilic or hydrophobic groups [113]. In some cases, the LCST of PNIPAM-containing nanomedicine can be even tuned by incorporating the inorganic materials, e.g., gold nanoparticles [114, 115].

Most of the investigated temperature-responsive systems are liposomes [116, 117] and micelles [118, 119] that exhibit the LCST of >40°C. The thermoresponsiveness of the liposomes usually comes from the phase transition of their constituent lipids and resultant conformational changes in the lipid bilayers. Phase transition temperatures of various types of lipids have been measured. The liposomes formulated by the lipids with the appropriate phase transition temperature are able to respond to the elevated temperatures resulting in the temperature-dependent drug release [120]. Dipalmitoylphosphatidylcholine (DPPC), one of the mostly used lipids for hyperthermia/thermosensitivity [121, 122], has a gel-to-liquid crystalline transition temperature (T_g) of around 41°C. Many studies using DPPC as a major component of liposomes indicate that the lipid undergoes a phase transition and conformational change in response to hyperthermia, which compromises the integrity of the liposomes, resulting in the cargo release. In addition, the temperature-dependent site-specific targeting can be achieved by the combined use of the thermo-sensitive liposomes with the regional hyperthermia, allowing for the controlled release of the entrapped drugs in the restricted heated tissues [116, 122]. Thermo-responsive liposomes are perhaps the most advanced stimuli-responsive nanomedicine, as evidenced by their use in several clinical trials/applications. For example, the DOX-loaded thermo-responsive liposomes (TSLs) (ThermoDox, Celsion Corporation), in association with hyperthermia or radiofrequency ablation, are investigated in the phase II clinical trials for treating breast cancer and colorectal liver metastasis and in the phase III clinical trials for treating hepatocellular carcinoma.

1.2.2.2 Light-responsive nanomedicines

Light is an attractive external stimulus because it can be easily and conveniently controlled in a spatial and temporal manner. Moreover, the light's wavelength ranging from the ultraviolet (UV) and visible to the near-infrared (NIR) regions and light's intensity can be readily adjusted. Therefore, light-responsive nanomedicines have recently received a lot of attention particularly in the spatiotemporal and/or on-demand drug delivery and therapy. Polymeric micelles and liposomes are two most widely studied light-responsive nanocarriers [24]. Other organic nanoparticles, e.g., nanogels for topical diagnose/therapy [123, 124], and inorganic nanoparticles, e.g., silica nanoparticles and gold nanoparticles [125–127], have also been employed. Though the light-responsive nanomedicines are diverse, they must contain at least one type of photo-responsive groups or chromophores, as the key component, to harvest light. In response to light, these groups may undergo the photo-isomerization usually accompanying with a change in the hydrophilicity/hydrophobicity balance, photo-induced cleavage, photo-induced cross-linking, and/or photo-oxidation [128].

A variety of photo-responsive groups have been discovered, such as azobenzene, spirobenzopyran, triphenylmethane, and cinnamonyl. Under UV–vis light, these photo-responsive groups undergo reversible structural changes. For example, the azobenzene undergoes a *trans* to *cis* photo-isomerization of its nitrogen double bond upon the UV irradiation, while this conversion can be reversed under the visible light [128, 129]. The photo-responsive groups may be encapsulated within the nanocarriers or conjugated to the surface of the nanoparticles.

For the light-responsive polymeric nanomedicines, e.g., polymeric micelles, the major functional building blocks, i.e., the light-responsive polymers, are usually amphiphilic polymers, containing one hydrophilic block, such as poly-(ethylene oxide) (PEO) and poly(acrylic acid), and one hydrophobic block bearing the photochromic groups. For instance, the UV/NIR-mediated cleavage of the photochromic *o*-nitrobenzyl- and coumarin-incorporated copolymers was sufficient to induce the light responsiveness of the assembled micellar nanomedicines [127, 130, 131]. Azagarsamy et al. prepared two *o*-nitrobenzyl and coumarin methylester based units [132], which could be selectively and effectively cleaved by different wavelengths of light and showed the wavelength-dependent photo-degradation of the nanoparticles and release of the loaded proteins. The on-demand manipulation of the cargo release rate in real time is of great interest to drug delivery and tissue engineering. In addition, the photo-induced rearrangements and isomerization of nanoparticle building blocks are also used in the light-responsive nanomedicine [133–135].

However, the UV-vis light is not capable of deep penetration through the body because of their absorption by the skin, blood, and tissue. Thus, the conventional light-responsive drug delivery is only applicable to the parts of the body that can be directly illuminated (such as the eye and skin). In contrast, the NIR light can penetrate much deeper (up to several inches). By using photosensitive groups responding to longer wavelengths or exploiting two-photon technology, it becomes possible to replace the UV–visible light with an NIR laser (700–1,000 nm) for deeper tissue penetration, lower light scattering, and minimal nonspecific tissue damage [136, 137], which makes the NIR-responsive system extremely promising for noninvasive diagnoses and treatments [138]. Furthermore, the NIR-absorbing plasmonic materials' light-to-heat conversion capacity may be used to design the light/thermo responsive nanosystems to trigger and facilitate the drug release. For example, DOX-loaded hollow gold nanospheres showed the accelerated drug release when irradiated at 808 nm, allowing for the enhanced anticancer activity and reduced systemic toxicity compared with the free drug [139].

1.2.2.3 Magnetic field-responsive nanomedicines

Magnetic field is another widely used external stimulus in nanomedicine and considered one of the best external stimuli because of its negligible potential damage to the body. Magnetic nanoparticles with various sizes (usually <10 nm) and morphologies, such as magnetite (Fe_3O_4) or maghemite (Fe_2O_3) [140], are extensively explored in biomedical applications including drug and gene delivery, magnetic resonance imaging (MRI), magnetic hyperthermia, magnetic transfection, and cell and protein isolation and purification [141]. Magnetic field-responsive nanomedicines are typically fabricated via the encapsulation/incorporation of the therapeutic payloads along with the magnetic nanoparticles in a pharmaceutically stable formulation. In the presence of an external magnetic field, the magnetic nanoparticles are capable of delivering therapeutics specifically to the intended site. In order to reduce the aggregation caused by the hydrophobicity of the magnetic nanoparticles and improve their biocompatibility, the magnetic nanoparticles can be surface-modified with hydrophilic biocompatible molecules, such as glutamic acid, dextran, pullulan, citric acid, and silane.

Magnetic field-responsive nanomedicine has showed great potential in tumor targeting and drug delivery. Zhou et al. prepared the PEGylated iron oxide nanoparticles which showed the tumor-targeted MRI and photothermal anticancer activity in a mouse tumor model [142]. Magnetic field-responsive nanocarriers have been also used in the delivery of nucleic acids, mostly siRNA and genes [143, 144]. Nucleic acids are usually condensed with cationic nanomaterials containing magnetic nanoparticles, e.g., iron oxide doped tricalcium phosphate (Fe_2O_3@TCP) nanoparticles, to form the magnetic complexes [145] for gene transfection in a magnetic field, called magnetofection, a highly efficient transfection technique. Moreover, magnetic field-responsive genetic nanocomplexes may be used for pathogen detection. For example, a magneto-DNA nanoparticle was developed for rapid detection and phenotyping of bacteria [146].

When magnetic nanoparticles are placed in an oscillating or alternating magnetic field (AMF), due to hysteresis loss and/or Néel relaxation, the heat can be generated. Therefore, magnetic nanoparticles can be used as an induced external heat source for magnetic hyperthermia [147].

1.2.2.4 Ultrasound-responsive nanomedicines

Ultrasound plays an important role in modern biomedical applications, particularly as a diagnostic tool for bioimaging at low frequencies and as a therapeutic tool for the removal of masses and treatment of cancer at high frequencies. The ultrasound-based applications are appealing mainly because of their non-invasiveness, absence of ionizing radiations, and controllable penetration depth. In addition, the characteristics of ultrasound such as the frequency, duty cycles, and exposure time can be easily tuned. Therefore, the use of ultrasound is a simple and effective way for spatiotemporally controlling the release of cargoes including drugs, genes and imaging agents, at the desired site and lowering the risk of nonspecific damage to the healthy tissues [148]. Another benefit is that, under ultrasound, the permeability of biological barriers, such as cell membranes and blood brain barrier (BBB), can be significantly increased, via ultrasound's capability of formation of cavitation bubbles and increasing temperature, thus enhancing the drug uptake [108, 149].

Drug delivery and release with ultrasound relies on the thermal and/or mechanical effects generated by acoustic cavitation or radiation forces. Development of the drug-loaded ultrasound contrast agents (i.e., ultrasound-responsive drug delivery systems) is one of the major applications of ultrasound. Among the ultrasound-responsive systems, the microbubble-containing systems are a major one which, consisted of a gas core (perfluorocarbons, nitrogen, or air) and a shell (lipids, polymers, or proteins), help to enhance the specificity and sensitivity of ultrasonic imaging [150]. However, intravenous injection leads to the accumulation of microbubbles in the blood pool due to their large size (2–8 μm). As a result, the applications of microbubbles have been limited to the diseases of the cardiovascular system [150]. In order to broaden the applications of the contrast agents, the nanoscale ultrasound-responsive systems including nanobubbles [151–153], nanodroplets [154], liposomes [155], nanemulsion [156], and micelles [157], have been developed for extravascular ultrasonic imaging and targeted delivery of small molecules, proteins, and nucleic acids.

1.3 Current Status and Future Perspectives

The design of nanocarriers/nanomedicines that are able to target the disease sites and deliver drugs in an on-demand and controllable manner in response to endogenous or exogenous stimuli, represents an advanced, "smart" drug delivery strategy compared to the first generation of nanomedicine, mainly the plain liposomes and nanoparticles. The variety of stimuli able to trigger the cargo release at the right place and right time and the diversity of emerging stimuli-responsive materials, allows great flexibility in the preparation and application of stimuli-responsive nanomedicine. Although many kinds of stimuli-responsive systems have been reported, only a few of them enter the clinical trials (Table 1.3) due to several unsolved issues/challenges.

First of all, the complexity of nanosystems in terms of preparation, components, and architectures, and the difficulty of scale-up are likely to hamper the translation of the stimuli-responsive nanomedicine from the bench to bedside. As a general rule, the simpler and easier a nanomedicine and its preparation are, the higher its chance of reaching the clinic is. Therefore, it is essential to "keep it simple, stupid" in both the elements and processes that involved in the preparation of the stimuli-responsive nanomedicine. The thermosensitive liposomes, thermoDox, which is already in the clinical trial, may be a good example.

Second, biocompatibility/safety is another key factor influencing the applications of the stimuli-responsive nanomedicine. The synthetic/semi-synthetic materials are widely used in nanopreparations. It is easy to control the molecular structure and size to produce high-quality nanomaterials. But these materials may have poor immune tolerance and biodegradability. Poor immune tolerance may directly cause undesirable inflammatory reactions and/or accumulation of nanomedicine in the MPS, resulting in the nonspecific biodistribution and off-target toxicity, while poor biodegradability may lead to the slow and incomplete clearance/elimination through the elimination organs, mainly liver and kidney, increasing the risk of the long-term toxicity. It is imperative to systematically evaluate the biocompatibility/safety of stimuli-responsive nanomaterials both in vitro and in vivo.

Table 1.3 Examples of stimuli-responsive nanomedicines in clinical trials

Trade names	Stimuli	Carriers	Active agents	Targets	Clinical status	References
Thermo DOX	Temperature	Liposomes	DOX & HIFU	Pediatric refractory solid tumors	Phase I	NCT02536183
			DOX & hyperthermia	Recurrent chest-wall breast cancer	Phase I/II	NCT00826085
			DOX & HIFU	Liver tumors	Phase I	NCT02181075
			DOX & RFA	Hepatocellular carcinoma	Phase III	NCT02112656
				Primary and metastatic liver tumors	Phase I	NCT00441376
				Hepatocellular carcinoma	Phase III	NCT00617981
TM-DOX	Magnetic	Iron and carbon particles	DOX & nanoparticles	Hepatocellular carcinoma	Phase I/II	NCT00054951
				Liver metastasis	Phase I/II	NCT00041808
Magnetic Nanoparticle	Magnetic	Magnetic nanoparticles	Magnetic nanoparticles	Prostate cancer	Early Phase I	NCT02033447

Abbreviations: HIFU, high-intensity focused ultrasound; RFA, radiofrequency ablation.

Third, a remarkable and robust stimulus is still challenging. For external stimuli, the specificity, intensity and penetration depth need to be well controlled to minimize the negative impact of the nanomedicine and stimuli on the surrounding healthy tissues. This may be improved through the development of new stimuli-generating devices/sources. For internal stimuli that are usually disease- and individual-dependent, are less controllable. Therefore, the identification and quantification of the disease-associated stimuli are key. A robust specific stimulus would secure the success of the stimuli-responsive strategy.

Fourth, many current ideas focusing on the novelty in the nanomaterial and nanocarrier design show less potential in their clinical translation. More efforts should be made towards the development of the clinically acceptable/"friendly" systems. The clinically relevant disease models are highly recommended to evaluate the stimuli-responsive nanomedicine.

Finally, the limitations of current drug delivery and therapy strategies and complexity of the diseases require the development of multifunctional drug delivery systems to address various unmet needs of the currently used therapeutics [172]. In these multifunctional nanocarriers, the "smart" stimuli-responsive strategy would "cooperatively" work with other strategies for optimal therapeutic effects. For example, to have better therapeutic outcomes, the stimuli-sensitive imaging-guided drug delivery systems, i.e., the theranostics, that are capable of monitoring and tracking diseases, therapeutics, and treatment outcomes, have been investigated [173].

As evidenced by recent advances, the stimuli-responsive nanomedicine has shown great potential for effective and smart drug delivery. In the future, a simple, multifunctional, and biocompatible stimuli-responsive nanomedicine with high chance of clinical translation will be a critical need.

Acknowledgments

The work was supported by the National Cancer Institute of the National Institutes of Health (R15CA213103).

References

1. Farokhzad, O. C., and R. Langer, Nanomedicine: Developing smarter therapeutic and diagnostic modalities. *Advanced Drug Delivery Reviews*, 2006. **58**(14): pp. 1456–1459.
2. Kim, B. Y., J. T. Rutka, and W. C. Chan, Nanomedicine. *New England Journal of Medicine*, 2010. **363**(25): pp. 2434–2443.
3. Allen, T. M., and P. R. Cullis, Drug delivery systems: Entering the mainstream. *Science*, 2004. **303**(5665): pp. 1818–1822.
4. Torchilin, V. P., Recent advances with liposomes as pharmaceutical carriers. *Nature Reviews Drug Discovery*, 2005. **4**(2): pp. 145–160.
5. Torchilin, V. P., Micellar nanocarriers: Pharmaceutical perspectives. *Pharmaceutical Research*, 2007. **24**(1): pp. 1–16.
6. Gillies, E. R., and J. M. Frechet, Dendrimers and dendritic polymers in drug delivery. *Drug Discovery Today*, 2005. **10**(1): pp. 35–43.
7. Müller, R. H., M. Radtke, and S. A. Wissing, Solid lipid nanoparticles (SLN) and nanostructured lipid carriers (NLC) in cosmetic and dermatological preparations. *Advanced Drug Delivery Reviews*, 2002. **54**: pp. S131–S155.
8. Elsabahy, M., and K. L. Wooley, Design of polymeric nanoparticles for biomedical delivery applications. *Chemical Society Reviews*, 2012. **41**(7): pp. 2545–2561.
9. Smith, A. M., et al., Bioconjugated quantum dots for in vivo molecular and cellular imaging. *Advanced Drug Delivery Reviews*, 2008. **60**(11): pp. 1226–1240.
10. Ghosh, P., et al., Gold nanoparticles in delivery applications. *Advanced Drug Delivery Reviews*, 2008. **60**(11): pp. 1307–1315.
11. Bianco, A., K. Kostarelos, and M. Prato, Applications of carbon nanotubes in drug delivery. *Current Opinion in Chemical Biology*, 2005. **9**(6): pp. 674–679.
12. Chari, R. V., M. L. Miller, and W. C. Widdison, Antibody–drug conjugates: An emerging concept in cancer therapy. *Angewandte Chemie International Edition*, 2014. **53**(15): pp. 3796–3827.
13. Zhu, L., and R. I. Mahato, Lipid and polymeric carrier–mediated nucleic acid delivery. *Expert Opinion on Drug Delivery*, 2010. **7**(10): pp. 1209–1226.
14. Harris, J. M., and R. B. Chess, Effect of pegylation on pharmaceuticals. *Nature Reviews Drug Discovery*, 2003. **2**(3): pp. 214–221.

15. Pasut, G., and F. M. Veronese, State of the art in PEGylation: The great versatility achieved after forty years of research. *Journal of Controlled Release*, 2012. **161**(2): pp. 461–472.
16. Fang, J., H. Nakamura, and H. Maeda, The EPR effect: Unique features of tumor blood vessels for drug delivery, factors involved, and limitations and augmentation of the effect. *Advanced Drug Delivery Reviews*, 2011. **63**(3): pp. 136–151.
17. Zhu, L., and V. P. Torchilin, Stimulus-responsive nanopreparations for tumor targeting. *Integrative Biology*, 2013. **5**(1): pp. 96–107.
18. Ling, D., et al., Multifunctional tumor pH-sensitive self-assembled nanoparticles for bimodal imaging and treatment of resistant heterogeneous tumors. *Journal of the American Chemical Society*, 2014. **136**(15): pp. 5647–5655.
19. De La Rica, R., D. Aili, and M. M. Stevens, Enzyme-responsive nanoparticles for drug release and diagnostics. *Advanced Drug Delivery Reviews*, 2012. **64**(11): pp. 967–978.
20. Saito, G., J. A. Swanson, and K.-D. Lee, Drug delivery strategy utilizing conjugation via reversible disulfide linkages: Role and site of cellular reducing activities. *Advanced Drug Delivery Reviews*, 2003. **55**(2): pp. 199–215.
21. Perche, F., et al., Hypoxia-responsive copolymer for siRNA delivery. *RNA Imaging: Methods and Protocols*, 2016. **1372**: pp. 139–162.
22. Thambi, T., et al., Hypoxia-responsive polymeric nanoparticles for tumor-targeted drug delivery. *Biomaterials*, 2014. **35**(5): pp. 1735–1743.
23. Al-Ahmady, Z., and K. Kostarelos, Chemical components for the design of temperature-responsive vesicles as cancer therapeutics. *Chemical Reviews*, 2016. **116**(6): pp. 3883–3918.
24. Alvarez-Lorenzo, C., L. Bromberg, and A. Concheiro, Light-sensitive intelligent drug delivery systems. *Photochemistry and Photobiology*, 2009. **85**(4): pp. 848–860.
25. Sun, C., J. S. Lee, and M. Zhang, Magnetic nanoparticles in MR imaging and drug delivery. *Advanced Drug Delivery Reviews*, 2008. **60**(11): pp. 1252–1265.
26. NIKOLAEV, A., et al., Ultrasonic nanomedicine in the aspect of cancer therapy. *Nanostructures, Nanomaterials, and Nanotechnologies to Nanoindustry*, 2014: Chapter 22: pp. 335–348.
27. Sarmento, B., et al., Oral bioavailability of insulin contained in polysaccharide nanoparticles. *Biomacromolecules*, 2007. **8**(10): pp. 3054–3060.

28. Garbern, J. C., et al., Delivery of basic fibroblast growth factor with a pH-responsive, injectable hydrogel to improve angiogenesis in infarcted myocardium. *Biomaterials*, 2011. **32**(9): pp. 2407–2416.

29. Xu, R., et al., An injectable nanoparticle generator enhances delivery of cancer therapeutics. *Nature Biotechnology*, 2016. **34**(4): pp. 414–418.

30. Jiang, L., et al., Overcoming drug-resistant lung cancer by paclitaxel loaded dual-functional liposomes with mitochondria targeting and pH-response. *Biomaterials*, 2015. **52**: pp. 126–139.

31. Zhu, L., P. Kate, and V. P. Torchilin, Matrix metalloprotease 2-responsive multifunctional liposomal nanocarrier for enhanced tumor targeting. *ACS Nano*, 2012. **6**(4): pp. 3491–3498.

32. Zhang, J., et al., Multifunctional envelope-type mesoporous silica nanoparticles for tumor-triggered targeting drug delivery. *Journal of the American Chemical Society*, 2013. **135**(13): pp. 5068–5073.

33. Gajanayake, T., et al., A single localized dose of enzyme-responsive hydrogel improves long-term survival of a vascularized composite allograft. *Science Translational Medicine*, 2014. **6**(249): pp. 249ra110–249ra110.

34. Li, J., et al., Redox-sensitive micelles self-assembled from amphiphilic hyaluronic acid-deoxycholic acid conjugates for targeted intracellular delivery of paclitaxel. *Biomaterials*, 2012. **33**(7): pp. 2310–2320.

35. Kurtoglu, Y. E., et al., Poly (amidoamine) dendrimer–drug conjugates with disulfide linkages for intracellular drug delivery. *Biomaterials*, 2009. **30**(11): pp. 2112–2121.

36. Shim, M. S., and Y. Xia, A reactive oxygen species (ROS)-responsive polymer for safe, efficient, and targeted gene delivery in cancer cells. *Angewandte Chemie International Edition*, 2013. **52**(27): pp. 6926–6929.

37. Lee, M., et al., Hypoxia-inducible VEGF gene delivery to ischemic myocardium using water-soluble lipopolymer. *Gene Therapy*, 2003. **10**(18): pp. 1535–1542.

38. Perche, F., et al., Hypoxia-Targeted siRNA delivery. *Angewandte Chemie International Edition*, 2014. **53**(13): pp. 3362–3366.

39. Karimi, M., et al., pH-Sensitive stimulus-responsive nanocarriers for targeted delivery of therapeutic agents. *Wiley Interdisciplinary Reviews: Nanomedicine and Nanobiotechnology*, 2016. **8**(5): pp. 696–716.

40. Engin, K., et al., Extracellular pH distribution in human tumours. *International Journal of Hyperthermia*, 1995. **11**(2): pp. 211–216.

41. Mura, S., J. Nicolas, and P. Couvreur, Stimuli-responsive nanocarriers for drug delivery. *Nature Materials*, 2013. **12**(11): 991–1003.

42. Gao, W., J. M. Chan, and O. C. Farokhzad, pH-responsive nanoparticles for drug delivery. *Molecular Pharmaceutics*, 2010. **7**(6): pp. 1913–1920.

43. Hoffman, A. S., Stimuli-responsive polymers: Biomedical applications and challenges for clinical translation. *Advanced Drug Delivery Reviews*, 2013. **65**(1): pp. 10–16.

44. Liu, J., et al., pH-sensitive nano-systems for drug delivery in cancer therapy. *Biotechnology Advances*, 2014. **32**(4): pp. 693–710.

45. Lee, E. S., K. Na, and Y. H. Bae, Doxorubicin loaded pH-sensitive polymeric micelles for reversal of resistant MCF-7 tumor. *Journal of Controlled Release*, 2005. **103**(2): pp. 405–418.

46. Yin, H., et al., Physicochemical characteristics of pH-sensitive poly(L-histidine)-b-poly (ethylene glycol)/poly (L-lactide)-b-poly (ethylene glycol) mixed micelles. *Journal of Controlled Release*, 2008. **126**(2): pp. 130–138.

47. Kim, D., et al., Doxorubicin-loaded polymeric micelle overcomes multidrug resistance of cancer by double-targeting folate receptor and early endosomal pH. *Small*, 2008. **4**(11): pp. 2043–2050.

48. Luo, S., et al., Amphiphilic block copolymers bearing six-membered ortho ester ring in side chains as potential drug carriers: Synthesis, characterization, and in vivo toxicity evaluation. *Journal of Biomaterials Science, Polymer Edition*, 2014. **25**(10): pp. 965–984.

49. Lavasanifar, A., J. Samuel, and G. S. Kwon, Poly (ethylene oxide)-block-poly (L-amino acid) micelles for drug delivery. *Advanced Drug Delivery Reviews*, 2002. **54**(2): pp. 169–190.

50. Xu, R., et al., An injectable nanoparticle generator enhances delivery of cancer therapeutics. *Nature Biotechnology*, 2016. pp. 414–418.

51. Rim, H. P., et al., pH-tunable calcium phosphate covered mesoporous silica nanocontainers for intracellular controlled release of guest drugs. *Angewandte Chemie International Edition*, 2011. **50**(38): pp. 8853–8857.

52. Rasmussen, J. W., et al., Zinc oxide nanoparticles for selective destruction of tumor cells and potential for drug delivery applications. *Expert Opinion on Drug Delivery*, 2010. **7**(9): pp. 1063–1077.

53. Yuan, L., et al., Preparation of pH-responsive mesoporous silica nanoparticles and their application in controlled drug delivery. *The Journal of Physical Chemistry C*, 2011. **115**(20): pp. 9926–9932.

54. Lu, Y., et al., Transformable liquid-metal nanomedicine. *Nature Communications*, 2015. **6**: pp. 10066–10075.
55. He, Q., et al., A pH-responsive mesoporous silica nanoparticles-based multi-drug delivery system for overcoming multi-drug resistance. *Biomaterials*, 2011. **32**(30): pp. 7711–7720.
56. Lu, Y., et al., Bioresponsive materials. *Nature Reviews Materials*, 2016. **1**: pp. 16075–16091.
57. Egeblad, M., and Z. Werb, New functions for the matrix metalloproteinases in cancer progression. *Nature Reviews Cancer*, 2002. **2**(3): pp. 161–174.
58. Vandenbroucke, R. E., and C. Libert, Is there new hope for therapeutic matrix metalloproteinase inhibition? *Nature Reviews Drug Discovery*, 2014. **13**(12): pp. 904–927.
59. Overall, C. M., and O. Kleifeld, Validating matrix metalloproteinases as drug targets and anti-targets for cancer therapy. *Nature Reviews Cancer*, 2006. **6**(3): pp. 227–239.
60. Zhu, L., et al., Enhanced anticancer activity of nanopreparation containing an MMP2-sensitive PEG-drug conjugate and cell-penetrating moiety. *Proceedings of the National Academy of Sciences of the United States of America*, 2013. **110**(42): pp. 17047–17052.
61. Tu, Y., and L. Zhu, Enhancing cancer targeting and anticancer activity by a stimulus-sensitive multifunctional polymer-drug conjugate. *Journal of Controlled Release*, 2015. **212**: pp. 94–102.
62. Yao, Q., et al., Building stable MMP2-responsive multifunctional polymeric micelles by an all-in-one polymer–lipid conjugate for tumor-targeted intracellular drug delivery. *ACS Applied Materials & Interfaces*, 2017. **9**(38): pp. 32520–32533.
63. Yao, Q., et al., Improving tumor specificity and anticancer activity of dasatinib by dual-targeted polymeric micelles. *ACS Applied Materials & Interfaces*, 2017. **9**(42): pp. 36642–36654.
64. Dai, Z., Q. Yao, and L. Zhu, MMP2-sensitive PEG-lipid copolymers: A new type of tumor-targeted P-glycoprotein inhibitor. *ACS Applied Materials & Interfaces*, 2016. **8**(20): pp. 12661–12673.
65. Jiang, T., et al., Tumor imaging by means of proteolytic activation of cell-penetrating peptides. *Proceedings of the National Academy of Sciences of the United States of America*, 2004. **101**(51): pp. 17867–17872.
66. Olson, E. S., et al., Activatable cell penetrating peptides linked to nanoparticles as dual probes for in vivo fluorescence and MR

imaging of proteases. *Proceedings of the National Academy of Sciences*, 2010. **107**(9): pp. 4311–4316.

67. Olson, E. S., et al., In vivo characterization of activatable cell penetrating peptides for targeting protease activity in cancer. *Integrative Biology*, 2009. **1**(5–6): pp. 382–393.
68. Ouyang, M., et al., Simultaneous visualization of protumorigenic Src and MT1-MMP activities with fluorescence resonance energy transfer. *Cancer Research*, 2010. **70**(6): pp. 2204–2212.
69. Aguilera, T. A., et al., Systemic in vivo distribution of activatable cell penetrating peptides is superior to that of cell penetrating peptides. *Integrative Biology*, 2009. **1**(5–6): pp. 371–381.
70. Li, J., Z. Ge, and S. Liu, PEG-sheddable polyplex micelles as smart gene carriers based on MMP-cleavable peptide-linked block copolymers. *Chemical Communications*, 2013. **49**(62): pp. 6974–6976.
71. Zhu, L., et al., Matrix metalloproteinase 2-sensitive multifunctional polymeric micelles for tumor-specific co-delivery of siRNA and hydrophobic drugs. *Biomaterials*, 2014. **35**(13): pp. 4213–4222.
72. Giustarini, D., et al., Glutathione, glutathione disulfide, and S-glutathionylated proteins in cell cultures. *Free Radical Biology and Medicine*, 2015. **89**: pp. 972–981.
73. Schafer, F. Q., and G. R. Buettner, Redox environment of the cell as viewed through the redox state of the glutathione disulfide/glutathione couple. *Free Radical Biology and Medicine*, 2001. **30**(11): pp. 1191–1212.
74. Estrela, J. M., et al., Glutathione in metastases: From mechanisms to clinical applications. *Critical Reviews in Clinical Laboratory Sciences*, 2016. **53**(4): pp. 253–267.
75. Traverso, N., et al., Role of glutathione in cancer progression and chemoresistance. *Oxidative Medicine and Cellular Longevity*, 2013. **2013**: p. 972913–972922.
76. Li, X., et al., Targeting mitochondrial reactive oxygen species as novel therapy for inflammatory diseases and cancers. *Journal of Hematology & Oncology*, 2013. **6**(1): pp. 1–19.
77. Lee, S. H., et al., Current progress in reactive oxygen species (ROS)-responsive materials for biomedical applications. *Advanced Healthcare Materials*, 2013. **2**(6): pp. 908–915.
78. Wen, X., et al., Deconvoluting the role of reactive oxygen species and autophagy in human diseases. *Free Radical Biology and Medicine*, 2013. **65**: pp. 402–410.

79. Brown, D. I., and K. K. Griendling, Regulation of signal transduction by reactive oxygen species in the cardiovascular system. *Circulation Research*, 2015. **116**(3): pp. 531–549.

80. Cheng, R., et al., Glutathione-responsive nano-vehicles as a promising platform for targeted intracellular drug and gene delivery. *Journal of Controlled Release*, 2011. **152**(1): pp. 2–12.

81. Huo, M., et al., Redox-responsive polymers for drug delivery: From molecular design to applications. *Polymer Chemistry*, 2014. **5**(5): pp. 1519–1528.

82. Meng, F., W. E. Hennink, and Z. Zhong, Reduction-sensitive polymers and bioconjugates for biomedical applications. *Biomaterials*, 2009. **30**(12): pp. 2180–2198.

83. Zhao, M., et al., Redox-responsive nanocapsules for intracellular protein delivery. *Biomaterials*, 2011. **32**(22): pp. 5223–5230.

84. Miyata, K., et al., Block catiomer polyplexes with regulated densities of charge and disulfide cross-linking directed to enhance gene expression. *Journal of the American Chemical Society*, 2004. **126**(8): pp. 2355–2361.

85. Meng, F., W. E. Hennink, and Z. Zhong, Reduction-sensitive polymers and bioconjugates for biomedical applications. *Biomaterials*, 2009. **30**(12): pp. 2180–2198.

86. Wen, H.-Y., et al., Rapidly disassembling nanomicelles with disulfide-linked PEG shells for glutathione-mediated intracellular drug delivery. *Chemical Communications*, 2011. **47**(12): pp. 3550–3552.

87. Ma, N., et al., Dual redox responsive assemblies formed from diselenide block copolymers. *Journal of the American Chemical Society*, 2009. **132**(2): pp. 442–443.

88. Yang, J., et al., Platinum (IV)-coordinate polymers as intracellular reduction-responsive backbone-type conjugates for cancer drug delivery. *Biomaterials*, 2011. **32**(34): pp. 9136–9143.

89. Levine, M. N., and R. T. Raines, Trimethyl lock: A trigger for molecular release in chemistry, biology, and pharmacology. *Chemical Science*, 2012. **3**(8): pp. 2412–2420.

90. Napoli, A., et al., Oxidation-responsive polymeric vesicles. *Nature Materials*, 2004. **3**(3): pp. 183–189.

91. Mathew, A. P., et al., Stimuli-regulated smart polymeric systems for gene therapy. *Polymers*, 2017. **9**(4): pp. 152–177.

92. Broaders, K. E., S. Grandhe, and J. M. Fréchet, A biocompatible oxidation-triggered carrier polymer with potential in therapeutics. *Journal of the American Chemical Society*, 2010. **133**(4): pp. 756–758.

93. Noh, J., et al., Amplification of oxidative stress by a dual stimuli-responsive hybrid drug enhances cancer cell death. *Nature Communications*, 2015. **6**: pp. 6907–6915.

94. Liu, X., et al., Fusogenic reactive oxygen species triggered charge-reversal vector for effective gene delivery. *Advanced Materials*, 2016. **28**(9): pp. 1743–1752.

95. Wang, M., et al., Reactive oxygen species-responsive protein modification and its intracellular delivery for targeted cancer therapy. *Angewandte Chemie*, 2014. **126**(49): pp. 13662–13666.

96. Hwang, J.-W., et al., Antioxidant activity and protective effect of anthocyanin oligomers on H_2O_2-triggered G2/M arrest in retinal cells. *Journal of Agricultural and Food Chemistry*, 2012. **60**(17): pp. 4282–4288.

97. Apel, K., and H. Hirt, Reactive oxygen species: Metabolism, oxidative stress, and signal transduction. *Annual Review of Plant Biology*, 2004. **55**: pp. 373–399.

98. Kim, H. A., R. I. Mahato, and M. Lee, Hypoxia-specific gene expression for ischemic disease gene therapy. *Advanced Drug Delivery Reviews*, 2009. **61**(7): pp. 614–622.

99. Harris, A. L., Hypoxia—a key regulatory factor in tumour growth. *Nature Reviews Cancer*, 2002. **2**(1): pp. 38–47.

100. Rhim, T., D. Y. Lee, and M. Lee, Hypoxia as a target for tissue specific gene therapy. *Journal of Controlled Release*, 2013. **172**(2): pp. 484–494.

101. Vaupel, P., and A. Mayer, Hypoxia in cancer: Significance and impact on clinical outcome. *Cancer and Metastasis Reviews*, 2007. **26**(2): pp. 225–239.

102. Brown, J. M., and W. R. Wilson, Exploiting tumour hypoxia in cancer treatment. *Nature Reviews Cancer*, 2004. **4**(6): pp. 437–447.

103. Sengupta, S., et al., Temporal targeting of tumour cells and neovasculature with a nanoscale delivery system. *Nature*, 2005. **436**(7050): pp. 568–572.

104. Wilson, W. R., and M. P. Hay, Targeting hypoxia in cancer therapy. *Nature Reviews Cancer*, 2011. **11**(6): pp. 393–410.

105. Thambi, T., J. H. Park, and D. S. Lee, Hypoxia-responsive nanocarriers for cancer imaging and therapy: Recent approaches and future perspectives. *Chemical Communications*, 2016. **52**(55): pp. 8492–8500.

106. Koppolu, B., et al., Temperature-sensitive polymer-coated magnetic nanoparticles as a potential drug delivery system for targeted

therapy of thyroid cancer. *Journal of Biomedical Nanotechnology*, 2012. **8**(6): pp. 983–990.

107. Sershen, S., et al., Temperature-sensitive polymer–nanoshell composites for photothermally modulated drug delivery. *Journal of Biomedical Materials Research Part A*, 2000. **51**(3): pp. 293–298.

108. Dromi, S., et al., Pulsed-high intensity focused ultrasound and low temperature–sensitive liposomes for enhanced targeted drug delivery and antitumor effect. *Clinical Cancer Research*, 2007. **13**(9): pp. 2722–2727.

109. Song, C. W., Effect of local hyperthermia on blood flow and microenvironment: A review. *Cancer Research*, 1984. **44**(10 Supplement): pp. 4721s–4730s.

110. Soni, G., and K. S. Yadav, High encapsulation efficiency of poloxamer-based injectable thermoresponsive hydrogels of etoposide. *Pharmaceutical Development and Technology*, 2014. **19**(6): pp. 651–661.

111. Yang, J., et al., A thermoresponsive biodegradable polymer with intrinsic antioxidant properties. *Biomacromolecules*, 2014. **15**(11): pp. 3942–3952.

112. Jeong, B., S. W. Kim, and Y. H. Bae, Thermosensitive sol–gel reversible hydrogels. *Advanced Drug Delivery Reviews*, 2012. **64**: pp. 154–162.

113. Cheng, C., et al., Thermo-triggered and biotinylated biotin-P (NIPAAm-co-HMAAm)-b-PMMA micelles for controlled drug release. *Journal of Biomedical Materials Research Part A*, 2009. **88**(3): pp. 814–822.

114. Wang, C., N. T. Flynn, and R. Langer, Controlled structure and properties of thermoresponsive nanoparticle–hydrogel composites. *Advanced Materials*, 2004. **16**(13): pp. 1074–1079.

115. Timko, B. P., et al., Near-infrared–actuated devices for remotely controlled drug delivery. *Proceedings of the National Academy of Sciences*, 2014. **111**(4): pp. 1349–1354.

116. Lindner, L. H., et al., Novel temperature-sensitive liposomes with prolonged circulation time. *Clinical Cancer Research*, 2004. **10**(6): pp. 2168–2178.

117. de Smet, M., et al., Temperature-sensitive liposomes for doxorubicin delivery under MRI guidance. *Journal of Controlled Release*, 2010. **143**(1): pp. 120–127.

118. Liu, S., Y. Tong, and Y.-Y. Yang, Incorporation and in vitro release of doxorubicin in thermally sensitive micelles made from

poly(N-isopropylacrylamide-co-N,N-dimethylacrylamide)-b-poly(D,L-lactide-co-glycolide) with varying compositions. *Biomaterials*, 2005. **26**(24): pp. 5064–5074.

119. Wei, H., et al., Thermo-sensitive polymeric micelles based on poly(N-isopropylacrylamide) as drug carriers. *Progress in Polymer Science*, 2009. **34**(9): pp. 893–910.
120. Kong, G., and M. Dewhirst, Review hyperthermia and liposomes. *International Journal of Hyperthermia*, 1999. **15**(5): pp. 345–370.
121. Lopez-Pinto, J., M. Gonzalez-Rodriguez, and A. Rabasco, Effect of cholesterol and ethanol on dermal delivery from DPPC liposomes. *International Journal of Pharmaceutics*, 2005. **298**(1): pp. 1–12.
122. Zhu, L., et al., Targeted delivery of methotrexate to skeletal muscular tissue by thermosensitive magnetoliposomes. *International Journal of Pharmaceutics*, 2009. **370**(1–2): pp. 136–143.
123. Nishimura, T., et al., A light sensitive self-assembled nanogel as a tecton for protein patterning materials. *Chemical Communications*, 2016. **52**(6): pp. 1222–1225.
124. Charati, M. B., et al., Light-sensitive polypeptide hydrogel and nanorod composites. *Small*, 2010. **6**(15): pp. 1608–1611.
125. Lee, K.-S., and M. A. El-Sayed, Gold and silver nanoparticles in sensing and imaging: Sensitivity of plasmon response to size, shape, and metal composition. *The Journal of Physical Chemistry B*, 2006. **110**(39): pp. 19220–19225.
126. Zhou, X., and J. Zhou, Improving the signal sensitivity and photostability of DNA hybridizations on microarrays by using dye-doped core-shell silica nanoparticles. *Analytical Chemistry*, 2004. **76**(18): pp. 5302–5312.
127. Agasti, S. S., et al., Photoregulated release of caged anticancer drugs from gold nanoparticles. *Journal of the American Chemical Society*, 2009. **131**(16): pp. 5728–5729.
128. Yan, Q., D. Han, and Y. Zhao, Main-chain photoresponsive polymers with controlled location of light-cleavable units: From synthetic strategies to structural engineering. *Polymer Chemistry*, 2013. **4**(19): pp. 5026–5037.
129. Rijcken, C., et al., Triggered destabilisation of polymeric micelles and vesicles by changing polymers polarity: An attractive tool for drug delivery. *Journal of Controlled Release*, 2007. **120**(3): pp. 131–148.
130. Zhao, Y., Light-responsive block copolymer micelles. *Macromolecules*, 2012. **45**(9): pp. 3647–3657.

131. Ghosh, H. N., J. B. Asbury, and T. Lian, Direct observation of ultrafast electron injection from coumarin 343 to TiO_2 nanoparticles by femtosecond infrared spectroscopy. *The Journal of Physical Chemistry B*, 1998. **102**(34): pp. 6482–6486.

132. Azagarsamy, M. A., and K. S. Anseth, Wavelength-controlled photocleavage for the orthogonal and sequential release of multiple proteins. *Angewandte Chemie International Edition*, 2013. **52**(51): pp. 13803–13807.

133. Plech, A., et al., Femtosecond laser near-field ablation from gold nanoparticles. *Nature Physics*, 2006. **2**(1): pp. 44–47.

134. Cavalleri, A., and M. Rini, Ultra-broadband femtosecond measurements of the photo-induced phase transition in VO_2: From the mid-IR to the hard X-rays. *Journal of the Physical Society of Japan*, 2006. **75**(1): pp. 011004–011004.

135. Paci, B., et al., Enhancement of photo/thermal stability of organic bulk heterojunction photovoltaic devices via gold nanoparticles doping of the active layer. *Nanoscale*, 2012. **4**(23): pp. 7452–7459.

136. Ntziachristos, V., C. Bremer, and R. Weissleder, Fluorescence imaging with near-infrared light: New technological advances that enable in vivo molecular imaging. *European Radiology*, 2003. **13**(1): pp. 195–208.

137. Sasaki, E., et al., Highly sensitive near-infrared fluorescent probes for nitric oxide and their application to isolated organs. *Journal of the American Chemical Society*, 2005. **127**(11): pp. 3684–3685.

138. Liu, G., W. Liu, and C.-M. Dong, UV-and NIR-responsive polymeric nanomedicines for on-demand drug delivery. *Polymer Chemistry*, 2013. **4**(12): pp. 3431–3443.

139. You, J., et al., Effective photothermal chemotherapy using doxorubicin-loaded gold nanospheres that target EphB4 receptors in tumors. *Cancer Research*, 2012. **72**(18): pp. 4777–4786.

140. Wahajuddin, S. A., Superparamagnetic iron oxide nanoparticles: Magnetic nanoplatforms as drug carriers. *International Journal of Nanomedicine*, 2012. **7**: pp. 3445–3471.

141. Torchilin, V., Multifunctional and stimuli-sensitive pharmaceutical nanocarriers. *European Journal of Pharmaceutics and Biopharmaceutics*, 2009. **71**(3): pp. 431–444.

142. Zhou, Z., et al., Iron/iron oxide core/shell nanoparticles for magnetic targeting MRI and near-infrared photothermal therapy. *Biomaterials*, 2014. **35**(26): pp. 7470–7478.

143. Jiang, S., et al., Lipidoid-coated iron oxide nanoparticles for efficient DNA and siRNA delivery. *Nano Letters*, 2013. **13**(3): pp. 1059–1064.

144. Hu, S. H., et al., Surfactant-free, lipo-polymersomes stabilized by iron oxide nanoparticles/polymer interlayer for synergistically targeted and magnetically guided gene delivery. *Advanced Healthcare Materials*, 2014. **3**(2): pp. 273–282.

145. Puddu, M., et al., Magnetically deliverable calcium phosphate nanoparticles for localized gene expression. *RSC Advances*, 2015. **5**(13): pp. 9997–10004.

146. Chung, H. J., et al., A magneto-DNA nanoparticle system for rapid detection and phenotyping of bacteria. *Nature Nanotechnology*, 2013. **8**(5): pp. 369–375.

147. Kumar, C. S., and F. Mohammad, Magnetic nanomaterials for hyperthermia-based therapy and controlled drug delivery. *Advanced Drug Delivery Reviews*, 2011. **63**(9): pp. 789–808.

148. Wu, J., and W. L. Nyborg, Ultrasound, cavitation bubbles and their interaction with cells. *Advanced Drug Delivery Reviews*, 2008. **60**(10): pp. 1103–1116.

149. Lentacker, I., et al., Understanding ultrasound induced sonoporation: Definitions and underlying mechanisms. *Advanced Drug Delivery Reviews*, 2014. **72**: pp. 49–64.

150. Ferrara, K., R. Pollard, and M. Borden, Ultrasound microbubble contrast agents: Fundamentals and application to gene and drug delivery. *Biomedical Engineering*, 2007. pp. 415–447.

151. Yin, T., et al., Tumor-penetrating codelivery of siRNA and paclitaxel with ultrasound-responsive nanobubbles hetero-assembled from polymeric micelles and liposomes. *Biomaterials*, 2014. **35**(22): pp. 5932–5943.

152. Xie, X., et al., Cell-penetrating peptide-siRNA conjugate loaded YSA-modified nanobubbles for ultrasound triggered siRNA delivery. *Colloids and Surfaces B: Biointerfaces*, 2015. **136**: pp. 641–650.

153. Cavalli, R., et al., New chitosan nanobubbles for ultrasound-mediated gene delivery: Preparation and in vitro characterization. *International Journal of Nanomedicine*, 2012. **7**: pp. 3309–3318.

154. Rapoport, N., Phase-shift, stimuli-responsive perfluorocarbon nanodroplets for drug delivery to cancer. *Wiley Interdisciplinary Reviews: Nanomedicine and Nanobiotechnology*, 2012. **4**(5): pp. 492–510.

155. Schroeder, A., J. Kost, and Y. Barenholz, Ultrasound, liposomes, and drug delivery: Principles for using ultrasound to control the release of drugs from liposomes. *Chemistry and Physics of Lipids*, 2009. **162**(1): pp. 1–16.

156. Rapoport, N. Y., et al., Controlled and targeted tumor chemotherapy by ultrasound-activated nanoemulsions/microbubbles. *Journal of Controlled Release*, 2009. **138**(3): pp. 268–276.

157. Zhang, H., et al., High intensity focused ultrasound-responsive release behavior of PLA-b-PEG copolymer micelles. *Journal of Controlled Release*, 2009. **139**(1): pp. 31–39.

158. Shim, W. S., et al., pH-and temperature-sensitive, injectable, biodegradable block copolymer hydrogels as carriers for paclitaxel. *International Journal of Pharmaceutics*, 2007. **331**(1): pp. 11–18.

159. Chen, S., et al., Temperature-responsive magnetite/PEO–PPO–PEO block copolymer nanoparticles for controlled drug targeting delivery. *Langmuir*, 2007. **23**(25): pp. 12669–12676.

160. Kono, K., et al., Highly temperature-sensitive liposomes based on a thermosensitive block copolymer for tumor-specific chemotherapy. *Biomaterials*, 2010. **31**(27): pp. 7096–7105.

161. Kono, K., et al., Multi-functional liposomes having temperature-triggered release and magnetic resonance imaging for tumor-specific chemotherapy. *Biomaterials*, 2011. **32**(5): pp. 1387–1395.

162. You, J., et al., Near-infrared light-sensitive liposomes for the enhanced photothermal tumor treatment by the combination with chemotherapy. *Pharmaceutical Research*, 2014. **31**(3): pp. 554–565.

163. Yuan, Q., et al., Photon-manipulated drug release from a mesoporous nanocontainer controlled by azobenzene-modified nucleic acid. *ACS Nano*, 2012. **6**(7): pp. 6337–6344.

164. Chang, Y. T., et al., Near-infrared light-responsive intracellular drug and siRNA release using Au nanoensembles with oligonucleotide-capped silica shell. *Advanced Materials*, 2012. **24**(25): pp. 3309–3314.

165. Ma, X., et al., A functionalized graphene oxide-iron oxide nano-composite for magnetically targeted drug delivery, photothermal therapy, and magnetic resonance imaging. *Nano Research*, 2012. **5**(3): pp. 199–212.

166. Lee, J. H., et al., On-demand drug release system for in vivo cancer treatment through self-assembled magnetic nanoparticles. *Angewandte Chemie*, 2013. **125**(16): pp. 4480–4484.

167. Lim, E. K., et al., pH-Triggered drug-releasing magnetic nanoparticles for cancer therapy guided by molecular imaging by MRI. *Advanced Materials*, 2011. **23**(21): pp. 2436–2442.

168. Paris, J. L., et al., Polymer-grafted mesoporous silica nanoparticles as ultrasound-responsive drug carriers. *ACS Nano*, 2015. **9**(11): pp. 11023–11033.

169. Paris, J. L., et al., Vectorization of ultrasound-responsive nanoparticles in placental mesenchymal stem cells for cancer therapy. *Nanoscale*, 2017. **9**(17): pp. 5528–5537.

170. Uesugi, Y., et al., An ultrasound-responsive nano delivery system of tissue-type plasminogen activator for thrombolytic therapy. *Journal of Controlled Release*, 2010. **147**(2): pp. 269–277.

171. Di, J., et al., Ultrasound-triggered regulation of blood glucose levels using injectable nano-network. *Advanced Healthcare Materials*, 2014. **3**(6): pp. 811–816.

172. Torchilin, V. P., Multifunctional, stimuli-sensitive nanoparticulate systems for drug delivery. *Nature Reviews Drug Discovery*, 2014. **13**(11): pp. 813–827.

173. Caldorera-Moore, M. E., W. B. Liechty, and N. A. Peppas, Responsive theranostic systems: Integration of diagnostic imaging agents and responsive controlled release drug delivery carriers. *Accounts of Chemical Research*, 2011. **44**(10): pp. 1061–1070.

Chapter 2

pH-Responsive Nanomedicine for Image-Guided Drug Delivery

Jong Hoon Choi,[a] Eunsoo Yoo,[a] Jung Hoon Kim,[b] and Dongin Kim[a]

[a]*Department of Pharmaceutical Sciences, College of Pharmacy, Texas A&M University, Reynolds Medical Building, Suite 159, Mail Stop 1114, College Station, Texas 77843-1114, USA*
[b]*Department of Radiation Oncology, Konyang University Hospital, Daejeon 302-701, Korea*

Dongin.Kim@pharmacy.tamhsc.edu

2.1 Introduction

Cancer is a second leading cause of death and current treatments which are applicable to cancers are still confined to chemotherapeutics followed by surgical debulking [1, 2]. There have been limited in the safety and efficacy of current anticancer therapeutics because of a narrow therapeutic window due to randomly distributed in the whole body [3]. This non-specific biodistribution may cause systemic toxicity to normal cells. Therefore, there is a great need to develop innovative systems

Stimuli-Responsive Nanomedicine
Edited by Lin Zhu

ISBN 978-981-4800-70-9 (Hardcover), 978-0-429-29529-4 (eBook)
www.jennystanford.com

to meet following conditions to overcome limitations: (1) noninvasive assessment of biodistribution of therapeutic agents, (2) quantification of localized accumulation and (3) monitoring of therapeutic efficacy [4, 5].

A real-time noninvasive imaging assessment of therapeutic response has the strong potential to control or suppress cancer because imaging property offers the quantification of nanomedicine at the intended disease site, thus the potential assurance of adequate treatment and elimination of undesirable delay leading to poor clinical outcomes. Currently, image-guided drug delivery system (IGDDS) has strong satisfaction for its request and there has been intensively researched to develop innovative IGDDS by using nanoparticles including organic or inorganic theranostic (therapy + diagnostic) systems [6–8]. Among them, liposomes, polymeric micelles, dendrimers, and carbon nanotubes represent major organic systems while gold nanoparticles, iron oxide nanoparticles, quantum dots, and silica nanoparticles are major inorganic systems [9–15]. The acidic microenvironment of cancer was discovered several decades ago and has been widely investigated in drug delivery. So far, the pH-sensitive nanoparticles have showed great potential in controlling cancer because they can specifically deliver the encapsulated agents into the tumor area without systemic release [16–18]. Combining these pH-sensitive advantages of nanoparticle with the theranostic property will have high chance to overcome cancer. In this chapter, we will introduce various examples of pH-responsive delivery systems and discuss their theranostic applications. Finally, we will address the summary and future direction of current pH-responsive theranostic systems.

2.2 pH-Sensitive Nanomedicine in Extracellular Region

The common state of pathology such as ischemia, infection, inflammation, and cancer involves acidic pH profiles [19–22]. Carcinogenic tissues have pH 5.7 for the least (average pH 6.8–7.0), and a normal blood pH 7.2–7.4 [23]. The average pH of the cancer region in various xenograft models showed pH 6.84 as shown in Fig. 2.1. This is because the formation of new blood vessels is

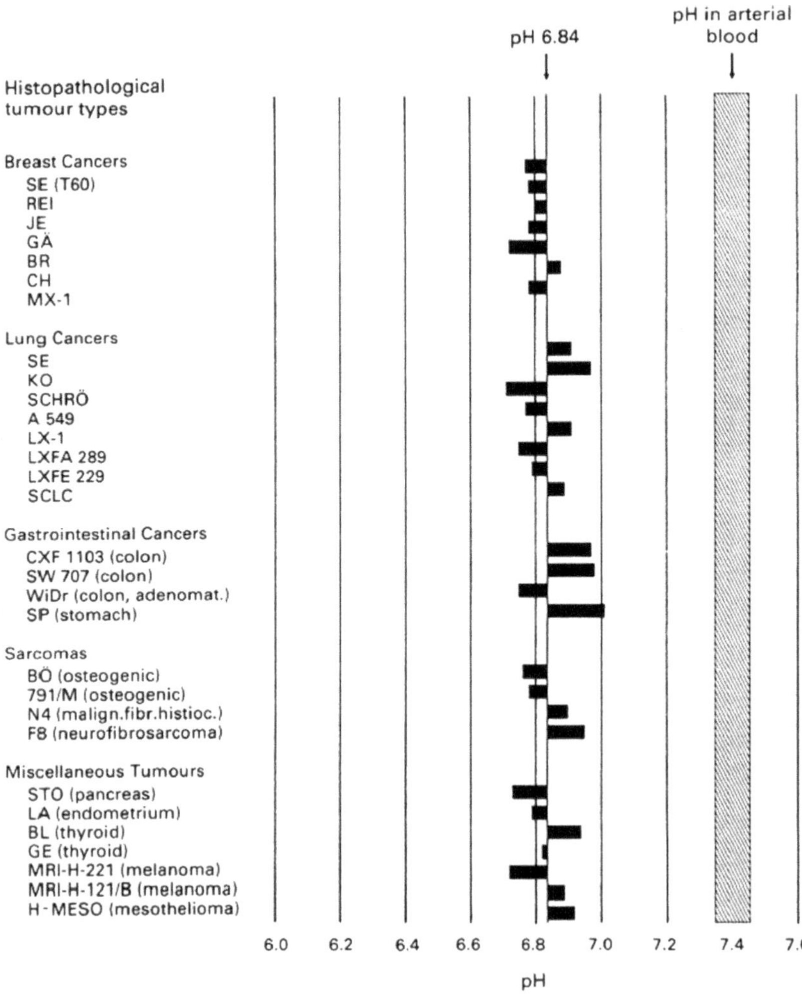

Figure 2.1 pH of cancer region in various xenograft models. Reprinted with permission from [19]. Copyright 1993 Nature Publishing Group.

often roughly organized, and the supply of nutrition and oxygen may not be sufficient to the growing population of cancer cells, that can result in an energy-deficient state of hypoxia, lactic acid production, and hydrolysis of ATP. Moreover, many tumors show the increase in their glycolysis that contributes to the increase of proton production. Many researchers have confirmed that many transporters in plasma membrane such as the sodium hydrogen exchanger (NHE-1), bicarbonate/chloride exchangers, and the monocarboxylate transporter (MCT) family [24–26] contributed to active extrusion of protons and weak acids. It is also significant to note that cell proliferation is initiated by cytoplasmic alkalization driven by the activated NHE-1 due to oncogenic signaling [27]. These transporters mentioned above and the transcriptional targets of HIF-1 are upregulated by hypoxic condition [28, 29]. Acidification is caused and strengthened by hypoxic condition and incomplete proton and acid clearance that happened due to a dysfunction of neovascularization in tumors. Contributing to the make-up of an acidic microenvironment, this deficient proton clearance provided a high interstitial pressure of tumors and short lymphatic drainage. Under such situation, the design of the drug delivery systems that are sensitive to tumoral pH is of great interest, to achieve the localized highly concentrated drugs in the tumor and make minimal unwanted effects.

Table 2.1 pH-sensitive polymers and chemical bonds

Charge reversal in acidic target region		**Acid-cleavable bond in acidic target region**	
Name	**Structure**	**Name**	**Structure**
PHis	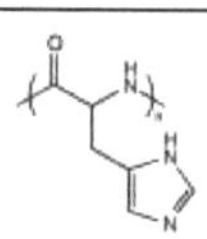	benzonic-imine	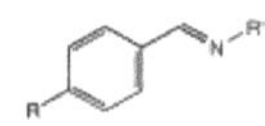
PSD	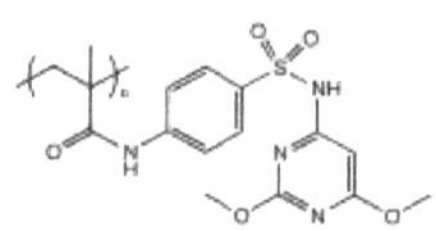	2,3-dimethyl-maleic amide	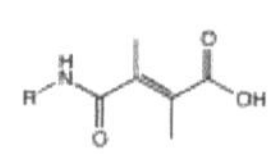
PDPA	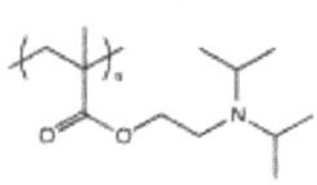	hydrazone	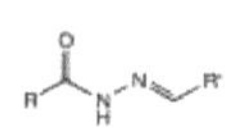
PVBA	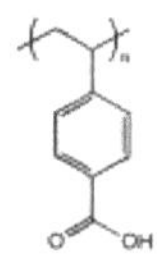	thiopropionate	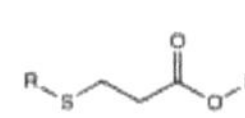
PMAA	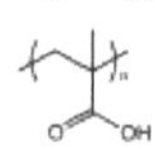	imine	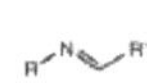
PGlu	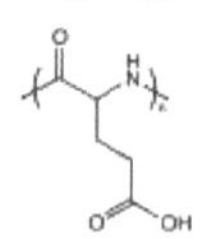	ortho ester	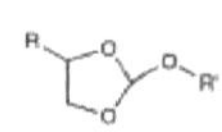

Charge reversal in acidic target region		Acid-cleavable bond in acidic target region	
Name	**Structure**	**Name**	**Structure**
PAsp	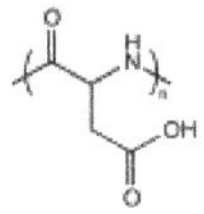	ketal	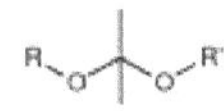
PAsp(DET)	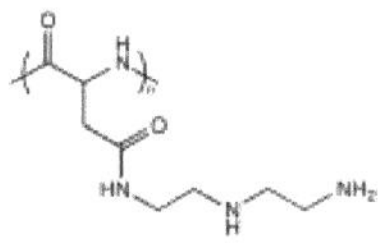	vinylether	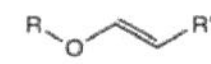
PAsp(DIP)	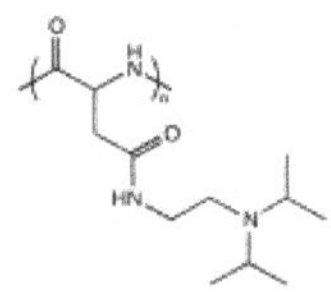	acetal	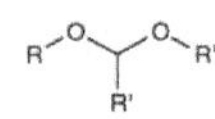
Polylysine	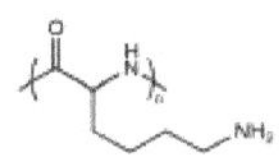	cyclic acetal	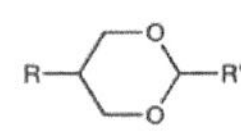
PVP	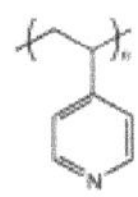	phosphoramidate	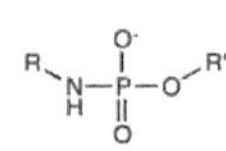

Fundamentally systematized by subtle pH gradient, pH-sensitive drug transport requires ultrasensitivity. Over the past decade, a variety of ultrasensitive pH-sensitive polymers have been researched to produce polymeric delivery carriers having a sensitivity to weakly acidic microenvironments. Table 2.1 represents pH-sensitive polymers and chemical bonds which are usually acid-cleavable and charge reversal in acidic condition.

A pKa of 6.15 is exhibited by the imidazole functionality in histidine. A pKa of 6.5 elevating to approximately 7.0 by linking to a hydrophilic polymer such as poly(ethylene) glycol (PEG) is exhibited by PHis homopolymers [30]. These pKa values of PHis pieces fit for constructing polymeric delivery carriers that are reactive to the acidic microenvironment of the extracellular region of the tumor. Figure 2.2 shows the pH gradients in the extracellular region of normal tissue and cancer. Bae et al. built pH-sensitive PEG-b-PHis polymeric micelles displaying the sensitivity to tumor tissues of slightly acidic pH. The protonating PHis cores could steadily disassemble the micelles designed at pH 8 when the pH was under 7.4. Because most solid tumors retain an acidic extracellular pH, the thorough optimization of the pH-responsive range of PEG-b-PHis micelles brought expectations for their application on solid tumor chemotherapy.

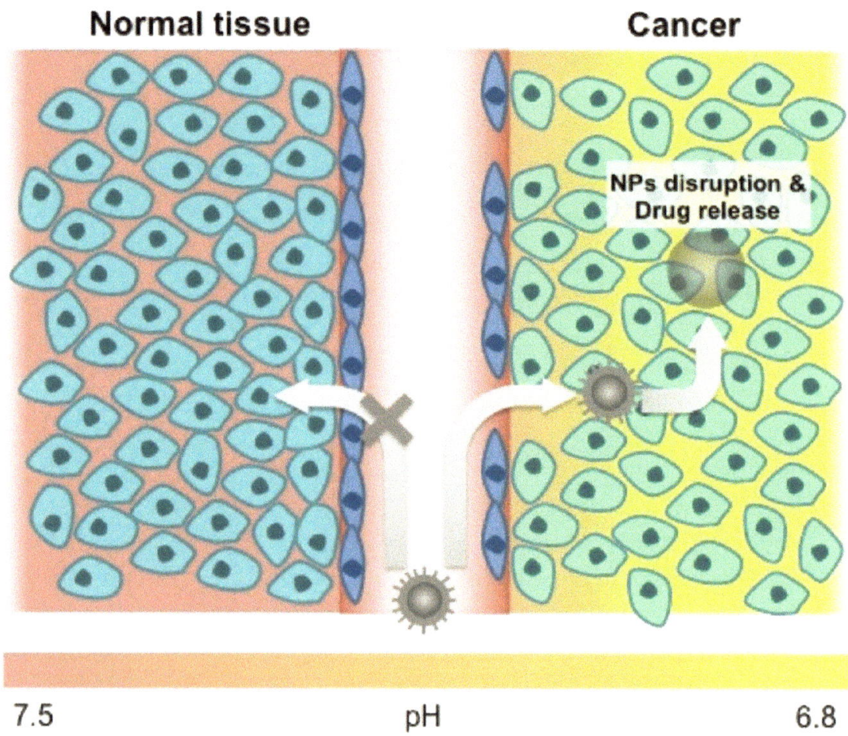

Figure 2.2 pH gradients in the extracellular region of the tumor and normal tissue.

Ding et al. recently presented N-Boc-histidine functionality on chain terminals of biocompatible and biodegradable PLGA-b-PEG-b-PLGA triblock copolymer [31]. Micelles built in aqueous media above the critical micelle concentration (CMC) exhibited a substantial difference in release profiles at pH 6.2 and pH 7.5 in vitro, while micelles of unmodified triblock copolymers had doxorubicin release characteristics presenting no sensitivity to pH.

Sulfonamide, a weak acid, easily loses a proton from its amide. The amide substitution groups substantially change the pKa of sulfonamide group within a range of 3–11 [32, 33]. Polysulfadimethoxine (PSD) is negatively charged at pH of 7.4 or neutral under pH of 7.0. Negatively charged biomacromolecules such as plasmid DNA can be conjugated with the cationic polyethyleneimine (PEI). Sequentially, they can be conjugated with pH-responsive PEG-b-PSD, a diblock copolymer, via dipole-dipole interactions [34]. PEI and PEG-b-PSD interacting electrostatically build shielding effect and leads to a low cytotoxicity and transfection of prepared nanoparticles at neutral pH. The removal of PEG-b-PSD at pH 6.6 allows high transfection efficiency due to the interaction between cationic PEI and cell membranes. The neutral charge of PSD in weakly acidic condition results in building TLP targeting, enhancing tumor localized movement of non-specific targeting ligands [35].

Conjugation of primary or secondary amines to diacrylates can produce poly (β-amino acid) (PAE) polymers [36]. PAE polymers that can attain pH-sensitive characteristics through their tertiary amine moieties may be used as vectors for gene delivery. In addition, the acidic microenvironment of the tumor (pH 6.8–7.2) can cause the micellization-disintegration shift of the PAE polymeric micelles [37]. Because of the pH-responsive micellar disintegration, the drug (DOX) uptake by tumor cells is elevated at pH of 6.4 compared to that at neutral pH. According to the results of in vivo anti-tumor study, the tumor growth was remarkably inhibited by the DOX-loaded micelles with the extended survival rates of tumor-bearing mice [38]. To enhance therapeutic efficiency, the hybrid micelles were prepared by using tumor-targeting AP peptide (CRKRLDRN) functionalized PEG-b-PLA block copolymer and pH-sensitive PEG-b-PAE block copolymer [39]. Co-delivery of black hole quencher (BHQ) and tetramethylrhodamine isothiocyanate (TRITC) by the PEG-b-PAE micelles also resulted in noninvasive in vivo imaging of the acidic tumor microenvironment [40]. Both in vivo and in vitro pH-sensitive imaging may be achieved through the adjustment of the critical pH range for fluorescence by changing the alkyl chain lengths. In another example, protoporphyrin IX was encapsulated in the pH-sensitive micelles to obtain simultaneous tumor imaging and tumor-targeted photodynamic therapy via tumoral pH-induced micellar disintegration [41].

A hydrophilic block and a methacrylate block that involves tertiary amine have been used to synthesize the block copolymers that exhibit tunable pH-sensitive micellization behavior. A series of ultrasensitive micellar nanoparticles were constructed to give off pH-responsive fluorescence in the pH range of pH 5.0–7.4. These nanoparticles could undergo a micellization-disintegration transition within a narrow pH range of 0.25 units [42]. Fluorescence quenching is attained by equal and associated contributions from the photo-induced electron transfer (PET) effect by deprotonation of tertiary amines in the micellar state and homo-FRET (self-quenching of various dyes within micellar cores). At mildly acidic pH, these two quenching agents diminishes by demicellarization leading to initiation of prominent emission. However, one limitation holds on this well-designed work that the originally nonfluorescent polymeric micelles will dissociate by high dilution (e.g., intravenous injection) into unimers at a concentration below the CMC, which initiates premature emission prior to reaching the pathological location. Moreover, complementary characteristics such as improved penetration depths and capability of 3D imaging may be provided by the application of pH-responsive fluorescence imaging with other responsive imaging modalities such as magnetic resonance imaging (MRI) [43–45]. It is also notable that despite well-documentation of biological responsiveness of small molecule-based MRI contrast agents in several reports, bio-responsive macromolecular MRI agents built upon polymeric assemblies or nanoparticles are far less researched [46–48]. The polymeric micelles that are core cross-linked (CCL) and labeled covalently with tetraxetan (Gd) and green-emitting fluorophores within pH-sensitive cores are built. This dual-modality magnetic resonance/fluorescence imaging system shows mildly acidic pH-responsiveness or enhanced signal intensities for both imaging modalities [49]. Both MR and fluorescence imaging have better performance with the CCL micelles than non-cross-linked diblock precursors because of their structural stability. Also, the dual-modality imaging system of selective enhancement MR/fluorescence imaging can be utilized in response to the intracellular microenvironments such as mildly acidic organelles.

Both imine and benzoic-imine bonds are subjects to hydrolysis under slightly acidic conditions. However, aliphatic imine bonds are unstable at physiological pH while benzoic-imine bonds are stable in both neutral and alkaline pH [50]. This benzoic-imine linkage containing PEG-b-C18 polymers are prepared by conjugating PEG with a terminal hydrophobic stearoyl moiety. The polymers have stability at physiological pH and cleavage upon decreasing pH to the pH characteristic of a tumor extracellular microenvironment. When the linkage is cleaved, the exposed primary amine moieties can significantly enhance cellular uptake. Therefore, the benzoic-imine motif is promisingly expected as the pH-sensitive materials for preparation of smart pH-responsive delivery systems.

To achieve the transformation of negatively charged carboxylate functionalities to positively charged primary amine ones, primary amines react with DMMA to form the amide bond containing polymers which have stability under neutral conditions and can be cleaved under slightly acidic pH conditions. Kataoka et al. initially proposed the strategy of charge conversion and prepared the charge conversion polymers (CCP) with citraconic amide [51]. Recently, Shen et al. compared anhydrides of varying chemical structures with pH-responsive cleavage feature of a series of synthesized primary amides and found that DMMA-based amide linkages dissociated most rapidly in the pH range of 6.0–7.0. To construct block copolymer micelles that attain a cascade of pH-sensitivity, Wang et al. synthesized two different types of pH-responsive linkages, hydrazine and DMMA-based amide linkages, with various stability towards acidity, leading to facilitated cellular internalization of nanoparticles and significant release of covalently conjugated DOX [52].

2.3 pH-Sensitive Nanomedicine in Intracellular Region

The delivery and release of drugs intracellularly are significant because most anti-cancer drugs are effective only within cells. The pH gradient exists between the extracellular environment (pH 7.4) and intracellular compartment such as the lysosomes and endosomes (pH 4.5–6.5) [53]. Figure 2.3 shows the pathways for drug release of pH-sensitive nanocarriers in the cancer cell. The pH-sensitive nanocarriers are endocytosed, followed by destabilization of endosome membrane, leading to drug release into cytoplasm. Since many drugs and drug carriers undergo endocytosis and trapped within lysosomes and endosomes, the assessment of pH gradient is important [54–56]. Endocytosis is described as internalization of macromolecules by cells into membrane-bound transport vesicles such as endosomes and lysosomes to create the following invagination and pocketing off of the plasma membrane [57]. The route of entry and involved proteins determine different fates of the internalized macromolecules. A good example is that internalized materials via clathrin-coated vesicles mature with acidification from early to late endosomes (pH 5.0–6.5). The materials will then be transferred to lysosomes, degradation terminal of the endocytic pathway, and also undergo several content mixing processes within late endosomes and lysosomes [58]. Retaining an internal acidic pH 4.5–5.0, lysosomes recruit numerous hydrolytic enzymes such as nucleases, proteases, phospholipases, esterases, and glycosidases to break down the entrapped molecules [59]. Thus, the drug delivery systems capable of exploiting the acidic pH of endosomes and lysosomes to avoid degradation and release considerable drug to the cytosol are valuable.

Several nanocarriers used for pH-sensitive drug delivery have been developed to function effectively within endosomes or lysosomes of cancer cells. Endosomal or lysosomal pH-sensitive polymer segments used in drug delivery and imaging can be categorized into two main classifications according to their pKa and chemical bonds cleavage, respectively, as demonstrated in Table 2.1. However, it is also noteworthy that it is not only tumor cells having acidic intracellular compartments for intracellular pH-responsive drug delivery but also normal cells. Therefore, the therapeutic effectiveness of tumor-targeted drug nanocarriers built upon pH-sensitive block copolymers considerably depends on their targeting capability towards tumor cells rather than normal cells.

The transitions between hydrophilic and hydrophobic states or between charged and neutral states are exhibited by block copolymers with a block having a pKa near or in the range of pH gradients when the pH varies from neutral to acidic pH in the endosomes or lysosomes of cancerous cells, leading to microstructural adjustment and reorganization. This can be applied to deliver anticancer drugs or imaging substances to the tumor. The double hydrophilic and amphiphilic pH-sensitive block copolymers are mainly discussed below.

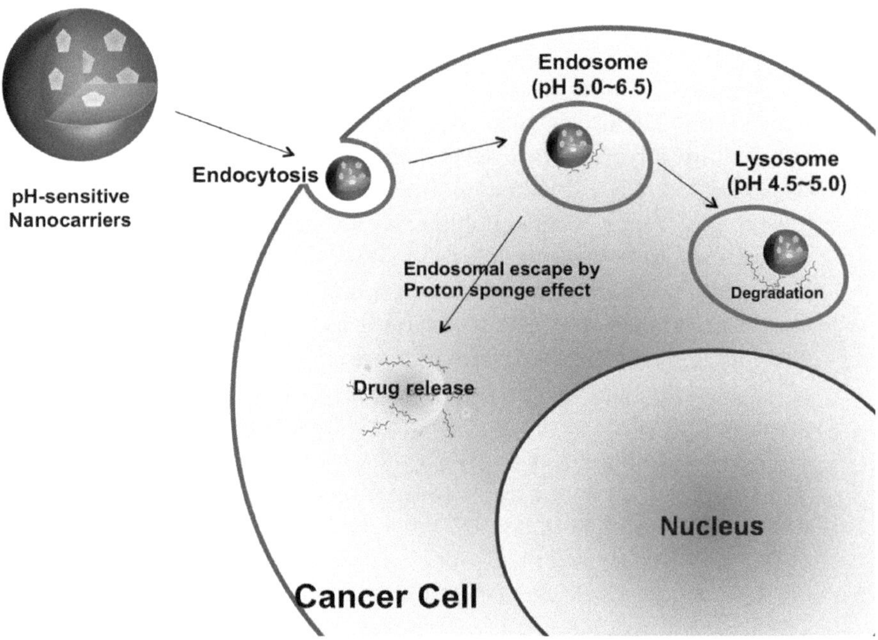

Figure 2.3 Schematic diagram of the working mechanism of pH-sensitive nanocarriers in the intracellular region.

Armes and Liu synthesized DHBCs (double hydrophilic block copolymers), that consisted of poly-(4-vinylbenzoic acid) (PVBA) and poly(2-N-(morpholine)ethyl methacrylate) (PMEMA), which exhibit "schizophrenic" micellization behavior due to the reversely modulated water solubility of one or both blocks by external stimuli [60, 61]. If the external conditions are properly met, they can self-assemble into micelles with varying microstructures in aqueous solution. Below pH 6, PVBA-core micelles were formed. However, above pH 6, the diblock copolymers could be dissolved as unimers. In alkaline media, the well-defined PMEMA-core micelles are formed when enough concentration of Na_2SO_4

is present. Aggregates of this zwitter-ionic diblock copolymer can be switched from PVBA-core micelles to PMEMA-core micelles by simply manipulating the pH of the solution. Furthermore, Armes et al. prepared poly(2-methacryloyloxyethyl phosphorylcholine)-b-poly(2-(diisopropylamino)ethyl methacrylate) block copolymers, PMPC-b-PDPA, that can spontaneously form vesicles through deprotonation of the PDPA block (pKa 6.3) when the solution pH increases [62]. They also used the vesicular assemblies to deliver the anticancer drug, DOX. During vesicle formation, DOX was encapsulated. Then, a retarded release was observed. At around neutral pH, DNA was also able to be encapsulated within the hydrophilic interior of vesicles formed. The vesicles release the encapsulated DNA as they dissolve completely into unimers at lower pH. Thus, these polymeric vesicles could successfully encapsulate GFP-encoding plasmid DNA and deliver intracellularly [63]. This new strategy allows better protection of DNA than the conventional electrostatic complexation approach because of its simple physical encapsulation nature.

When divalent metal ions such as Ca^{2+} (or Ba^{2+}, Sr^{2+}) are present, polyplex micelles can form from DHBCs, PEG-b-PMAA (PMAA: polymethacrylic acid) due to charge neutralization of the polyion block by metal ions [64]. After cross-linking of PEG-b-PMAA/Ca^{2+} micelles using 1,2-ethylenediamine, stable polymeric micelles with cross-linked ionic cores can be obtained. The ionization and swelling of cross-linked polyplex micelles can be actuated upon cellular entry. This feature is useful for the design of drug nanocarriers allowing for controlled loading and release of drugs. Moreover, polyplex micelles were also fabricated from PEG-b-PLys and anionic zinc porphyrin dendrimers, which can effectively accumulate in endosomes or lysosomes after endocytosis. Thereafter, the zinc porphyrin dendrimers can be promptly released into the cytoplasm upon photoirradiation. The polyplex micelles showed significantly higher in vivo photodynamic anticancer efficacy than clinically used Photofrin (poly(hematoporphyrin) esters, PHE) in the mouse models bearing lung adenocarcinoma cells [65].

Being high-efficiency gene delivery carriers, polyamine-containing water-soluble block copolymer based polyplex micelles have attracted great attention in the past decade. For gene delivery, endosomal escape is essential since the biotherapeutics such as plasmid DNA may undergo degradation or deactivation in endosomes/lysosomes. Termed as "proton sponge effect", polyamines when subjected to pH decrease in endosomes become protonated, acting as a pH buffer [66]. Destabilization of endosomes can be followed by the proton buffering effect in endosomes. For instance, Kataoka et al. designed an elegant ABC triblock copolymer, PEG-b-PMPA-b-PLys (PMPA: poly((3-morpholinopropyl)-aspartamine)), containing two blocks (PMPA and PLys) with different pKa values (6.2 and 9.4), respectively [67]. It is concluded that the PLys block can be employed to condense DNA at pH 7.4, while the PMPA block can disrupt endosomal membranes.

On the other hand, the design of block copolymers containing two blocks that have varying buffer capacities between pH 5.0 and pH 7.4 can be replaced by including only one block but with multiple pKa values in block copolymers. Thus, introducing a block exhibiting pH-responsive protonation in varied pH ranges, such as PEI, to the nanomaterials can also enhance their delivery efficiency [68, 69]. Diblock copolymers, PEG-b-PAsp(DET), contain PAsp(DET) segments with two pKa values of 9.1 and 6.3 owing to the presence of a distinctive gauche–anti conformational transition, which could induce endosomal disruption efficiently upon subjection to early endosomal pH. Furthermore, by minimizing the cumulative side toxicity due to polycations, the biodegradability of PAsp(DET) further facilitates safe and sustained gene expression [70].

Similarly, flower-type micelles formed from the triblock copolymer PVP-b-PEG-b-PCL were also explored for pH-responsive anticancer drug delivery, where PVP has a pKa of 5.0 [71]. Shuai et al. recently reported the diblock copolymer containing pH-sensitive segments, folate-PEG-b-poly(N-(N,N-diisopropyl-aminoethyl) aspartamide)-cholic acid (FA-PEG-b-PAsp(DIP)-CA) [72]. In aqueous solution, the diblock copolymer could self-assemble into micellar structures with CA cores, PAsp (DIP) inner shells, and PEG coronas. In this design, pH-responsive release under acidic lysosomal conditions could be observed in the inner shell, while negligible release at neutral pH was observed (e.g., in blood circulation). Folic acid moieties in micellar coronas can function for active tumor targeting. In addition to drug loading, negatively charged quantum dots can be embedded into micellar cores to make micelles fluorescent. The experiments showed very effective tumor suppression with PTX-loaded micelles.

Another series of pH-responsive block copolymers exhibit pH-triggered cleavage of chemical bonds in intracellular acidic endosomes or lysosomes. These linkages are relatively stable during blood circulation. Drug release, corona deshielding, cross-linker degradation, charge conversion, or the reversal of water solubility for one of the blocks can be led by pH-triggered cleavage of these covalent linkages within block copolymers and their gathering.

In acidic endosomal or lysosomal organelles (pH 4.5–6.5), prodrugs fabricated by conjugating anticancer drugs to the backbone of block copolymers through pH-cleavable linkages (e.g., hydrazone) can release parent drugs. (pH 4.5–6.5). Kataoka et al. first came up with this design by the conjugation of DOX onto polymer chains and synthesized PEG-b-P(Asp-Hyd-ADR) diblock copolymers [73]. Here (P(Asp-Hyd-ADR) refers to poly(aspartic acid hydrazone adriamycin)). pH-reactive polymeric micelles were fabricated through the self-assembly of PEG-b-P(Asp-Hyd-ADR), that were programmed to release DOX in endosomes or lysosomes and experience dynamic structural changes in pH variation responsively. These micelles do have several advantages including the capability of intracellular pH-triggered drug release, tumor penetration, and effective antitumor activity with little side toxicity. After further modification with folate (cancer cell targeting ligand) and fine adjustment of folate content on the micellar surface, the folate-conjugated micelles showed even lower in vivo side toxicity and higher antitumor activity, resulting in a broad therapeutic window from 7.50 to 26.21 mg/kg, which was 5-fold broader than that exhibited by the free drug.

Furthermore, Gong et al. conjugated PEG-b-P(Glu-Hyd-DOX) block copolymer onto the surface of hyperbranched polyester in order to synthesize structurally stable unimolecular micelles [74]. The targeted theranostic nanocarriers could be prepared by using cyclic RGD (cRGD) as a targeting moiety, and positron emission tomography (PET) isotope chelator as a PET imaging probe. To achieve both the tumor-targeted delivery and MRI contrast enhancement (based on SPIO nanoparticles), a hetero-functional amphiphilic triblock copolymer, FA-PEG114-b-P(Glu-Hyd-DOX)-b-PEG46-acrylate, was synthesized with the capability of forming the multifunctional pH-responsive polymeric nanocarriers [75]. The amphiphilic triblock copolymers can self-assemble into stable vesicles in aqueous solution. Longer FA-PEG114 segments were segregated into the outer hydrophilic layer providing active tumor targeting; shorter PEG46-acrylate segments were segregated into the inner hydrophilic layer, rendering the possibility of further core crosslinking via free radical polymerization of acrylate functionalities. Hydrophobic vesicle membranes were formed by conjugating DOX to the PGlu segment through hydrazone linkages, while hydrophilic SPIO nanoparticles were encapsulated into the aqueous core of stable vesicles to achieve sensitive MRI detection. These multifunctional SPIO/DOX-loaded vesicles provided excellent in vivo stability, pH-controlled drug release, as well as enhanced tumor MRI.

2.3.1 pH-Triggered Deshielding

For systemic administration, prolonged blood circulation and improved targeting are usually necessary, and surface shielding of nanocarriers with non-ionic and flexible hydrophilic polymers such as PEG and PHPMA is the most extensively employed approach which helps to avoid the clearance by the RES system, reduce interaction with proteins, and prevent aggregation. However, the endosomal escape capacity of these nanoparticles will be reduced by the incorporation of these motifs, which is very important for the intracellular delivery of some anticancer drugs and nucleic acids [76]. One solution for this issue is to design endosomal pH-triggered deshielding delivery systems. The neutral polymer segments can be used as the endosomal pH-cleavable chemical linkage in smart intracellular delivery system. Cyclic ortho ester, b-thiopropionate, hydrazone, and acetal, which are typical endosomal acid-labile chemical bonds, have been explored to achieve endosomal pH cleavable deshielding [77–80]. For example, Li and coworkers used a cyclic ortho ester linkage to prepare an acid-labile diblock copolymer (PEG-b-PDMAEMA), which could condense with pDNA to form polyplex nanoparticles with an acid-triggered reversible PEG shield. After 6 h-incubation at pH 5.0, in vitro transfection efficiency of the acid-labile copolymer greatly increased, approaching a level comparable to that of PDMAEMA. By assembling PLys and the lactosylated-PEG-b-siRNA conjugates bearing acid-labile b-thiopropionate linkages between PEG and siRNA, the pH-responsive polyplex micelles were formed. In a receptor-mediated manner, the polyplex micelles with clustered lactose moieties on the surface were transported into hepatoma cells, releasing siRNA into the cytosol responsively to acidification in endosomes. Using this micellar system, roughly 100-fold enhancement in the gene silencing activity compared to that of the free conjugate was achieved.

2.3.2 pH-Responsive Destabilization

One of the most important approaches to stabilize block copolymer micelles is core or shell crosslinking (CCL or SCL). In the micelles with the cross-linked core or shell as barrier, the loaded drugs are difficult to be released due to the high stability and crosslinked nature. As a solution for this difficulty, CCL or SCL micelles were prepared by integrating various pH-triggered degradable crosslinkers. For instance, the degradable bifunctional crosslinker from the living RAFT-end groups of PBA chains was introduced during the RAFT polymerization to synthesize the amphiphilic diblock copolymer, poly(2-hydroxyethyl acrylate)-b-poly(N-butyl acrylate) (PHEA-b-PBA), yielding a core-crosslinked micelle without significantly affecting the micellar size [81]. Compared to that under neutral pH conditions, hydrolysis of less than half of the crosslinkers in the micelles with high DOX loading capacities (60 wt%) was sufficient to release DOX at a much faster rate under acidic pH.

Additionally, Boyer and coworkers prepared amphiphilic block copolymers consisting of methanethiosulfonate (MTS) and pentafluorophenyl acrylate (PFP-A) pendant functionality. Different functional groups were introduced using the thiol/MTS exchange chemistry by reacting with thiol-containing compounds. Micelles with a diameter of 20 nm were yielded via self-assembly of the block copolymers in aqueous media. Upon addition of bifunctional amino compounds (crosslinkers) bearing an acid-cleavable linkage (ketal), activated esters in micellar cores were crosslinked to generate pH-sensitive CCL micelles. Using hydrophobic (Nile red) and hydrophilic (thiol-modified fluorescein isothiocyanate, FITC) dyes as probes, drug encapsulation and release were studied. Selective release of a single dye or the simultaneous release of both dyes in response to external stimuli could be achieved. A pH-responsive SCL micelle containing an acetal-containing crosslinker was prepared by Wooley et al. An UV-vis active probe was used within the pH-cleavable crosslinker in order to provide direct and sensitive detection of the crosslinker cleavage. Accelerated degradation could be obtained at lysosomal pH compared to that occurring at physiological pH [82].

Micellar dissociation or swelling and concomitant drug release can be achieved due to a pH-responsive solubility transition of the core-forming block when hydrophobic moieties are incorporated into the main chain of the core-forming segments bridged by acid-labile bonds. As the acid-labile functionalities, acetal and ortho ester groups are frequently explored. Linear-dendritic block copolymers consisting of PEG and PLys or polyester dendron on which highly acid-sensitive cyclic acetal linkages was used to link hydrophobic groups with the periphery were developed by Frechet et al. [83]. The copolymers were structured to form relatively stable micelles in aqueous solution at pH of 7.4, which could break into unimers following the loss of hydrophobic moieties upon acetal hydrolysis at acidic pH. The anticancer drug DOX was encapsulated into the cores of the micelles and exhibited cytotoxicity, comparable to that of the free drug, while empty micelles are nontoxic. The localization of DOX in intracellular organelles was in stark contrast to that exhibited by free DOX.

2.4 Classification of pH-Sensitive Nanomedicine

2.4.1 Polymer–Drug Conjugates

Polymer–drug conjugates consist of specific regions of the polymer backbone and drugs through covalent bond in order to facilitate both efficient delivery to the required intracellular site and availability within a specific period of time [84]. For pH-sensitive polymer–anticancer drug conjugates, various synthetic approaches have been adopted to include spacers such as acid-labile linkages between the polymer and drug in the polymer backbone or side-chain. pH-sensitive spacers located between polymer and drug allows polymer–drug conjugates to release drug in relatively acidic extracellular space either in a tumor, endosomes, or lysosomes after endocytosis as shown in Fig. 2.4 (polymer–drug conjugates).

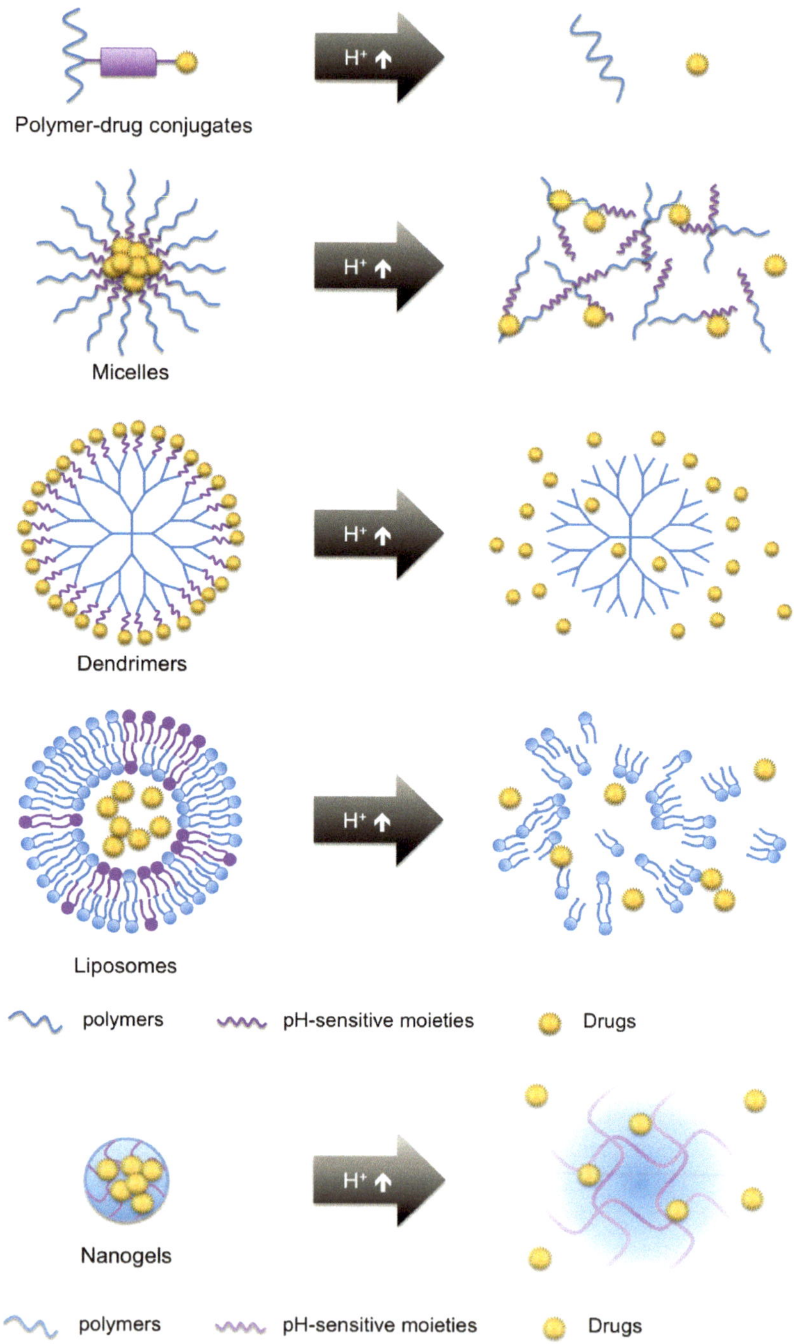

Figure 2.4 Schematic representation of pH-sensitive delivery systems.

2.4.2 Liposomes

Lipids are organic molecules which is consist of a hydrophilic head group and a hydrophobic tail region. Liposomes are classified based on their size and the number of lipid bilayers. Much like polymer micelles, liposomes were prepared via aggregation of double-layer structure of lipids in water. The hydrophobic tails packed in a core and the hydrophilic heads freely interacting with the surrounding water. Due to this feature, liposomes can be used as pharmaceutical carriers in a variety of applications such as controlled and targeted drugs/DNA/antigen delivery as shown in Fig. 2.4.

The components of liposome can be easily modified for drug releasing based on their response to certain environmental conditions such as pH and temperature. pH-sensitive liposomes are specifically designed to be destabilized and release the aqueous drugs in acidic condition of the endosomal system, even though these pH-sensitive liposomes are stable at physiological pH 7.4 as shown in Fig. 2.4 (Liposomes). The first generation of pH-sensitive liposomes contains dioleoylphosphatidylethanolamine (DOPE) which is the cone-shaped lipid.

2.4.3 Dendrimers

Dendrimers are highly branched, unique macromolecules with many arms radiating from an inner core [85]. They are highly monodispersed, water soluble, and have a high surface functionality; these make dendrimers attractive for drug delivery and other biomedical applications [86]. Due to their stepwise synthesis, they present highly regular star-like pattern, specific molecular weight, narrow size distribution, and well-defined number of peripheral groups. Poly(amidoamine) (PAMAM) dendrimers were the first to be synthesized and commercialized among the dendrimer family members. PAMAM dendrimers are synthesized initially from an alkyldiamine core, for example, ethylene diamine, which can react with methylacrylate monomers to produce a branched intermediate followed by transformation to the lowest generation of PAMAM dendrimers with $-NH_2$, -OH, or -COOH surface groups. The advantage of PAMAM dendrimers as drug carrier is that they can enhance the biodistribution of drugs via the EPR effect for targeting tumors [87]. The pH-responsive dendrimers can be fabricated by tailoring the surface functional groups of the surface of dendrimers. In addition, the anticancer drug can be attached via pH-sensitive linkages to form dendrimer-drug conjugates. The mechanism of drug release is the same as for polymer–drug conjugates. The linkages of pH-responsive dendrimers remain stable in the systemic circulation but are quickly hydrolyzed in an acidic microenvironment like the tumor tissue, thus releasing the incorporated drug inside the target cell, as shown in Fig. 2.4 (Dendrimers).

2.4.4 Polymeric Micelles

Polymeric micelles are spherical supramolecular aggregates ranging from 20 to 200 nm in size that has attracted considerable interest as potential drug delivery carriers due to their unique properties such as high solubility, high drug-loading capacity, and low toxicity [88]. Polymeric micelles are created by a spontaneous self-assembly of individual polymeric molecules that are synthetic amphiphilic di-block or tri-block copolymers comprised hydrophilic and hydrophobic blocks. Usually, micelles are spherically shaped structures with a hydrophobic core and hydrophilic shell. pH-responsive polymeric micellar systems are developed by using pH-sensitive linkages (hydrazone linkage) or attaching "titratable" groups such as amines or carboxylic acids to the block copolymers such that the micelle formation and stability can be controlled by the protonation of these groups. Particularly, at the tumoral acidic pH, sudden de-micellization of pH-responsive micelles engenders rapid drug release, but very slow release rate under physiological conditions (pH 7.4), as shown in Fig. 2.4 (Micelles).

2.4.5 Nanogels

The pH-responsive hydrogels are composed of cross-linked polyelectrolyte polymers with pendant acidic and/or basic functional groups in their backbones. These groups can accept or release protons in aqueous media, responding to different pH. The pendant groups are able to ionize and develop the fixed charges on the polymer network at the appropriate pH. The fixed charges will produce electrostatic repulsive forces which are responsible for pH-dependent swelling or de-swelling of the hydrogel, by which the drug release can be controlled [89]. Acidic (anionic) pendant groups of hydrogels can be ionized at a pH above the polymer's pKa and the ions produce a large osmotic swelling force for the hydrogels, as shown in Fig. 2.4 (nanogels). On the other hand, basic (cationic) pendant groups of hydrogels can be ionized (swollen) below the pKa. According to these approaches, the anticancer drug could be released rapidly in the acidic microenvironment of the endosomes/lysosomes or tumor tissues.

2.4.6 Others

Recently, pH-sensitive hybrid nanocarriers composed of organic and inorganic materials have been synthesized and explored as new platforms for diagnostic and therapeutic purposes. Mono dispersed hollow silica nanoparticles (SiO_2 HNPs) were introduced by Deng et al. Chitosan was used to conjugate with SiO_2 HNPs via the cross-linking reaction with (3-glycidyloxypropyl) trimethoxysilane [90]. The resulting SiO_2 HNPs that have a pH-sensitive polyelectrolyte layer are able to be further engineered by the TNF-α antibody in order to for targeted delivery of drugs to breast cancer cells.

2.5 Summary and Future Direction

Although the image-guided technique has shown a lot of promise in drug delivery and targeting, especially for its excellence in investigating the biodistribution and pharmacokinetics of therapeutics or their delivery systems, and in assisting the fulfillment of personalized medicine, the current stage of pH-sensitive IGDDS is very early. One of the challenges is the difficulty and complexity of integration of various functions within one system. In order to have ideal therapeutic effects, a pH-sensitive IGDDS should include efficient, specific delivery of theranostic agents without or negligible systemic toxicity. Moreover, the accurate measurement of disease characteristic/status and therapeutic outcome should be obtained in a noninvasive and spatiotemporal way. Specifically, pH-sensitive nanoparticles for IGDDS need to integrate the quantitative, noninvasive imaging and therapeutic agents within one entity with the sensitivity to the desirable pH. Thus, as a future direction, it is necessary to fully understand the pH status of the physiological and pathological conditions and basic principles in the design and clinical application of pH-sensitive theranostic nanoparticles. Many factors including biocompatibility, pharmacokinetics, targetability, and cost-effectiveness need to be optimized for advanced theranostic applications. In addition, the future directions of the pH-sensitive IGDDS will include its immunotherapeutic application by monitoring and manipulating immune components in the body. For example, it may be used for either imaging immune components or triggering or boosting immune components as an anticancer immunotherapy drug or both. It will be beneficial in minimizing the current chemotherapy issues such as systemic toxicity and drug resistance because of targeting immune cells.

In conclusion, the pH-sensitive nanomedicine will continue to grow and make its way from bench to bedside for specific, effective, and accurate diagnosis and therapy.

References

1. Kehoe, S., Hook, J., Nankivell, M., Jayson, G. C., Kitchener, H., Lopes, T., Luesley, D., Perren, T., Bannoo, S., Mascarenhas, M., Dobbs, S., Essapen, S., Twigg, J., Herod, J., McCluggage, G., Parmar, M., Swart, A.-M. (2015), Primary chemotherapy versus primary surgery for newly diagnosed advanced ovarian cancer (CHORUS): An open-label, randomised, controlled, non-inferiority trial, *The Lancet*, **386**, pp. 249–257.
2. Siegel, R. L., Miller, K. D., Jemal, A. (2015), Cancer statistics, *CA: A Cancer Journal for Clinicians*, **65**, pp. 5–29.
3. Panowski, S., Bhakta, S., Raab, H., Polakis, P., Junutula, J. R. (2014), Site-specific antibody drug conjugates for cancer therapy, *mAbs*, **6**, pp. 34–45.
4. Alqawlaq, S., Sivak, J. M., Huzil, J. T., Ivanova, M. V., Flanagan, J. G., Beazely, M. A., Foldvari, M. (2014), Preclinical development and ocular biodistribution of gemini-DNA nanoparticles after intravitreal and topical administration: Towards non-invasive glaucoma gene therapy, *Nanomedicine: Nanotechnology, Biology and Medicine*, **10**, pp. 1637–1647.

5. Rizzo, L. Y., Theek, B., Storm, G., Kiessling, F., Lammers, T. (2013), Recent progress in nanomedicine: Therapeutic, diagnostic and theranostic applications, *Current Opinion in Biotechnology*, **24**, pp. 1159–1166.
6. Chen, F., Hong, H., Zhang, Y., Valdovinos, H. F., Shi, S., Kwon, G. S., Theuer, C. P., Barnhart, T. E., Cai, W. (2013), In vivo tumor targeting and image-guided drug delivery with antibody-conjugated, radiolabeled mesoporous silica nanoparticles, *ACS Nano*, **7**, pp. 9027–9039.
7. Hollis, C. P., Weiss, H. L., Leggas, M., Evers, B. M., Gemeinhart, R. A., Li, T. (2013), Biodistribution and bioimaging studies of hybrid paclitaxel nanocrystals: Lessons learned of the EPR effect and image-guided drug delivery, *Journal of Controlled Release*, **172**, pp. 12–21.
8. Ranjan, A., Jacobs, G. C., Woods, D. L., Negussie, A. H., Partanen, A., Yarmolenko, P. S., Gacchina, C. E., Sharma, K. V., Frenkel, V., Wood, B. J., Dreher, M. R. (2012), Image-guided drug delivery with magnetic resonance guided high intensity focused ultrasound and temperature sensitive liposomes in a rabbit Vx2 tumor model, *Journal of Controlled Release*, **158**, pp. 487–494.
9. Chakravarty, R., Goel, S., Hong, H., Chen, F., Valdovinos, H. F., Hernandez, R., Barnhart, T. E., Cai, W. (2015), Hollow mesoporous silica nanoparticles for tumor vasculature targeting and PET image-guided drug delivery, *Nanomedicine*, **10**, pp. 1233–1246.
10. Guo, J., Hong, H., Chen, G., Shi, S., Zheng, Q., Zhang, Y., Theuer, C. P., Barnhart, T. E., Cai, W., Gong, S. (2013), Image-guided and tumor-targeted drug delivery with radiolabeled unimolecular micelles, *Biomaterials*, **34**, pp. 8323–8332.
11. Huang, J., Li, Y., Orza, A., Lu, Q., Guo, P., Wang, L., Yang, L., Mao, H. (2016), Magnetic nanoparticle facilitated drug delivery for cancer therapy with targeted and image-guided approaches, *Advanced Functional Materials*, **26**, pp. 3818–3836.
12. Langereis, S., Hijnen, N., Strijkers, G., Nicolay, K., Grüll, H. (2013), Research spotlight: Multifunctional liposomes for MRI and image-guided drug delivery, *Therapeutic Delivery*, **5**, pp. 21–24.
13. Lorenzato, C., Cernicanu, A., Meyre, M.-E., Germain, M., Pottier, A., Levy, L., de Senneville, B. D., Bos, C., Moonen, C., Smirnov, P. (2013), MRI contrast variation of thermosensitive magnetoliposomes triggered by focused ultrasound: A tool for image-guided local drug delivery, *Contrast Media & Molecular Imaging*, **8**, pp. 185–192.

14. Taratula, O., Schumann, C., Naleway, M. A., Pang, A. J., Chon, K. J., Taratula, O. (2013), A multifunctional theranostic platform based on phthalocyanine-loaded dendrimer for image-guided drug delivery and photodynamic therapy, *Molecular Pharmaceutics*, **10**, pp. 3946–3958.

15. Ye, F., Barrefelt, Å., Asem, H., Abedi-Valugerdi, M., El-Serafi, I., Saghafian, M., Abu-Salah, K., Alrokayan, S., Muhammed, M., Hassan, M. (2014), Biodegradable polymeric vesicles containing magnetic nanoparticles, quantum dots and anticancer drugs for drug delivery and imaging, *Biomaterials*, **35**, pp. 3885–3894.

16. Gerweck, L. E., Seetharaman, K. (1996), pH gradient in tumor versus normal tissue: Potential exploitation for the treatment of cancer, *Cancer Research*, **56**, pp. 1194–1198.

17. Kim, D., Lee, E. S., Park, K., Kwon, I. C., Bae, Y. H. (2008), Doxorubicin loaded pH-sensitive Micelle: Antitumoral efficacy against ovarian A2780/DOXR tumor, *Pharmaceutical Research*, **25**, pp. 2074–2082.

18. Lee, E. S., Gao, Z., Kim, D., Park, K., Kwon, I. C., Bae, Y. H. (2008), Super pH-sensitive multifunctional polymeric micelle for tumor pHe specific TAT exposure and multidrug resistance, *Journal of Controlled Release*, **129**, pp. 228–236.

19. Volk, T., Jähde, E., Fortmeyer, H. P., Glüsenkamp, K. H., Rajewsky, M. F. (1993), pH in human tumour xenografts: Effect of intravenous administration of glucose, *British Journal of Cancer*, **68**, pp. 492–500.

20. Ganta, S., Devalapally, H., Shahiwala, A., Amiji, M. (2008), A review of stimuli-responsive nanocarriers for drug and gene delivery, *Journal of Controlled Release*, **126**, pp. 187–204.

21. Gaucher, G., Dufresne, M.-H., Sant, V. P., Kang, N., Maysinger, D., Leroux, J.-C. (2005), Block copolymer micelles: Preparation, characterization and application in drug delivery, *Journal of Controlled Release*, **109**, pp. 169–188.

22. Stubbs, M., McSheehy, P. M. J., Griffiths, J. R., Bashford, C. L. (2000), Causes and consequences of tumour acidity and implications for treatment, *Molecular Medicine Today*, **6**, pp. 15–19.

23. Engin, K., Leeper, D. B., Cater, J. R., Thistlethwaite, A. J., Tupchong, L., McFarlane, J. D. (1995), Extracellular pH distribution in human tumours, *International Journal of Hyperthermia*, **11**, pp. 211–216.

24. Gillies, R. J., Raghunand, N., Karczmar, G. S., Bhujwalla, Z. M. (2002), MRI of the tumor microenvironment, *Journal of Magnetic Resonance Imaging*, **16**, pp. 430–450.

25. Griffiths, J. R. (1991), Are cancer cells acidic?, *British Journal of Cancer*, **64**, pp. 425–427.

26. Madshus, I. H. (1988), Regulation of intracellular pH in eukaryotic cells, *Biochemical Journal*, **250**, pp. 1–8.

27. Cardone, R. A., Casavola, V., Reshkin, S. J. (2005), The role of disturbed pH dynamics and the Na^+/H^+ exchanger in metastasis, *Nature Reviews Cancer*, **5**, pp. 786–795.

28. Parks, S. K., Chiche, J., Pouyssegur, J. (2013), Disrupting proton dynamics and energy metabolism for cancer therapy, *Nature Reviews Cancer*, **13**, pp. 611–623.

29. Pouyssegur, J., Dayan, F., Mazure, N. M. (2006), Hypoxia signalling in cancer and approaches to enforce tumour regression, *Nature*, **441**, pp. 437–443.

30. Lee, E. S., Shin, H. J., Na, K., Bae, Y. H. (2003), Poly(l-histidine) PEG block copolymer micelles and pH-induced destabilization, *Journal of Controlled Release*, **90**, pp. 363–374.

31. Chang, G., Li, C., Lu, W., Ding, J. (2010), N-boc-histidine-capped PLGA-PEG-PLGA as a smart polymer for drug delivery sensitive to tumor extracellular pH, *Macromolecular Bioscience*, **10**, pp. 1248–1256.

32. Kang, S. I., Bae, Y. H. (2002), pH-Induced solubility transition of sulfonamide-based polymers, *Journal of Controlled Release*, **80**, pp. 145–155.

33. Park, S. Y., Bae, Y. H. (1999), Novel pH-sensitive polymers containing sulfonamide groups, *Macromolecular Rapid Communications*, **20**, pp. 269–273.

34. Sethuraman, V. A., Na, K., Bae, Y. H. (2006), pH-responsive sulfonamide/PEI system for tumor specific gene delivery: An in vitro study, *Biomacromolecules*, **7**, pp. 64–70.

35. Sethuraman, V. A., Lee, M. C., Bae, Y. H. (2008), A biodegradable pH-sensitive micelle system for targeting acidic solid tumors, *Pharmaceutical Research*, **25**, pp. 657–666.

36. Lynn, D. M., Langer, R. (2000), Degradable poly(β-amino esters): Synthesis, characterization, and self-assembly with plasmid DNA, *Journal of the American Chemical Society*, **122**, pp. 10761–10768.

37. Kim, M. S., Hwang, S. J., Han, J. K., Choi, E. K., Park, H. J., Kim, J. S., Lee, D. S. (2006), pH-responsive PEG-Poly(β-amino ester) block copolymer micelles with a sharp transition, *Macromolecular Rapid Communications*, **27**, pp. 447–451.

38. Ko, J., Park, K., Kim, Y.-S., Kim, M. S., Han, J. K., Kim, K., Park, R.-W., Kim, I.-S., Song, H. K., Lee, D. S., Kwon, I. C. (2007), Tumoral acidic extracellular pH targeting of pH-responsive MPEG-poly(β-amino ester) block copolymer micelles for cancer therapy, *Journal of Controlled Release*, **123**, pp. 109–115.
39. Wu, X. L., Kim, J. H., Koo, H., Bae, S. M., Shin, H., Kim, M. S., Lee, B.-H., Park, R.-W., Kim, I.-S., Choi, K., Kwon, I. C., Kim, K., Lee, D. S. (2010), Tumor-targeting peptide conjugated pH-responsive micelles as a potential drug carrier for cancer therapy, *Bioconjugate Chemistry*, **21**, pp. 208–213.
40. Ko, J. Y., Park, S., Lee, H., Koo, H., Kim, M. S., Choi, K., Kwon, I. C., Jeong, S. Y., Kim, K., Lee, D. S. (2010), pH-sensitive nanoflash for tumoral acidic pH imaging in live animals, *Small*, **6**, pp. 2539–2544.
41. Koo, H., Lee, H., Lee, S., Min, K. H., Kim, M. S., Lee, D. S., Choi, Y., Kwon, I. C., Kim, K., Jeong, S. Y. (2010), In vivo tumor diagnosis and photodynamic therapy via tumoral pH-responsive polymeric micelles, *Chemical Communications*, **46**, pp. 5668–5670.
42. Zhou, Z., Shen, Y., Tang, J., Fan, M., Van Kirk, E. A., Murdoch, W. J., Radosz, M. (2009), Charge-reversal drug conjugate for targeted cancer cell nuclear drug delivery, *Advanced Functional Materials*, **19**, pp. 3580–3589.
43. Choi, K. Y., Liu, G., Lee, S., Chen, X. (2012), Theranostic nanoplatforms for simultaneous cancer imaging and therapy: Current approaches and future perspectives, *Nanoscale*, **4**, pp. 330–342.
44. Khemtong, C., Kessinger, C. W., Gao, J. (2009), Polymeric nanomedicine for cancer MR imaging and drug delivery, *Chemical Communications*, **28**, pp. 3497–3510.
45. Kievit, F. M., Zhang, M. (2011), Cancer nanotheranostics: Improving imaging and therapy by targeted delivery across biological barriers, *Advanced Materials*, **23**, pp. H217–H247.
46. Barrett, T., Kobayashi, H., Brechbiel, M., Choyke, P. L. (2006), Macromolecular MRI contrast agents for imaging tumor angiogenesis, *European Journal of Radiology*, **60**, pp. 353–366.
47. Moats, R. A., Fraser, S. E., Meade, T. J. (1997), A "smart" magnetic resonance imaging agent that reports on specific enzymatic activity, *Angewandte Chemie International Edition in English*, **36**, pp. 726–728.
48. Tang, J., Sheng, Y., Hu, H., Shen, Y. (2013), Macromolecular MRI contrast agents: Structures, properties and applications, *Progress in Polymer Science*, **38**, pp. 462–502.

49. Hu, J., Liu, T., Zhang, G., Jin, F., Liu, S. (2013), Synergistically enhance magnetic resonance/fluorescence imaging performance of responsive polymeric nanoparticles under mildly acidic biological milieu, *Macromolecular Rapid Communications*, **34**, pp. 749–758.

50. Ding, C., Gu, J., Qu, X., Yang, Z. (2009), Preparation of multifunctional drug carrier for tumor-specific uptake and enhanced intracellular delivery through the conjugation of weak acid labile linker, *Bioconjugate Chemistry*, **20**, pp. 1163–1170.

51. Lee, Y., Fukushima, S., Bae, Y., Hiki, S., Ishii, T., Kataoka, K. (2007), A protein nanocarrier from charge-conversion polymer in response to endosomal pH, *Journal of the American Chemical Society*, **129**, pp. 5362–5363.

52. Du, J.-Z., Du, X.-J., Mao, C.-Q., Wang, J. (2011), Tailor-made dual pH-sensitive polymer doxorubicin nanoparticles for efficient anticancer drug delivery, *Journal of the American Chemical Society*, **133**, pp. 17560–17563.

53. Mellman, I., Fuchs, R., Helenius, A. (1986), Acidification of the endocytic and exocytic pathways, *Annual Review of Biochemistry*, **55**, pp. 663–700.

54. Bareford, L. M., Swaan, P. W. (2007), Endocytic mechanisms for targeted drug delivery, *Advanced Drug Delivery Reviews*, **59**, pp. 748–758.

55. Iversen, T.-G., Skotland, T., Sandvig, K. (2011), Endocytosis and intracellular transport of nanoparticles: Present knowledge and need for future studies, *Nano Today*, **6**, pp. 176–185.

56. Sahay, G., Alakhova, D. Y., Kabanov, A. V. (2010), Endocytosis of nanomedicines, *Journal of Controlled Release*, **145**, pp. 182–195.

57. Conner, S. D., Schmid, S. L. (2003), Regulated portals of entry into the cell, *Nature*, **422**, pp. 37–44.

58. Luzio, J. P., Rous, B. A., Bright, N. A., Pryor, P. R., Mullock, B. M., Piper, R. C. (2000), Lysosome-endosome fusion and lysosome biogenesis, *Journal of Cell Science*, **113**, pp. 1515–1524.

59. Kornfeld, S., Mellman, I. (1989), The biogenesis of lysosomes, *Annual Review of Cell and Developmental Biology*, **5**, pp. 483–525.

60. Cölfen, H. (2001), Double-hydrophilic block copolymers: Synthesis and application as novel surfactants and crystal growth modifiers, *Macromolecular Rapid Communications*, **22**, pp. 219–252.

61. Liu, S., Armes, S. P. (2003), Synthesis and aqueous solution behavior of a pH-responsive schizophrenic diblock copolymer, *Langmuir*, **19**, pp. 4432–4438.

62. Du, J., Tang, Y., Lewis, A. L., Armes, S. P. (2005), pH-sensitive vesicles based on a biocompatible zwitterionic diblock copolymer, *Journal of the American Chemical Society*, **127**, pp. 17982–17983.

63. Lomas, H., Canton, I., MacNeil, S., Du, J., Armes, S. P., Ryan, A. J., Lewis, A. L., Battaglia, G. (2007), Biomimetic pH sensitive polymersomes for efficient DNA encapsulation and delivery, *Advanced Materials*, **19**, pp. 4238–4243.

64. Bronich, T. K., Keifer, P. A., Shlyakhtenko, L. S., Kabanov, A. V. (2005), Polymer micelle with cross-linked ionic core, *Journal of the American Chemical Society*, **127**, pp. 8236–8237.

65. Nishiyama, N., Nakagishi, Y., Morimoto, Y., Lai, P.-S., Miyazaki, K., Urano, K., Horie, S., Kumagai, M., Fukushima, S., Cheng, Y., Jang, W.-D., Kikuchi, M., Kataoka, K. (2009), Enhanced photodynamic cancer treatment by supramolecular nanocarriers charged with dendrimer phthalocyanine, *Journal of Controlled Release*, **133**, pp. 245–251.

66. Mintzer, M. A., Simanek, E. E. (2009), Nonviral vectors for gene delivery, *Chemical Reviews*, **109**, pp. 259–302.

67. Fukushima, S., Miyata, K., Nishiyama, N., Kanayama, N., Yamasaki, Y., Kataoka, K. (2005), PEGylated polyplex micelles from triblock catiomers with spatially ordered layering of condensed pDNA and buffering units for enhanced intracellular gene delivery, *Journal of the American Chemical Society*, **127**, pp. 2810–2811.

68. Boussif, O., Lezoualc'h, F., Zanta, M. A., Mergny, M. D., Scherman, D., Demeneix, B., Behr, J. P. (1995), A versatile vector for gene and oligonucleotide transfer into cells in culture and in vivo: Polyethylenimine, *Proceedings of the National Academy of Sciences*, **92**, pp. 7297–7301.

69. Ziebarth, J. D., Wang, Y. (2010), Understanding the protonation behavior of linear polyethylenimine in solutions through Monte Carlo simulations, *Biomacromolecules*, **11**, pp. 29–38.

70. Itaka, K., Ishii, T., Hasegawa, Y., Kataoka, K. (2010), Biodegradable polyamino acid-based polycations as safe and effective gene carrier minimizing cumulative toxicity, *Biomaterials*, **31**, pp. 3707–3714.

71. Van Butsele, K., Cajot, S., Van Vlierberghe, S., Dubruel, P., Passirani, C., Benoit, J.-P., Jérôme, R., Jérôme, C. (2009), pH-responsive flower-type micelles formed by a biotinylated poly(2-vinylpyridine)-block-poly(ethylene oxide)-block-poly(ε-caprolactone) triblock copolymer, *Advanced Functional Materials*, **19**, pp. 1416–1425.

72. Wang, W., Cheng, D., Gong, F., Miao, X., Shuai, X. (2012), Design of multifunctional micelle for tumor-targeted intracellular drug release and fluorescent imaging, *Advanced Materials*, **24**, pp. 115–120.

73. Bae, Y., Fukushima, S., Harada, A., Kataoka, K. (2003), Design of environment-sensitive supramolecular assemblies for intracellular drug delivery: Polymeric micelles that are responsive to intracellular pH change, *Angewandte Chemie International Edition*, **42**, pp. 4640–4643.

74. Xiao, Y., Hong, H., Javadi, A., Engle, J. W., Xu, W., Yang, Y., Zhang, Y., Barnhart, T. E., Cai, W., Gong, S. (2012), Multifunctional unimolecular micelles for cancer-targeted drug delivery and positron emission tomography imaging, *Biomaterials*, **33**, pp. 3071–3082.

75. Yang, X., Grailer, J. J., Rowland, I. J., Javadi, A., Hurley, S. A., Matson, V. Z., Steeber, D. A., Gong, S. (2010), Multifunctional stable and pH-responsive polymer vesicles formed by heterofunctional triblock copolymer for targeted anticancer drug delivery and ultrasensitive MR imaging, *ACS Nano*, **4**, pp. 6805–6817.

76. Otsuka, H., Nagasaki, Y., Kataoka, K. (2003), PEGylated nanoparticles for biological and pharmaceutical applications, *Advanced Drug Delivery Reviews*, **55**, pp. 403–419.

77. Knorr, V., Allmendinger, L., Walker, G. F., Paintner, F. F., Wagner, E. (2007), An acetal-based PEGylation reagent for pH-sensitive shielding of DNA polyplexes, *Bioconjugate Chemistry*, **18**, pp. 1218–1225.

78. Lin, S., Du, F., Wang, Y., Ji, S., Liang, D., Yu, L., Li, Z. (2008), An acid-labile block copolymer of PDMAEMA and PEG as potential carrier for intelligent gene delivery systems, *Biomacromolecules*, **9**, pp. 109–115.

79. Oishi, M., Sasaki, S., Nagasaki, Y., Kataoka, K. (2003), pH-responsive oligodeoxynucleotide (ODN)–poly(ethylene glycol) conjugate through acid-labile β-thiopropionate linkage: Preparation and polyion complex micelle formation, *Biomacromolecules*, **4**, pp. 1426–1432.

80. Walker, G. F., Fella, C., Pelisek, J., Fahrmeir, J., Boeckle, S., Ogris, M., Wagner, E. (2005), Toward synthetic viruses: Endosomal pH-triggered deshielding of targeted polyplexes greatly enhances gene transfer in vitro and in vivo, *Molecular Therapy*, **11**, pp. 418–425.

81. Chan, Y., Wong, T., Byrne, F., Kavallaris, M., Bulmus, V. (2008), Acid-labile core cross-linked micelles for pH-triggered release of antitumor drugs, *Biomacromolecules*, **9**, pp. 1826–1836.

82. Duong, H. T. T., Marquis, C. P., Whittaker, M., Davis, T. P., Boyer, C. (2011), Acid degradable and biocompatible polymeric nanoparticles for

the potential codelivery of therapeutic agents, *Macromolecules*, **44**, pp. 8008–8019.

83. Gillies, E. R., Jonsson, T. B., Fréchet, J. M. J. (2004), Stimuli-responsive supramolecular assemblies of linear-dendritic copolymers, *Journal of the American Chemical Society*, **126**, pp. 11936–11943.
84. Duncan, R. (2006), Polymer conjugates as anticancer nanomedicines, *Nature Reviews Cancer*, **6**, pp. 688–701.
85. Tomalia, D. A., Fréchet, J. M. J. (2002), in *Dendrimers and Other Dendritic Polymers*, John Wiley & Sons, Ltd, pp. 1–44.
86. Esfand, R., Tomalia, D. A. (2001), Poly(amidoamine) (PAMAM) dendrimers: From biomimicry to drug delivery and biomedical applications, *Drug Discovery Today*, **6**, pp. 427–436.
87. Sato, N., Kobayashi, H., Hiraga, A., Saga, T., Togashi, K., Konishi, J., Brechbiel, M. W. (2001), Pharmacokinetics and enhancement patterns of macromolecular MR contrast agents with various sizes of polyamidoamine dendrimer cores, *Magnetic Resonance in Medicine*, **46**, pp. 1169–1173.
88. Kataoka, K., Harada, A., Nagasaki, Y. (2001), Block copolymer micelles for drug delivery: Design, characterization and biological significance, *Advanced Drug Delivery Reviews*, **47**, pp. 113–131.
89. Kost, J., Lapidot, S. A., Kumar, N., Chaubal, M., Domb, A. J., Kumar, R., Majeti, N. V. (2000), in *Kirk-Othmer Encyclopedia of Chemical Technology*, John Wiley & Sons, Inc.
90. Deng, Z., Zhen, Z., Hu, X., Wu, S., Xu, Z., Chu, P. K. (2011), Hollow chitosan silica nanospheres as pH-sensitive targeted delivery carriers in breast cancer therapy, *Biomaterials*, **32**, pp. 4976–4986.

Chapter 3

Enzyme-Responsive Nanomedicine

Hong Wu, SongYan Guo, and Tie Hong Yang

School of Pharmacy, Air Force Military Medical University,
169 Changle West Road, Xi'an, Shaanxi 710032, China

wuhong@fmmu.edu.cn

3.1 Introduction

Responsive nanomaterials have made tremendous growth in the fundamental research and applications in diverse fields, including drug and gene delivery, bioimaging, biosensor, diagnostics, and tissue engineering. Among a variety of internal and external stimuli such as pH, temperature, light, ultrasound, and magnetic fields that have been reported in biomedical sciences, the disease-associated enzymes have recently emerged to be a promising trigger for the design of responsive nanomaterials [1].

Enzymes are highly efficient biological catalysts that play the central roles in various biochemical processes. Enzyme-catalyzed reactions usually occur with high selectivity and efficiency toward

Stimuli-Responsive Nanomedicine
Edited by Lin Zhu

ISBN 978-981-4800-70-9 (Hardcover), 978-0-429-29529-4 (eBook)
www.jennystanford.com

enzymes' substrates under mild physiological or pathological conditions. Enzymes are involved in all biological and metabolic processes, serving as the prime protagonists in the chemistry of living organisms at a molecular level. Moreover, enzymes can be a diagnostic biomarker since the dysregulation of various enzymes is a characteristic feature of numerous diseases [2]. Even in the same host, the pattern of the enzyme activity and expression may be different in different tissues. The expression of certain enzymes like protease, esterase and glycosidase in pathological tissues suffering from inflammation or cancer is often higher compared with that in normal tissues [3]. These enzymes may serve as local stimuli for certain diseases. For example, esterases produced by macrophages are elevated at the site of inflammation. Myeloperoxidase is highly associated with cardiovascular and neurological diseases [4, 5]. Therefore, the disease-specific or tissue-associated enzymes when undergo the abnormal expression can be potential stimuli/targets for the nanocarriers that are programmed to deliver drugs via an enzyme-responsive manner.

The enzyme-triggered structural transformation of polymers can induce the tuning and modulation of various noncovalent interactions (electrostatic interactions, hydrogen bonding, van der Waals forces, hydrophobic interactions, etc.), leading to the self-assembly, disintegration, and morphological transition of polymers and their assemblies. Enzyme reactive polymer-based nanocarriers generally possess an enzyme-sensitive moiety which is an unstable connection in the polymer main chain or side groups. They can be also triggered by the enzyme reaction products or intermediates as a trigger signal input. Both enzyme-triggered disintegration and enzyme-driven self-assembly of synthetic polymers have been investigated. It is worth noting that these enzyme-responsive structural transformation or transition can be utilized for many biomedical applications such as drug delivery, controlled release, imaging, and diagnosis. So far, many tumor-associated enzymes have been successfully exploited in the nanomedicine for enzyme-sensitive cancer therapy [6, 7].

Figure 3.1 presents the typical strategies of enzyme-sensitive release of the cargoes from nanocarriers [2]. Some conventional polymeric nanoparticles or inorganic nanoparticles can be modified with the motifs that can be cleaved via specific enzymatic degradation. For examples, the proteases can trigger drug release when the drug is linked to the nanocarrier via a peptide and the glycosidases can trigger drug release when the nanocarrier is made of polysaccharides (Fig. 3.1a). The polymer-stabilized liposomes are programmed to respond to lipases for controlled drug release since the lipases can hydrolyze phospholipids, the major liposome building blocks (Fig. 3.1b). Moreover, the hydrolase is used to control the assembly or disassembly of some inorganic nanoparticles for the diagnostic purposes (Fig. 3.1c).

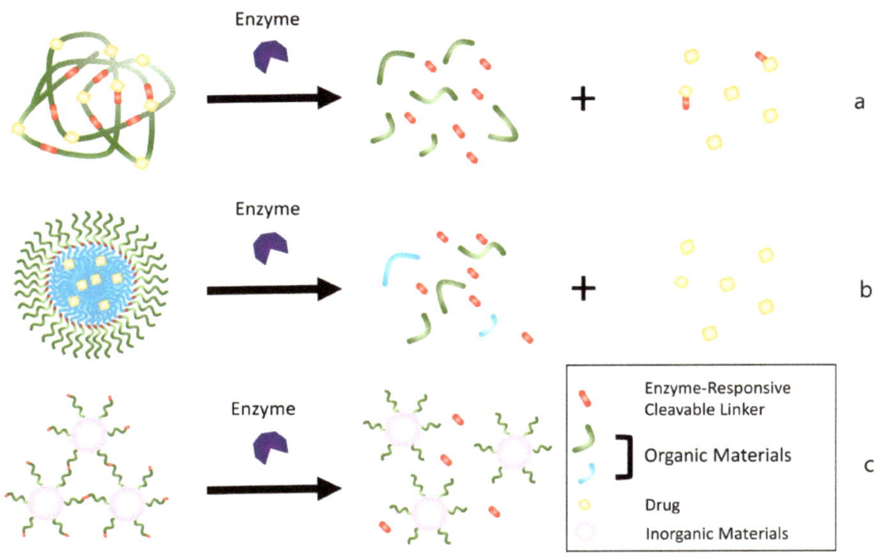

Figure 3.1 Representative enzyme-responsive strategies for drug delivery [2].

In this chapter, the stimuli-responsive drug carriers or cancer diagnostics in response to the widely investigated enzymes including proteases, phospholipases, and oxidoreductases are reviewed. The enzyme-driven self-assembly of synthetic polymers, especially hydrogel formation and disintegration systems, are also briefly described.

3.2 Nanomedicines Based on Enzyme-Triggered Cleavage/Degradation

3.2.1 Protease-Responsive Nanomedicines

Proteases, as a subclass of hydrolases, are strongly associated with cancer development and progression, especially with cancer invasion and metastasis, because of their ability to degrade the extracellular matrix and proteins [3, 8, 9]. To date, proteases are the most widely studied enzymes in cancer therapeutics and diagnostics.

3.2.1.1 Matrix metalloproteinase (MMP)-responsive nanomedicines

Alterations in the composition and expression of local enzymes such as matrix metalloproteinases have been considered the biomarkers for diagnosis and prognosis in many types and stages of cancer and also provide an opportunity for the delivery of drugs and imaging agents to pathological sites via an enzyme-triggered mechanism. MMPs are overexpressed in the tumor microenvironment compared to the normal tissue in most cancer patients [10]. In the MMP family, overexpression of MMP2 and MMP9 has been widely recognized as part of the invasion, progression and metastasis of most human tumors [11]. MMPs are proteolytic enzymes, and their basic mechanism of action, the degradation of proteins, regulates various cell behaviors in the tumor. Thus, MMPs have been widely employed to achieve enzyme-triggered site-specific drug release [7]. Many types of MMP-activated drugs have been developed, and an overview of these drugs is given in Table 3.1.

MMP-sensitive substrates, including proteins, peptides, and polymers, have been designed and used for delivery of drugs and imaging reagents [26, 27]. Jiang et al. developed an MMP2/MMP9 activated cell-penetrating peptide-modified nanoparticlulate drug delivery system containing Cy5, gadolinium, or both as the fluorescent probe [18, 19]. The reported activatable cell penetrating peptide (ACPP) was composed of a polycationic CPP sequence, an MMP-sensitive peptide linker (XPLGLAG), and a polyanionic inhibitory domain (poly Glu, E9), forming a hairpin structure

(Fig. 3.2). The ACPP surface-decorated nanoparticles showed increased tumor specificity. In the presence of MMPs, the polyanionic shielding domain could be dissociated from the ACPP at the MMP-sensitive cleavable site, resulting in the exposure of the polycationic peptides and cellular uptake of their cargoes.

Table 3.1 Examples of MMP-responsive drug delivery systems/ nanomedicines

Enzymes	Nanocarriers and drugs	Cleavable linkers	Disease models	Ref.
MMP 2	PTX-loaded NP	GPLGVRGC	Ovarian cancer cells (SKOV-3)	[12]
MMP 2	QD with cell penetrating peptide (CPP) and substrate-PEG	GGGGPLGVRGGGGK	Melanoma cells (MDA-MB-435)	[13]
MMP 2	Galactosylated liposomes with NOAC	GPLGIAGQ	Hepatocellular carcinoma cells (HepG2)	[14]
MMP 2	Nanoparticles with NIR fluorophore, CPP and TAMRA-labeled cleavable PEG	GK (TAMRA) GPLGVRGC	Fibrosarcoma cells (HT1080) and subcutaneous tumors in mice	[15]
MMP 2	PEG, PEI and lipid conjugate-based micelles for siRNA and PTX codelivery	GPLGIAGQ	Lung cancer cells (A549)	[16]
MMP 2 and 9	Anthrax toxin PA and LFn domains fused to ADP-ribosylation domain of pseudomonas exotoxin A	GPLGMLSQ	Fibrosarcoma cells (HT1080), breast cancer (MDA-MB-231) and melanoma (A2058) cells	[17]

(Conitnued)

Table 3.1 (*Continued*)

Enzymes	Nanocarriers and drugs	Cleavable linkers	Disease models	Ref.
MMP 2 and 9	A CPP with a polyanionic inhibitory domain (poly Glu, E9) and payloads	XPLGLAG	Fibrosarcoma cells (HT-1080)	[18, 19]
MMP 2, 8, 9 and 14	CPP with Cy5, CPP with paclitaxel	PLGLAG	Fibrosarcoma (HT1080), melanoma (B16F10), cervical (Hep2), prostate (PC3), colon (HCT)	[20–23]
All MMPs	Liposomes carrying nucleic acids shielded by a removable PEG-shell	GGGVPLSLYSGGGG	Fibrosarcoma cells (HT1080)	[20]
All MMPs	Gold NPs with quenched NIR fluorophore	PLGVRGC	Mice bearing SCC7 tumors	[24]
All MMPs	PEG-MSNPs	CGPQGIWGQGCR	Fibroblasts (3T3-J2)	[25]

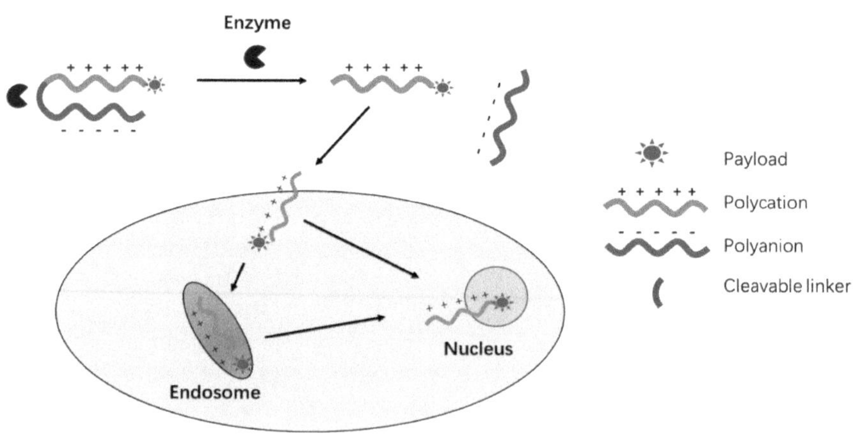

Figure 3.2 Schematic illustration of enzyme-responsive activatable CPPs.

Zhu et al. developed the nanoparticles possessing key features for siRNA and drug delivery, including the tumor targeting triggered by the upregulated tumoral MMP2 and enhanced cell internalization after the MMP2-mediated exposure of the previously hidden PEI. In the tumor, approximately 14.4% of the total cells internalized both siRNA and PTX after the administration of PEG-pp-PEI-PE/PTX/siRNA, around 2.4-fold higher than the internalization efficiency by its non-sensitive counterpart (6%) [16]. The combination of the MMP2-sensitive moiety with other functional moieties significantly increased the tumor-targeting effect as well as tumor cell internalization of the nanocarriers [26]. On the basis of these polymers, a novel multifunctional stimuli-sensitive nanocarrier was developed. The nanocarrier accumulated in the tumor and specifically targeted cancer cells responding to the upregulated extracellular MMP2, which facilitated the enhanced cellular internalization by a cell penetrating moiety (TATp) that became exposed and active when the nanocarrier reached the tumor microenvironment [26] (Fig. 3.3).

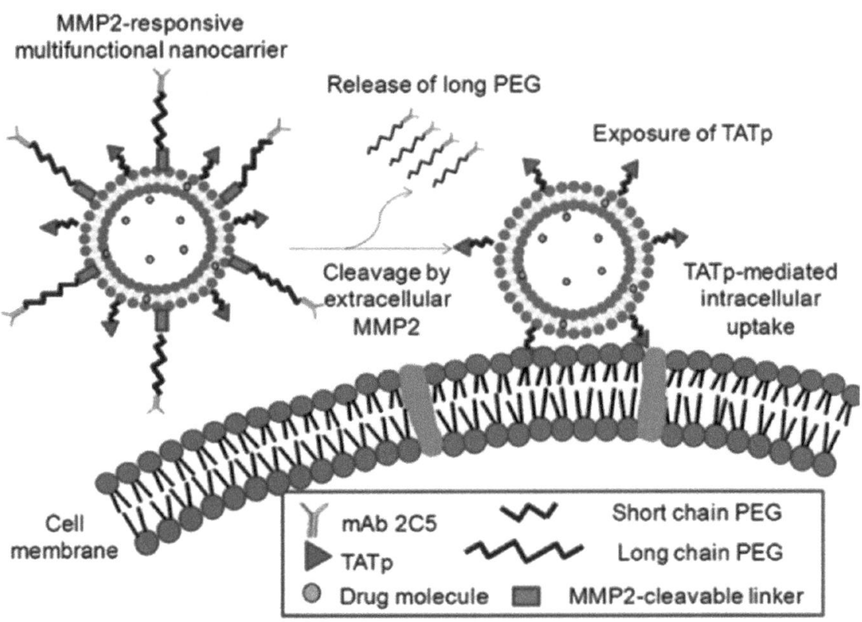

Figure 3.3 MMP2-responsive multifunctional liposomal nanocarriers and its drug delivery strategy [26].

Besides cancer therapeutics, the use of enzyme sensitivity in diagnosis offers a wide range of applications. In recent years, the targeted probes and activatable 'smart' sensors have become very popular. For instance, the Gd-based targeted molecular MRI probes that were activatable via the enzymatic cleavage have been intensively studied. To enhance the visualization of malignant tissues, these studies have focused on the use of MMPs as stimuli for the activation of the molecular probes because the elevated levels of MMPs, particularly MMP 2 and 9, are highly associated with the tumor growth and invasion [28–32]. Several approaches to ensure sufficient accumulation of the molecular imaging probes in the tumor have been reported. Lepage et al. designed two different MRI contrast agents which, upon cleavage by MMP 2 and 7, underwent a decrease in their solubility, leading to an accumulation of the imaging probes in the areas of high MMP concentration [33]. Other studies have implemented the opposite principle. The MMP-induced cleavage of the Gd-based MRI contrast agents led to an amphiphilic-to-hydrophilic rather than a hydrophilic-to-hydrophobic transition and an increased solubility. The washout kinetics of the contrast enhancement and the MMP expression and activity in the tumor could be interpreted [36]. In another study, a molecular imaging probe able to shift from a low-relaxivity state to a highly surface-charged, aggregative state in response to the MMP9 cleavage was created [15]. The MRI results in mice showed the improved accumulation and prolonged retention of the contrast agents in the tumor even after several days [18]. Moreover, the MMP-cleavable ACPP-dendrimers (ACPPDs) as MRI molecular imaging probes could detect the tumors and their metastases, including lymph node metastasis. The Gd-loaded ACPPs significantly improved the accuracy of diagnosis and staging for small invasive malignant tumor [34].

3.2.1.2 Cathepsin B (CTB)-responsive nanomedicines

Cathepsin is a kind of cysteine aspartic acid-specific protease and, as popular targets for early drug design, significantly influences the drug efficacy against cancer [35]. The natural milieu of these enzymes is the lysosomal compartment, where a low pH ensures optimal catalytic activity of the acid protease [36]. Cathepsin B, a cysteine protease, is recognized as a promising biomarker for cancer diagnosis and prognosis [37–40]. Malignant tumors are usually characterized by large amount and/or high activity of CTB, indicating that CTB may be the ideal target for the design of anticancer drugs [41]. Boc-Val-Cit- and Cbz-Val-Cit-, two CTB substrates, are the popularly used CTB-sensitive linkers in the nanomedicine. Table 3.2 summarizes the recent research on cathepsin B-responsive nanomedicine.

Table 3.2 Examples of cathepsin B-responsive drug delivery systems/ nanomedicines

Enzymes	Nanocarriers and drugs	Cleavable linkers	Disease models	Ref
Cathepsin B	Gemcitabine with vitamin E succinate	succinate	Bxpc-3 cells	[42]
	Floxuridine conjugated with 5′-monoamino acid and dipeptide	5′-monoamino acid and dipeptide	Capan-2 cells	[43]
	Doxorubicin and p-aminobenzylcarbonyl	Z-Phe-Lys	Hepatocellular carcinoma	[44]
	Doxorubicin conjugated with PAMAM-HPMA	GFLG	Lymphoma	[45]
	Doxorubicin conjugated with HPMA	GFLG	OVCAR-3	[46]
	PTX and GEM conjugated with HPMA	GFLG	A2780	[47]

The tetrapeptide, Gly-Phe-Leu-Gly (GFLG), has been found to be one of the most effective spacers in response to CTB. Zhang and his coworkers developed a novel multifunctional self-immolative linker which was applied in a camptothecin (CPT) prodrug. The $mPEG_{2000}$ was conjugated to the drug through the "click" chemistry, by which drug solubility was significantly improved, resulting in the nanoparticle formation. Both the CTB sensitive moiety and the PEG chain contributed to the decreased side toxicity in vivo [48]. As CTB is only found in the intracellular compartment, the drug exerts its therapeutic effect only after cellular internalization of the nanoparticles via endocytosis. This strategy can increase the therapeutic effects and reduce side effects of the drug as drug release only occurs in a localized/intracellular microenvironment. For example, Omelyanenko and his co-workers developed an HPMA-DOX-OV-TL16 conjugate (P-(GFLG)-DOX-Ab, 270,000 Da, 3.3 wt% DOX) recognized by the OA3 antigen which plays an important role in membrane transport and/or signal transduction through its multimembrane-spanning domain structure [54]. The P-(GFLG)-DOX-Ab could rapidly enter the OVCAR-3 cells and was located at the endolysosomes. Then, the nanoconjugate released DOX via the CTB-induced cleavage at the cleavable GFLG spacer, followed by drug diffusion through the endolysosomal membrane and drug accumulation in the cell nucleus [46]. The experimental data showed that the HPMA-DOX conjugate had a higher intracellular DOX concentration compared with free DOX in the multidrug resistant (MDR) A2780/AD cells [49]. Etrych et al. developed a PAMAM-based (D-NH_2)-Gly-Phe-Leu-Gly-HPMA-DOX dendrimer (D2-GFLG-P-DOX, 215,000 Da, DOX: 9.2 wt%) with a pH-sensitive hydrazone bond. The semi-telechelic HPMA copolymers with a molecular weight below the renal threshold were conjugated onto the dendrimer via an enzyme-sensitive GFLG oligopeptide. The in vitro study using the phosphate buffer at pH 5.0 or 7.4 at 37°C (for the hydrazone conjugates) and in a CTB-containing (5×10^{-7} M) phosphate buffer at 37°C (for the amide conjugates) indicated that the presence of CTB increased the rate of DOX release [45].

3.2.1.3 Legumain-responsive nanomedicines

Legumain is a highly conserved lysosomal/vascuolar cysteine protease, originally identified in legumes [50]. Under physiological conditions, legumain is primarily expressed in kidney tubuli and contributes to renal tubular reabsorption [51, 52]. The results of several studies have indicated that legumain expression is upregulated in a variety of different solid tumor types, with the level that is highly correlated with the malignancy [53–55]. Like cancer cells, several other cell types have been also reported to express high levels of legumain, especially the tumor-associated macrophages (TAMs) [56, 57]. In addition, the legumain expression was observed in both the cellular plasma and cell surface mainly due to the stress, such as hypoxia and starvation [58]. Based on these changes, legumain represents a good target for the design of the anticancer therapeutics. Several delivery systems in response to the legumain expression have been constructed and used as anticancer drug carriers [57–59]. Table 3.3 provides a summary of recent studies on legumain-responsive therapeutics.

Table 3.3 Examples of legumain-responsive drug delivery systems

Enzymes	Nanocarriers and drugs	Cleavable linkers	Disease models	Ref.
Legumain	Auristatin conjugated with di-desmethylauristatin E and monomethylauristatin E	AAN	MDA-MB-435 cells	[60]
	Auristatin/inhibitor of integrin $\alpha_v\beta_3$ with monomethylauristatin E	AAN	MDA-MB-435 cells	[61]
	Liposomes loaded with Doxorubicin	AAN-CBZ	4T1 cells	[62]
	N-succinyl-β-alanyl-L-alanyl-L-asparaginyl-L-leucyl-doxorubicin	LNAA-Suc	Various cells	[63]
	Colchicine conjugated with Suc-AANV	Suc-AANV	HEK293 cells	[64]
	Etoposide conjugated with Cbz-AAN-AMC	Cbz-AAN-AMC	HEK293 cells	[65]

Several groups have reported prodrugs of dolastatins or related molecules. The dolastatins are exceptionally potent pentapeptide antineoplastic agents (in the pM range) that bind tubulin close to the Vinca-binding site and disrupt microtubule function. However, their side effects are considerable. Liu et al. described the construction of a novel liposome-based drug delivery platform via the incorporation of an internalizing TAT moiety and legumain-based tumor targeting moiety [62]. This strategy made the liposomes remain stable until they were identified and cleaved by active legumain in the tumor. Through conjugation of the asparagine containing tripeptide (AAN) to didesmethylauristatin E (DDAE) or monomethylauristatin E (MMAE), a legumain-activatable prodrug was developed [60]. Researchers have developed several analogues that showed considerable activity against legumain-transfected MDA-MB-435 cells rather than the control cells. This improvement of activity was dependent on the legumain catalysis. Liu et al. developed MMAE based drug molecule mimetics through the (alanine–alanine–asparagine) tripeptide linker with a cap structure of integrin $\alpha_v\beta_3$ inhibitor [61]. Adding the inhibitors to a prodrug allows it to bind to cell surface integrin $\alpha_v\beta_3$ before activation by legumain. In vivo studies showed that the MMAE integrin-binding prodrug is more effective than the MMAE. Several legumain-cleavable peptide–doxorubicin conjugates were synthesized by Wu et al. [63]. However, these prodrugs containing a succinyl group decreased their cell permeability. DOX was therefore prevented from entering the cell until it had been activated by legumain in the tumor microenvironment. In the tumor microenvironment, legumain is highly expressed by the TAMs rather than the tumor cells [60, 63, 66]. However, the activated drug was suggested to have a greater antitumor efficacy through a "bystander effect" upon both tumor and stromal cells, rather than only selectively deleting the target-producing cells from the tumor. The legumain-sensitive DOX prodrug, LEG-3, that comprised DOX bound to a tetrapeptide (Leu-Asn-Ala-Ala) with an end cap of a succinyl group, was the most effective form [57]. It was shown that LEG-3 possessed higher anti-tumor efficacy than free DOX in mouse syngeneic tumor models and human tumor xenografts, including a drug-resistant prostate cancer model [63].

Another legumain activatable prodrug with etoposide as the active component was synthesized by Stern et al. [65]. The prodrug released the etoposide upon the enzymatic cleavage by legumain and showed an efficient cell growth inhibition on the legumain-overexpressed 293 HEK-Leg cells. The design of the prodrug able to be activated by legumain will be a promising approach for cancer therapy.

3.2.2 Esterase-Responsive Nanomedicines

Esterase is an enzyme that able to hydrolyze the ester-containing compounds into acids and alcohols. Many attempts have been made to explore various esterases residing in the endolysosomes of cancer cells for drug delivery.

3.2.2.1 Phospholipase A-responsive nanomedicines

The level of human secretory phospholipase A2 (sPLA2) increases in pathological tissues due to its secretion from a number of cells under pathological conditions such as inflammation and cancer [67]. Moreover, it has been found that sPLA2 was highly expressed in the evading zone of tumors, which worked as a part of the defense mechanism of the host [68]. The sPLA2-IIA subtype, primarily responding to anionic phospholipid membranes, showed a binding force to a negatively charged lipid membrane several times stronger than to a zwitterionic lipid membrane [69]. It has been suggested that with the stimulation by inflammatory cytokines, the carcinoma cells, hepatocytes or inflammatory cells might trigger the sPLA-IIA overexpression. In particular, the overexpression of sPLA2-IIA has been identified in a variety of cancer types including prostate, colon, breast and pancreatic cancer [70]. Although a large number of studies have shown that sPLA2 is present and upregulated in cancerous tissue, it is still being debated whether it plays a pro- or anti-tumorigenic role in human cancer development [70, 71].

The elevated sPLA2 levels in cancerous tissues suggest that sPLA2 may be a potential target for therapeutic intervention. For example, the sPLA2-degradable nanomaterials that carry cytotoxic compounds may be used to target the tumor, as they may accumulate in the leaky cancerous tissue and release cargo via the enzymatic degradation.

Interestingly, sPLA2 exhibits higher catalytic activity towards aggregated phospholipids such as liposomes and micelles than towards lipid monomers. Its activity is also highly dependent on the membrane charge, lipid composition, and physical state of the lipids. So, it is possible to prepare the liposomes to be more or less susceptible to the sPLA2 hydrolysis by appropriate selection of the liposomal phospholipids.

The drug-loaded liposomes made of phospholipids have been utilized as an enzyme-responsive drug delivery system, either sensitive to PLA2 exclusively or to PLA2 and other tumor associated enzyme (e.g., α-amylase) concurrently [72, 73]. In these designs, the liposomes could not only encapsulate the molecule of interest but also work as a prodrug when the highly cytotoxic lyso lipids (antitumor ether lipids or AELs) were formed after the enzymatic degradation of the lipids in the carrier. After loading to the designed liposomes, for example, the haemolytic side effect of AELs could be minimized. By this way, the safety and drug specificity can be both improved because PLA2 is found at high levels only in the tumor but not in the bloodstream. To demonstrate this concept, the liposomes made of (R)-1-O-hexadecyl-2-palmitoyl-sn-glycero-3-phosphocholine (1-O-DPPC) and phospholipid-PEG derivatives (1-O-DPPE-PEG) were exposed to the PLA2 from snake venom, and the release of calcein from non-hydrolysable liposomes was measured as an indicator for membrane permeation. The nanocarriers were stable in the bloodstream due to the steric repulsion of PEG molecules but subject to the enzymatic digestion, which destabilized the liposomes and generated AELs and fatty acids. After enzymatic activation, the AELs could permeate the cell membrane to enhance the uptake of drugs encapsulated in the liposomes [2, 74, 75].

PLA2 is also an interesting diagnostic biomarker and the PLA2 level is monitored on a regular basis for the diagnosis of cancer. A novel colorimetric gold nanoparticle-based strategy for measuring the PLA2 concentration and activity was recently developed. In this approach, a residue non-linear helix-loop-helix polypeptide was encapsulated in liposomes. This polypeptide was able to complementarily bind with a second peptide fixed on gold nanoparticles and the binding led to a folding-dependent nanoparticle bridging/aggregation. The peptide-loaded liposomes could be digested by PLA2 and release the encapsulated peptides, which induced nanoparticles' aggregation and an obvious spectral redshift [76].

In addition to the readily detectable colorimetric shift of the nanoparticles, this assay strategy had an advantage that it exploited a natural PLA2 substrate, i.e., the lipid bilayer. This advantage is particularly important for phospholipases because they are interfacially activated enzymes, meaning that their catalytic activity toward individual lipid monomers is very low while their activity increases when they are associated with a lipid bilayer membrane. Using this strategy, the physiologically relevant enzyme concentrations could be detected and the enzyme activity could be monitored by measuring the length of the lag phase [76].

3.2.2.2 α-amylase-responsive nanomedicines

Amylase is an enzyme that contains three subtypes, α-, β-, and γ-amylase and able to catalyze the hydrolysis of starch into sugars. In particular, the α-amylase existing in the saliva of humans and some other mammals cleaves the 1,4 α-glycosidic bond and initiates the digestion process. Li and his co-workers developed a novel hydroxyethyl starch (HES)-based α-amylase- and redox-dual-responsive system for delivery of PTX. The resultant HES nanoparticles could be degraded into smaller sizes (around 40 nm) by α-amylase and release the active PTX under a high redox potential [77]. Kim and co-workers developed another strategy for the esterase-triggered release of drugs from mesoporous silica nanoparticles (MSNPs). The porous channels of the MSNPs were blocked by cyclodextrin (CD) gatekeepers attached to the nanoparticle surface through a lipase cleavable linker. The hydrolysis of the ester moiety in the linker by the esterase removed the bulky CDs from the nanoparticle surface, releasing molecules entrapped in the pores. The drug release could also be triggered by α-amylase, which catalyzed the digestion of the CD gatekeepers, thus providing a delivery system responsive toward two key enzymes involved in the pathology of acute pancreatitis [78].

3.2.3 Oxidoreductase-Responsive Nanomedicines

Oxidoreductases have been widely exploited as a promising target for drug delivery systems and biosensors due to their fundamental role in oxidative environments which are generated by many diseases including diabetes and cancer [79, 80]. Moreover, they are utilized as diagnostic tools, for example, the detection of glucose by glucose oxidase (GOx) or as labels for immune detection with horseradish peroxidase [81].

GOx is one of the mostly investigated oxidoreductases, which has been used as a "gate" in the responsive, controlled release systems or as a diagnostic tool in detecting glucose [82]. An early example of the oxidoreductase-responsive nanomaterials was the polymeric vesicles that encapsulated GOx [83]. In the work, the nanocarriers encapsulating the enzyme allowed the permeation of the substrate glucose.

Furthermore, Gu et al. applied GOx as the trigger to develop a nano-network that could respond to the blood sugar level for closed-loop insulin delivery [67]. For example, the chitosan-based positively charged nanoparticles and alginate-based negatively charged nanoparticles formed a new nanoparticle with a hydrophobic core of insulin, ketal-modified dextran and enzymes (GOx and catalase) through the electrostatic interaction. In the hyperglycemic area, the glucose was converted to gluconic acid by the GOx loaded in the nanoparticles, leading to dissociation of the nanocarrier and subsequently insulin release. An in vitro pulsatile insulin release was observed in response to glucose concentrations. The in vivo results on streptozotocin (STZ)-induced diabetic mice indicated that a single injection of the nanoparticles could stabilize blood glucose levels within the normoglycemic range for up to 10 days. The use of the glucose-sensitive biodegradable nanoparticles might realize the self-regulated chronic diabetes management [7].

Moreover, the enzymatic reaction intermediates or products can also be utilized to trigger the self-assembly/disassembly of synthetic polymers. For example, the GOx-mediated self-destructing amphiphilic ABA triblock polymeric vesicles, PEG-*b*-PPS-*b*-PEG, were reported, which contained the hydrophobic B block of poly(propylene sulfide) (PPS) and two outer hydrophilic A block of PEGs. The triblock PEG-*b*-PPS-*b*-PEG copolymer could initially self-assemble into vesicles. In the presence of GOx and glucose, GOx transformed glucose to glucose lactone, by which hydrogen peroxide was generated as a side product. In turn, the hydrogen peroxide oxidized the PPS thioethers into sulfoxides and sulfones. The process induced conversion of the PEG-PPS-PEG from an initially amphiphilic polymer to a hydrophilic polymer, which triggered the dissolution of the vesicles. The approach might be used for drug delivery if the destabilization of the polymeric vesicles could trigger the release of the loaded therapeutic molecules [2, 83].

3.2.4 Other Enzyme-Responsive Nanomedicines

Since the expression and activity of specific enzymes are imbalanced in the pathological region in many diseases, these dysregulated enzymes can be used as the triggers for selective activation of drug delivery systems. Recently, the penicillin G amidase (PGA) was exploited to cleave the end groups of phenyl acetamide to destabilize their assembled micelles [84]. When micelles were incubated with the PGA, the monomeric hybrids could be detected, which was accompanied with the altered fluorescence of the encapsulated Nile Red dyes and rapid disappearance of the amphiphilic hybrids in a HPLC analysis. Moreover, the fully degraded tetra-amine hybrids could only be detected at high PGA concentration of 1.4 μM after 24 h incubation, while they were not detectable at low PGA concentration of 0.14 μM.

The azoreductase has received extensive attention in the treatment of colon diseases because it is produced by the microbial flora present in the human colon. Rao et al. prepared an azoreductase-responsive assembly by covalent insertion of an azobenzene linker into an amphiphilic diblock copolymer, poly (ethylene glycol)-*b*-poly(styrene) (PEG-*b*-PS) [85]. The azobenzene-linked PEG-*b*-PS copolymer was obtained by an atom transfer radical polymerization (ATRP)-based macroinitiator approach, which was able to assemble into micelles in water but dissociate to two segments (PEG and PS) when exposed to azoreductase. Due to the azoreductase selectivity and sensitivity, the azoreductase responsive micelles might have potential as a colon-targeted drug delivery system.

Although significant progress regarding stimuli-responsive tumor targeting has been made over the past few decades, the nanosystems responding to bacterial enzymes are seldom reported. Recently, a bacterial strain-selective polymeric vesicle has been constructed for delivery of antibiotics [86]. Liu et al. introduced side chain enzymatic cleavage mechanism to the polymeric vesicles containing PEG-*b*-PP and PEG-*b*-PC, and chose PGA and β-lactamase

(Bla) as target enzymes. Side chain moieties in hydrophobic blocks of PEG-*b*-PP and PEG-*b*-PC were subjected to enzymatic cleavage of terminal groups and subsequent self-immolative cleavage, both affording poly(2-aminoethyl methacrylate) scaffolds. As a result, the polymeric vesicles underwent self-immolative structural rearrangement and morphological transitions, leading to the sustained release of antibiotics. So, the strategy would satisfy the enhanced stability, reduced side effects, and bacterial strain-selective drug release. It is known that Bla is the main cause of bacterial resistance to β-lactam antibiotics. In a further study, the Bla-degradable polymeric vesicles were used for the methicillin-resistant S. aureus (MRSA)-triggered release of antibiotics, resulting in the in vitro inhibition of MRSA growth and enhanced wound healing in a murine model.

3.3 Nanomedicines Based on Enzyme-Triggered Polymeric Assemblies

Enzyme-catalyzed reactions exhibit many advantages like high efficiency, high selectivity, and specificity. An additional advantage is that enzyme-catalyzed reactions can be conducted in vitro. This allows for chemical synthesis in a nature-mimicking manner and construction of artificial systems. The design flexibility and applications of the enzyme-responsive nanomedicine can be broadened and improved by the integration of enzyme-catalyzed reactions with responsive polymers. Manipulation of the chemical and physical properties of polymers using enzymes may initiate the brand new applications, including the enzyme-induced surface activation and the in vivo formation of highly regulated nanostructures [87]. The previous sections mainly deal with enzyme-triggered disintegration and structural reorganization of polymeric assemblies and nanoparticles. In this section, the enzyme-driven self-assembly of synthetic polymers and enzyme-responsive formation and disintegration of polymeric hydrogel systems are focused and their applications in the controlled drug release, biological catalysis, and imaging will be discussed.

Synthetic polymers possessing enzyme-cleavable phosphate functionalities exhibit the triggered self-assembly and aggregation characteristics. Amir et al. have demonstrated that the polymerization of vinyl monomers with the cleavable enzymatic substrates generated the water-soluble double-hydrophilic block copolymers. Upon enzymatic activation (dephosphorylation), the diblock copolymers became amphiphilic and self-assembled into a well-defined core–shell nanostructure, due to the increased hydrophobicity of the terminal groups (Fig. 3.4) [87]. More sophisticated enzyme-triggered self-assembling systems can be constructed from polymer–peptide hybrid copolymers upon site-specific modification of the peptide block. A PEG–peptide conjugate containing a (TV) 5-peptide (five repeats of alternating threonine (T) and valine (V) diads) aggregator domain was synthesized. In the presence of acid phosphatase (AP), the hydrolysis of threonine phosphate ester moieties could turn on the self-assembly process, forming the nicely organized fibrillar nanostructures.

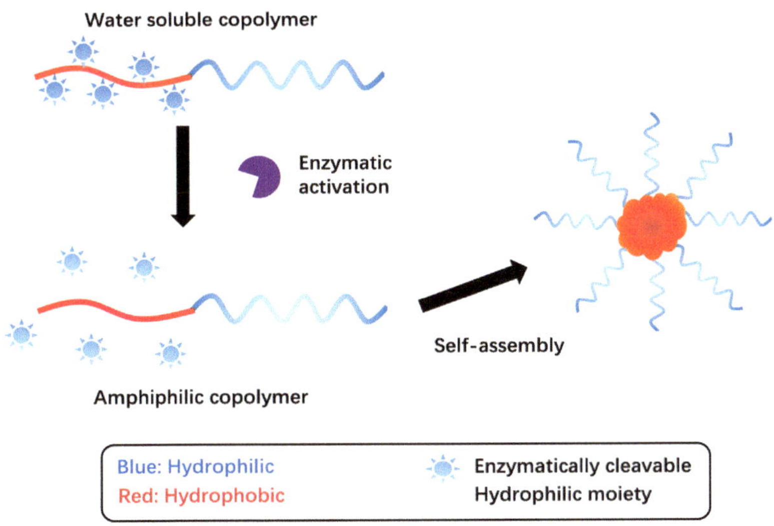

Figure 3.4 Schematic representation of enzymatic transformation of water-soluble diblock copolymers to amphiphilic self-assembling copolymers [91].

Furthermore, by taking advantage of highly specific and selective enzyme-catalyzed reactions, a variety of enzyme-responsive "sol-to-gel" and "gel-to-sol" transition systems based on synthetic polymers have been developed. The nature of stimuli-responsive in situ gel formation and disintegration endows these polymers with a broad applicability, like the controlled drug release, cell internalization, and tissue engineering. The enzyme-triggered gelation systems based on the transglutaminase (TGase) and the proteins or peptides which contain the g-carboxamide/glutamine residues and amine-containing residues, were reported [88]. The in situ formation of chemically cross-linked hydrogels could typically occur within a few minutes. It is known that the horseradish peroxidase (HRP) can catalyze the coupling of phenol or aniline derivatives in the presence of H_2O_2. For example, the H_2O_2/HRP-triggered rapid gelation occurred within ~1 min at pH 6.5 when the poly (aspartic acid) functionalized with tyramine, tyrosine, or aminophenol moieties were used in the formulation [89]. The enzyme-triggered cross-linking of tyramine-functionalized dextran could also trigger the in situ formation of hydrogels [90]. In the presence of HRP and H_2O_2, the hydrogel could be rapidly formed under physiological conditions when the polymer concentration was above ~2.5 wt%. Based on the similar strategy, a variety of cross-linked hydrogels in response to the enzyme-catalyzed reactions have been developed. The polymers include the gelatin-PEG-tyramine conjugates, tyramine-conjugated copolymer, sugar beet pectin, phenol moieties functionalized chitosan and dextran, hyaluronic acid-modified dextran-tyramine conjugates, and saccharide-peptide conjugates containing tyrosine [91]. All these polymers could undergo the HRP/H_2O_2–triggered sol-to-gel transitions under mild conditions. In addition, other enzyme-responsive in situ formation of cross-linked hydrogels have been also studied. Since the enzyme-triggered self-assembly processes in the above examples are highly selective and substrate-specific and happen in aqueous medium under mild conditions, the reported polymers and their hydrogels are excellent nanomaterials/nanocarriers for enzyme-responsive drug and gene delivery [91].

3.4 Conclusions

In this chapter, we review recent advances in the field of the enzyme-responsive nanomedicine and discuss the enzyme-driven self-assembly and enzyme-triggered disintegration or structural reorganization and their applications in drug delivery and targeting.

Although enzyme-responsive nanomedicines have made great progress in diagnosis, drug delivery, and treatment, there is a long way to go for their commercialization. In fact, the development of enzyme-responsive nanomedicine is still at its very early stage. There is still plenty of room for the design of enzyme-responsive polymers/biomolecules or their conjugates by taking advantage of the sequence-specific and programmed self-assembling nature of biomolecules such as peptides, proteins, and DNA. Moreover, the applications and risks of the current enzyme-responsive polymeric assemblies, nanoparticles, and hydrogels should be extensively evaluated under both in vitro and in vivo conditions. The microenvironments within specific cells, tissues, and organs are quite complex. Although enzymatic reactions are highly selective and substrate-specific, the presence of the same type of target enzymes or other homologous enzymes in healthy tissues or other intracellular organelles, even at relatively lower concentrations, will result in the nonspecific drug accumulation and release and incur side effects. So, the selection of the target enzymes is very important and the development of the nanomedicine which is stable in the non-target sites but shows enzymatic responsiveness only in the target tissues and cells, is highly desirable. In addition, the emerging field of enzyme-responsive nanomedicine calls for more revolutionary ideas and novel designing concepts, such as multi-responsiveness. Compared to the current enzyme-responsive nanomaterials which are mainly responsive to one type of enzyme, the integration of multi-enzyme mediated cascade reactions in nanomedicine will provide them with multi-responsiveness which are more fascinating and may broaden their applications. In addition, more stimuli will be found along with the progress in biology, chemistry, and physics, which are favorable for drug delivery. The combination of stimuli sensitivity with other strategies may result in the multi-component/-functional nanomedicine which will be of great interest.

Acting as a novel category of "smart" medicine, enzyme-responsive nanomedicine will continuously receive intense investigations and more exciting progresses in this field will be seen in the near future.

Acknowledgements

This work was supported by the National Nature Science Foundation of China (grant number 31771087, 81571786, 31671015 and 81271687).

References

1. J. Hu, G. Zhang, S. Liu, Enzyme-responsive polymeric assemblies, nanoparticles and hydrogels, *Chem. Soc. Rev.,* 41(18) (2012) 5933–5949.
2. R. de la Rica, D. Aili, M. M. Stevens, Enzyme-responsive nanoparticles for drug release and diagnostics, *Adv. Drug Deliv. Rev.*, 64(11) (2012) 967–978.
3. C. Lopez-Otin, J. S. Bond, Proteases: Multifunctional enzymes in life and disease, *J. Biol. Chem.,* 283(45) (2008) 30433–30437.
4. S. Aime, C. Cabella, S. Colombatto, S. Geninatti Crich, E. Gianolio, F. Maggioni, Insights into the use of paramagnetic Gd(III) complexes in MR-molecular imaging investigations, *J. Magn. Reson. Imaging*, 16(4) (2002) 394–406.
5. E. Rodriguez, M. Nilges, R. Weissleder, J. W. Chen, Activatable magnetic resonance imaging agents for myeloperoxidase sensing: Mechanism of activation, stability, and toxicity, *J. Am. Chem. Soc.*, 132(1) (2010) 168–177.
6. T. L. Andresen, D. H. Thompson, T. Kaasgaard, Enzyme-triggered nanomedicine: Drug release strategies in cancer therapy, *Mol. Membr. Biol.*, 27(7) (2010) 353–363.
7. Q. Hu, P. S. Katti, Z. Gu, Enzyme-responsive nanomaterials for controlled drug delivery, *Nanoscale*, 6(21) (2014) 12273–12286.
8. C. Lopez-Otin, L. M. Matrisian, Emerging roles of proteases in tumour suppression, *Nat. Rev. Cancer*, 7(10) (2007) 800–808.
9. C. Lopez-Otin, T. Hunter, The regulatory crosstalk between kinases and proteases in cancer, *Nat. Rev. Cancer*, 10(4) (2010) 278–292.
10. J. Westermarck, V. M. Kahari, Regulation of matrix metalloproteinase expression in tumor invasion, *FASEB J.*, 13(8) (1999) 781–792.
11. A. M. Mansour, J. Drevs, N. Esser, F. M. Hamada, O. A. Badary, C. Unger, I. Fichtner, F. Kratz, A new approach for the treatment of malignant melanoma: Enhanced antitumor efficacy of an

albumin-binding doxorubicin prodrug that is cleaved by matrix metalloproteinase 2, *Cancer Res.*, 63(14) (2003) 4062–4066.

12. E. Gullotti, J. Park, Y. Yeo, Polydopamine-based surface modification for the development of peritumorally activatable nanoparticles, *Pharm. Res.*, 30(8) (2013) 1956–1967.
13. H. Mok, K. H. Bae, C. H. Ahn, T. G. Park, PEGylated and MMP-2 specifically dePEGylated quantum dots: Comparative evaluation of cellular uptake, *Langmuir*, 25(3) (2009) 1645–1650.
14. T. Terada, M. Iwai, S. Kawakami, F. Yamashita, M. Hashida, Novel PEG-matrix metalloproteinase-2 cleavable peptide-lipid containing galactosylated liposomes for hepatocellular carcinoma-selective targeting, *J. Control. Release*, 111(3) (2006) 333–342.
15. E. Schellenberger, F. Rudloff, C. Warmuth, M. Taupitz, B. Hamm, J. Schnorr, Protease-specific nanosensors for magnetic resonance imaging, *Bioconjug. Chem.*, 19(12) (2008) 2440–2445.
16. L. Zhu, F. Perche, T. Wang, V. P. Torchilin, Matrix metalloproteinase 2-sensitive multifunctional polymeric micelles for tumor-specific co-delivery of siRNA and hydrophobic drugs, *Biomaterials*, 35(13) (2014) 4213–4222.
17. S. Liu, S. Netzel-Arnett, H. Birkedal-Hansen, S. H. Leppla, Tumor cell-selective cytotoxicity of matrix metalloproteinase-activated anthrax toxin, *Cancer Res.*, 60(21) (2000) 6061–6067.
18. E. S. Olson, T. Jiang, T. A. Aguilera, Q. T. Nguyen, L. G. Ellies, M. Scadeng, R. Y. Tsien, Activatable cell penetrating peptides linked to nanoparticles as dual probes for in vivo fluorescence and MR imaging of proteases, *Proc. Natl. Acad. Sci. U. S. A.*, 107(9) (2010) 4311–4316.
19. T. Jiang, E. S. Olson, Q. T. Nguyen, M. Roy, P. A. Jennings, R. Y. Tsien, Tumor imaging by means of proteolytic activation of cell-penetrating peptides, *Proc. Natl. Acad. Sci. U. S. A.*, 101(51) (2004) 17867–17872.
20. H. Hatakeyama, H. Akita, E. Ito, Y. Hayashi, M. Oishi, Y. Nagasaki, R. Danev, K. Nagayama, N. Kaji, H. Kikuchi, Y. Baba, H. Harashima, Systemic delivery of siRNA to tumors using a lipid nanoparticle containing a tumor-specific cleavable PEG-lipid, *Biomaterials*, 32(18) (2011) 4306–4316.
21. G. Gu, H. Xia, Q. Hu, Z. Liu, M. Jiang, T. Kang, D. Miao, Y. Tu, Z. Pang, Q. Song, L. Yao, H. Chen, X. Gao, J. Chen, PEG-co-PCL nanoparticles modified with MMP-2/9 activatable low molecular weight

protamine for enhanced targeted glioblastoma therapy, *Biomaterials*, 34(1) (2013) 196–208.

22. E. S. Olson, T. A. Aguilera, T. Jiang, L. G. Ellies, Q. T. Nguyen, E. H. Wong, L. A. Gross, R. Y. Tsien, In vivo characterization of activatable cell penetrating peptides for targeting protease activity in cancer, *Integr. Biol. (Camb.)*, 1(5–6) (2009) 382–393.
23. T. A. Aguilera, E. S. Olson, M. M. Timmers, T. Jiang, R. Y. Tsien, Systemic in vivo distribution of activatable cell penetrating peptides is superior to that of cell penetrating peptides, *Integr. Biol. (Camb.)*, 1(5–6) (2009) 371–381.
24. S. Lee, E. J. Cha, K. Park, S. Y. Lee, J. K. Hong, I. C. Sun, S. Y. Kim, K. Choi, I. C. Kwon, K. Kim, C. H. Ahn, A near-infrared-fluorescence-quenched gold-nanoparticle imaging probe for in vivo drug screening and protease activity determination, *Angew. Chem. Int. Ed. Engl.*, 47(15) (2008) 2804–2807.
25. N. Singh, A. Karambelkar, L. Gu, K. Lin, J. S. Miller, C. S. Chen, M. J. Sailor, S. N. Bhatia, Bioresponsive mesoporous silica nanoparticles for triggered drug release, *J. Am. Chem. Soc.*, 133(49) (2011) 19582–19585.
26. L. Zhu, P. Kate, V. P. Torchilin, Matrix metalloprotease 2-responsive multifunctional liposomal nanocarrier for enhanced tumor targeting, *ACS Nano*, 6(4) (2012) 3491–3498.
27. C. Wong, T. Stylianopoulos, J. Cui, J. Martin, V. P. Chauhan, W. Jiang, Z. Popovic, R. K. Jain, M. G. Bawendi, D. Fukumura, Multistage nanoparticle delivery system for deep penetration into tumor tissue, *Proc. Natl. Acad. Sci. U. S. A.*, 108(6) (2011) 2426–2431.
28. A. Juuti, J. Lundin, S. Nordling, J. Louhimo, C. Haglund, Epithelial MMP-2 expression correlates with worse prognosis in pancreatic cancer, *Oncology*, 71(1–2) (2006) 61–68.
29. A. Talvensaari-Mattila, P. Paakko, T. Turpeenniemi-Hujanen, Matrix metalloproteinase-2 (MMP-2) is associated with survival in breast carcinoma, *Br. J. Cancer*, 89(7) (2003) 1270–1275.
30. R. Hirvonen, A. Talvensaari-Mattila, P. Paakko, T. Turpeenniemi-Hujanen, Matrix metalloproteinase-2 (MMP-2) in T(1–2)N0 breast carcinoma, *Breast. Cancer Res. Treat.*, 77(1) (2003) 85–91.
31. X. Xu, Y. Wang, Z. Chen, M. D. Sternlicht, M. Hidalgo, B. Steffensen, Matrix metalloproteinase-2 contributes to cancer cell migration on collagen, *Cancer Res.*, 65(1) (2005) 130–136.
32. R. Lebel, B. Jastrzebska, H. Therriault, M. M. Cournoyer, J. O. McIntyre, E. Escher, W. Neugebauer, B. Paquette, M. Lepage, Novel

solubility-switchable MRI agent allows the noninvasive detection of matrix metalloproteinase-2 activity in vivo in a mouse model, *Magn. Reson. Med.*, 60(5) (2008) 1056–1065.

33. M. Lepage, W. C. Dow, M. Melchior, Y. You, B. Fingleton, C. C. Quarles, C. Pepin, J. C. Gore, L. M. Matrisian, J. O. McIntyre, Noninvasive detection of matrix metalloproteinase activity in vivo using a novel magnetic resonance imaging contrast agent with a solubility switch, *Mol. Imaging*, 6(6) (2007) 393–403.
34. J. Dupuis, F. Harel, Q. T. Nguyen, Molecular imaging of the pulmonary circulation in health and disease, *Clin. Transl. Imaging*, 2(5) (2014) 415–426.
35. R. Loser, J. Pietzsch, Cysteine cathepsins: Their role in tumor progression and recent trends in the development of imaging probes, *Front. Chem.*, 3 (2015) 37.
36. J. Vandooren, G. Opdenakker, P. M. Loadman, D. R. Edwards, Proteases in cancer drug delivery, *Adv. Drug Deliv. Rev.*, 97 (2016) 144–155.
37. M. R. Emmert-Buck, M. J. Roth, Z. Zhuang, E. Campo, J. Rozhin, B. F. Sloane, L. A. Liotta, W. G. Stetler-Stevenson, Increased gelatinase A (MMP-2) and cathepsin B activity in invasive tumor regions of human colon cancer samples, *Am. J. Pathol.*, 145(6) (1994) 1285–1290.
38. M. A. Nouh, M. M. Mohamed, M. El-Shinawi, M. A. Shaalan, D. Cavallo-Medved, H. M. Khaled, B. F. Sloane, Cathepsin B: A potential prognostic marker for inflammatory breast cancer, *J. Transl. Med.*, 9 (2011) 1.
39. A. T. Chan, Y. Baba, K. Shima, K. Nosho, D. C. Chung, K. E. Hung, U. Mahmood, K. Madden, K. Poss, A. Ranieri, D. Shue, R. Kucherlapati, C. S. Fuchs, S. Ogino, Cathepsin B expression and survival in colon cancer: Implications for molecular detection of neoplasia, *Cancer Epidemiol. Biomarkers Prev.*, 19(11) (2010) 2777–2785.
40. A. Mitrovic, B. Mirkovic, I. Sosic, S. Gobec, J. Kos, Inhibition of endopeptidase and exopeptidase activity of cathepsin B impairs extracellular matrix degradation and tumour invasion, *Biol. Chem.*, 397(2) (2016) 165–174.
41. J. E. Koblinski, M. Ahram, B. F. Sloane, Unraveling the role of proteases in cancer, *Clin. Chim. Acta*, 291(2) (2000) 113–135.
42. Y. Fang, F. Du, Y. Xu, H. Meng, J. Huang, X. Zhang, W. Lu, S. Liu, J. Yu, Enhanced cellular uptake and intracellular drug controlled release of VESylated gemcitabine prodrug nanocapsules, *Colloids Surf. B Biointerfaces*, 128 (2015) 357–362.

43. Y. Tsume, G. L. Amidon, The feasibility of enzyme targeted activation for amino acid/dipeptide monoester prodrugs of floxuridine; cathepsin D as a potential targeted enzyme, *Molecules*, 17(4) (2012) 3672–3689.

44. G. M. Dubowchik, R. A. Firestone, Cathepsin B-sensitive dipeptide prodrugs. 1. A model study of structural requirements for efficient release of doxorubicin, *Bioorg. Med. Chem. Lett.*, 8(23) (1998) 3341–3346.

45. T. Etrych, J. Strohalm, P. Chytil, P. Cernoch, L. Starovoytova, M. Pechar, K. Ulbrich, Biodegradable star HPMA polymer conjugates of doxorubicin for passive tumor targeting, *Eur. J. Pharm. Sci.*, 42(5) (2011) 527–539.

46. V. Omelyanenko, C. Gentry, P. Kopeckova, J. Kopecek, HPMA copolymer-anticancer drug-OV-TL16 antibody conjugates. II. Processing in epithelial ovarian carcinoma cells in vitro, *Int. J. Cancer*, 75(4) (1998) 600–608.

47. R. Zhang, J. Yang, M. Sima, Y. Zhou, J. Kopecek, Sequential combination therapy of ovarian cancer with degradable N-(2-hydroxypropyl)m ethacrylamide copolymer paclitaxel and gemcitabine conjugates, *Proc. Natl. Acad. Sci. U. S. A.*, 111(33) (2014) 12181–12186.

48. X. Zhang, K. Tang, H. Wang, Y. Liu, B. Bao, Y. Fang, X. Zhang, W. Lu, Design, Synthesis, and biological evaluation of new cathepsin B-sensitive camptothecin nanoparticles equipped with a novel multifuctional linker, *Bioconjug. Chem.*, 27(5) (2016) 1267–1275.

49. V. Omelyanenko, P. Kopeckova, C. Gentry, J. Kopecek, Targetable HPMA copolymer-adriamycin conjugates. Recognition, internalization, and subcellular fate, *J. Control. Release*, 53(1–3) (1998) 25–37.

50. R. Xiang, Y. Luo, A. G. Niethammer, R. A. Reisfeld, Oral DNA vaccines target the tumor vasculature and microenvironment and suppress tumor growth and metastasis, *Immunol. Rev.*, 222 (2008) 117–128.

51. E. Dall, H. Brandstetter, Mechanistic and structural studies on legumain explain its zymogenicity, distinct activation pathways, and regulation, *Proc. Natl. Acad. Sci. U. S. A.*, 110(27) (2013) 10940–10945.

52. G. Miller, S. P. Matthews, T. Reinheckel, S. Fleming, C. Watts, Asparagine endopeptidase is required for normal kidney physiology and homeostasis, *FASEB J.*, 25(5) (2011) 1606–1617.

53. C. Liu, C. Sun, H. Huang, K. Janda, T. Edgington, Overexpression of legumain in tumors is significant for invasion/metastasis and a

candidate enzymatic target for prodrug therapy, *Cancer Res.*, 63(11) (2003) 2957–2964.

54. Y. J. Chen, S. C. Wu, C. Y. Chen, S. C. Tzou, T. L. Cheng, Y. F. Huang, S. S. Yuan, Y. M. Wang, Peptide-based MRI contrast agent and near-infrared fluorescent probe for intratumoral legumain detection, *Biomaterials*, 35(1) (2014) 304–315.
55. N. Li, Q. Liu, Q. Su, C. Wei, B. Lan, J. Wang, G. Bao, F. Yan, Y. Yu, B. Peng, J. Qiu, X. Yan, S. Zhang, F. Guo, Effects of legumain as a potential prognostic factor on gastric cancers, *Med. Oncol.*, 30(3) (2013) 621.
56. R. A. Reisfeld, The tumor microenvironment: A target for combination therapy of breast cancer, *Crit. Rev. Oncog.*, 18(1–2) (2013) 115–133.
57. Y. Lin, C. Wei, Y. Liu, Y. Qiu, C. Liu, F. Guo, Selective ablation of tumor-associated macrophages suppresses metastasis and angiogenesis, *Cancer Sci.*, 104(9) (2013) 1217–1225.
58. D. Liao, Z. Liu, W. Wrasidlo, T. Chen, Y. Luo, R. Xiang, R. A. Reisfeld, Synthetic enzyme inhibitor: A novel targeting ligand for nanotherapeutic drug delivery inhibiting tumor growth without systemic toxicity, *Nanomedicine*, 7(6) (2011) 665–673.
59. L. E. Edgington, M. Verdoes, A. Ortega, N. P. Withana, J. Lee, S. Syed, M. H. Bachmann, G. Blum, M. Bogyo, Functional imaging of legumain in cancer using a new quenched activity-based probe, *J. Am. Chem. Soc.*, 135(1) (2013) 174–182.
60. K. M. Bajjuri, Y. Liu, C. Liu, S. C. Sinha, The legumain protease-activated auristatin prodrugs suppress tumor growth and metastasis without toxicity, *ChemMedChem*, 6(1) (2011) 54–59.
61. Y. Liu, K. M. Bajjuri, C. Liu, S. C. Sinha, Targeting cell surface alpha(v)beta(3) integrin increases therapeutic efficacies of a legumain protease-activated auristatin prodrug, *Mol. Pharm.*, 9(1) (2012) 168–175.
62. Z. Liu, M. Xiong, J. Gong, Y. Zhang, N. Bai, Y. Luo, L. Li, Y. Wei, Y. Liu, X. Tan, R. Xiang, Legumain protease-activated TAT-liposome cargo for targeting tumours and their microenvironment, *Nat. Commun.*, 5 (2014) 4280.
63. W. Wu, Y. Luo, C. Sun, Y. Liu, P. Kuo, J. Varga, R. Xiang, R. Reisfeld, K. D. Janda, T. S. Edgington, C. Liu, Targeting cell-impermeable prodrug activation to tumor microenvironment eradicates multiple drug-resistant neoplasms, *Cancer Res.*, 66(2) (2006) 970–980.
64. R. L. Smith, O. A. Astrand, L. M. Nguyen, T. Elvestrand, G. Hagelin, R. Solberg, H. T. Johansen, P. Rongved, Synthesis of a novel legumain-

cleavable colchicine prodrug with cell-specific toxicity, *Bioorg. Med. Chem.*, 22(13) (2014) 3309–3315.

65. L. Stern, R. Perry, P. Ofek, A. Many, D. Shabat, R. Satchi-Fainaro, A novel antitumor prodrug platform designed to be cleaved by the endoprotease legumain, *Bioconjug. Chem.*, 20(3) (2009) 500–510.
66. Y. Luo, H. Zhou, J. Krueger, C. Kaplan, S. H. Lee, C. Dolman, D. Markowitz, W. Wu, C. Liu, R. A. Reisfeld, R. Xiang, Targeting tumor-associated macrophages as a novel strategy against breast cancer, *J. Clin. Invest.*, 116(8) (2006) 2132–2141.
67. Z. Gu, A. A. Aimetti, Q. Wang, T. T. Dang, Y. Zhang, O. Veiseh, H. Cheng, R. S. Langer, D. G. Anderson, Injectable nano-network for glucose-mediated insulin delivery, *ACS Nano*, 7(5) (2013) 4194–4201.
68. T. Abe, K. Sakamoto, H. Kamohara, Y. Hirano, N. Kuwahara, M. Ogawa, Group II phospholipase A2 is increased in peritoneal and pleural effusions in patients with various types of cancer, *Int. J. Cancer*, 74(3) (1997) 245–250.
69. C. Leidy, L. Linderoth, T. L. Andresen, O. G. Mouritsen, K. Jorgensen, G. H. Peters, Domain-induced activation of human phospholipase A2 type IIA: Local versus global lipid composition, *Biophys. J.*, 90(9) (2006) 3165–3175.
70. J. R. Graff, B. W. Konicek, J. A. Deddens, M. Chedid, B. M. Hurst, B. Colligan, B. L. Neubauer, H. W. Carter, J. H. Carter, Expression of group IIa secretory phospholipase A2 increases with prostate tumor grade, *Clin. Cancer Res.*, 7(12) (2001) 3857–3861.
71. J. Jiang, B. L. Neubauer, J. R. Graff, M. Chedid, J. E. Thomas, N. W. Roehm, S. Zhang, G. J. Eckert, M. O. Koch, J. N. Eble, L. Cheng, Expression of group IIA secretory phospholipase A2 is elevated in prostatic intraepithelial neoplasia and adenocarcinoma, *Am. J. Pathol.*, 160(2) (2002) 667–671.
72. A. Goel, S. S. Chauhan, Role of proteases in tumor invasion and metastasis, *Indian J. Exp. Biol.*, 35(6) (1997) 553–564.
73. R. Weissleder, C. H. Tung, U. Mahmood, A. Bogdanov, Jr., In vivo imaging of tumors with protease-activated near-infrared fluorescent probes, *Nat. Biotechnol.*, 17(4) (1999) 375–378.
74. A. H. Hansen, O. G. Mouritsen, A. Arouri, Enzymatic action of phospholipase A(2) on liposomal drug delivery systems, *Int. J. Pharm.*, 491(1–2) (2015) 49–57.
75. T. Tagami, Y. Ando, T. Ozeki, Fabrication of liposomal doxorubicin exhibiting ultrasensitivity against phospholipase A2 for efficient

pulmonary drug delivery to lung cancers, *Int. J. Pharm.*, 517(1–2) (2017) 35–41.

76. D. Aili, M. Mager, D. Roche, M. M. Stevens, Hybrid nanoparticle-liposome detection of phospholipase activity, *Nano Lett.*, 11(4) (2011) 1401–1405.

77. Y. Li, H. Hu, Q. Zhou, Y. Ao, C. Xiao, J. Wan, Y. Wan, H. Xu, Z. Li, X. Yang, Alpha-amylase- and redox-responsive nanoparticles for tumor-targeted drug delivery, *ACS Appl. Mater. Interfaces*, 9(22) (2017) 19215–19230.

78. C. Park, H. Kim, S. Kim, C. Kim, Enzyme responsive nanocontainers with cyclodextrin gatekeepers and synergistic effects in release of guests, *J. Am. Chem. Soc.*, 131(46) (2009) 16614–16615.

79. D. A. Butterfield, S. S. Hardas, M. L. Lange, Oxidatively modified glyceraldehyde-3-phosphate dehydrogenase (GAPDH) and Alzheimer's disease: Many pathways to neurodegeneration, *J. Alzheimers Dis.*, 20(2) (2010) 369–393.

80. J. K. Kundu, Y. J. Surh, Nrf2-Keap1 signaling as a potential target for chemoprevention of inflammation-associated carcinogenesis, *Pharm. Res.*, 27(6) (2010) 999–1013.

81. L. Q. Chen, X. E. Zhang, W. H. Xie, Y. F. Zhou, Z. P. Zhang, A. E. Cass, Genetic modification of glucose oxidase for improving performance of an amperometric glucose biosensor, *Biosens Bioelectron.*, 17(10) (2002) 851–857.

82. R. Mo, T. Jiang, J. Di, W. Tai, Z. Gu, Emerging micro- and nanotechnology based synthetic approaches for insulin delivery, *Chem. Soc. Rev.*, 43(10) (2014) 3595–3629.

83. A. Napoli, M. J. Boerakker, N. Tirelli, R. J. Nolte, N. A. Sommerdijk, J. A. Hubbell, Glucose-oxidase based self-destructing polymeric vesicles, *Langmuir*, 20(9) (2004) 3487–3491.

84. A. J. Harnoy, I. Rosenbaum, E. Tirosh, Y. Ebenstein, R. Shaharabani, R. Beck, R. J. Amir, Enzyme-responsive amphiphilic PEG-dendron hybrids and their assembly into smart micellar nanocarriers, *J. Am. Chem. Soc.*, 136(21) (2014) 7531–7534.

85. J. Rao, A. Khan, Enzyme sensitive synthetic polymer micelles based on the azobenzene motif, *J. Am. Chem. Soc.*, 135(38) (2013) 14056–14059.

86. Y. Li, G. Liu, X. Wang, J. Hu, S. Liu, Enzyme-responsive polymeric vesicles for bacterial-strain-selective delivery of antimicrobial agents, *Angew. Chem. Int. Ed. Engl.*, 55(5) (2016) 1760–1764.

87. R. J. Amir, S. Zhong, D. J. Pochan, C. J. Hawker, Enzymatically triggered self-assembly of block copolymers, *J. Am. Chem. Soc.*, 131(39) (2009) 13949–13951.

88. B. H. Hu, P. B. Messersmith, Rational design of transglutaminase substrate peptides for rapid enzymatic formation of hydrogels, *J. Am. Chem. Soc.*, 125(47) (2003) 14298–14299.

89. S. J. Sofia, A. Singh, D. L. Kaplan, Peroxidase-catalyzed crosslinking of functionalized polyaspartic acid polymers, *J. Macromol. Sci.*, 39(10) (2007) 31.

90. R. Jin, C. Hiemstra, Z. Zhong, J. Feijen, Enzyme-mediated fast in situ formation of hydrogels from dextran-tyramine conjugates, *Biomaterials*, 28(18) (2007) 2791–2800.

91. J. Hu, G. Zhang, S. Liu, Enzyme-responsive polymeric assemblies, nanoparticles and hydrogels, *Chem. Soc. Rev.*, 41(18) (2012) 5933–5949.

Chapter 4

Redox-Responsive Nanomedicine

Shi Du, Hui Xiong, and Jing Yao

State Key Laboratory of Natural Medicines,
China Pharmaceutical University,
24 Tongjiaxiang, Nanjing 210009, China

yaojing@cpu.edu.cn

It is known that only inside the pathological cells many small-molecule therapeutics (e.g., anti-cancer drugs, photosensitizers, and anti-oxidants) and biotherapeutics (e.g., peptide and protein drugs, and siRNA) can exert therapeutic effects. Redox-sensitive nanosystems have been widely investigated and provide a promising drug delivery strategy in treating various diseases like cancer, inflammation, and so on. Redox-responsive drug delivery systems are inspired by the unique oxidation-reduction microenvironment inside the pathological cells. Usually, drug release may be triggered by the elevated redox potential and/or reactive oxygen species (ROS) concentration, which realizes the controlled drug release inside the diseased cells and avoids the possible burst release before drugs reach the lesions. In this

Stimuli-Responsive Nanomedicine
Edited by Lin Zhu

ISBN 978-981-4800-70-9 (Hardcover), 978-0-429-29529-4 (eBook)
www.jennystanford.com

chapter, we will review the mechanisms and advances of redox-responsive nanosized drug delivery systems which are able to control drug release in response to glutathione and/or reactive oxygen species.

4.1 Introduction

Nanotechnology-based drug delivery systems (nanoDDS) have emerged as a suitable strategy for overcoming pharmacokinetic limitations associated with conventional drug formulations [1, 2]. Although proven advantages in solubilizing poorly soluble drugs, prolonging drug circulation time, and reducing drug degradation [3], nanoDDS are in need of improvement. One of the dilemmas of the traditional nanoDDS is how to maintain the balance between drug release and system stability. As we know, the pathological microenvironment usually shows unique or abnormal characteristics compared with normal tissues, which provides great opportunities for the development of stimuli-sensitive drug delivery systems [4]. The biological redox status has been investigated for decades and its importance in developing redox-responsive nanoDDS has been emphasized recently as it is implicated that an increased level of glutathione/glutathione disulfide (GSH/GSSG) or reactive oxygen species (ROS) exists in numerous pathologies [5].

Although many molecules contribute to the highly reducing profile of the intracellular compartment, GSH/GSSG is considered the major redox buffer and accounts for the greatest amount among all reducing molecules. Often the total concentrations of GSH and/or GSSG are used to estimate the redox environment of the cell [6]. As a consequence, there exist a plethora of drug delivery systems that have taken advantage of the remarkable difference of GSH existing between the intracellular and extracellular environment of solid tumors. The intracellular concentration of GSH is about 1–10 mM, while the GSH concentration in the extracellular fluid in the tissue is only about 2–20 μM [7]. In addition, the GSH concentration in the tumor mass is also markedly higher than that in the normal tissue, which is the physiological basis of the reducing environment in tumors [7] (Fig. 4.1).

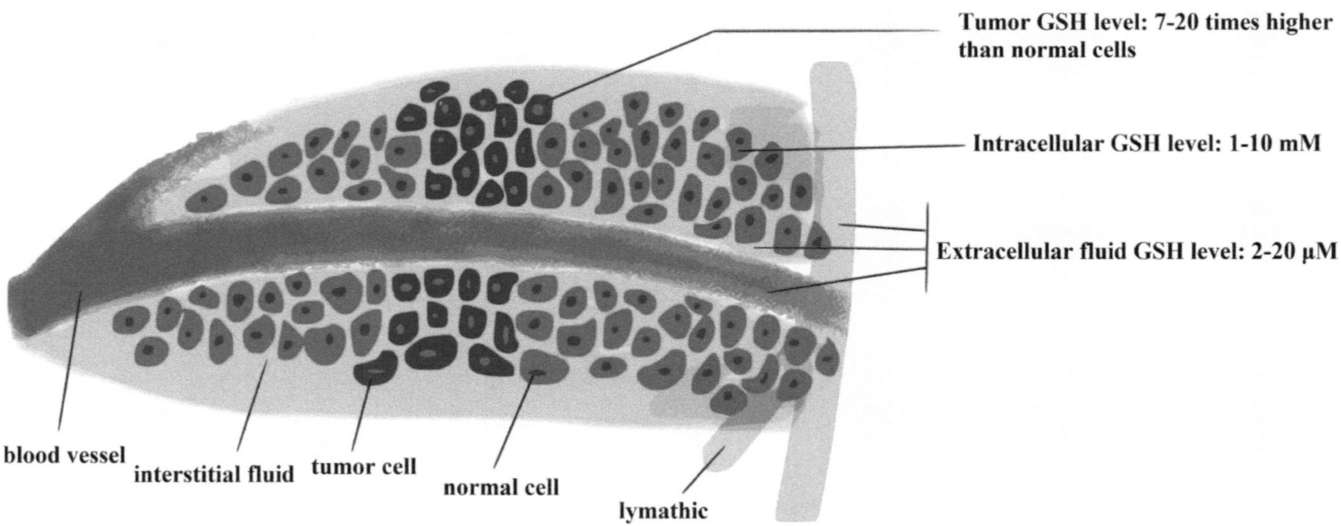

Figure 4.1 Schematic illustration of the intracellular and extracellular difference of GSH levels.

The reason behind such a concentration gradient of GSH is the rapid enzymatic degradation of glutathione in human plasma and more vigorous metabolism of tumor cells than normal cells. The GSH/GSSG redox couple imparts cells its oxidative property and is kept under a reductive condition by NADPH and glutathione reductase [8]. Other redox couples also contribute towards the intracellular levels of GSH, namely nicotinamide adenine dinucleotide ($NADH/NAD^+$), nicotinamide adenine dinucleotide phosphate ($NADPH/NADP^+$) and thioredoxin(red)/thioredoxin(ox) [9]. In addition to the cytosolic reducing conditions, the redox potential in the endosome is modulated by a specific reducing enzyme, gamma interferon-inducible lysosomal thiol reductase (GILT) in the co-presence of reducing agents such as cysteine but not GSH [10], while the lysosome's reductive environment is controlled by high amounts of cysteine, like thiols which keep low-mass iron in the reduced state in the acidic internal compartment [11]. Studies have shown that molecules like quinone reductase that are overexpressed in cancer cells, also contribute towards their increased redox potential. This attribute has been explored using substrates like quinone lipids to formulate liposomes for drug delivery [12].

In addition to the systems that respond to the concentration gradient of GSH for delivery of cargoes, the nanoDDS that are sensitive to ROS are also being developed. With the poly (propylene sulfide) containing nanomaterials emerged in 2004 [13], various ROS-responsive nanoDDS have been developed and been another integral part of redox-responsive nanoDDS. ROS include hydroxyl radicals (OH^-), hydrogen peroxides (H_2O_2), peroxynitrites ($ONOO^-$), and superoxides (O^{2-}) [14]. They are persistently generated, transformed, consumed and play crucial roles in various physiological processes such as cellular growth, signaling, immune response, and senescence [15]. However, the imbalance in the ROS production leads to the oxidative stress and inflammatory events, which damages the tissues and organs. Therefore, the imbalance is usually correlated with the generation and progression of various diseases, including cancer, diabetes, and cardiovascular and neurodegenerative diseases [16]. Among all the ROS species, H_2O_2 is the most common one in tumors and at higher concentrations when compared to others. Also, elevated generation rates of H_2O_2 are typically found in most types of cancer cells (up to ~0.5 × 10^{-4} nmol/cell/h) [17], monocytes and stimulated polymorphonuclear leukocytes (up to ~1.5 × 10^{-4} nmol/cell/h) [18]. The significance of ROS in physiological and pathological conditions, particularly high ROS concentration in inflammatory and cancer tissues, have motivated the construction of clever chemical tools such as ROS-responsive nanovehicles [4, 19].

This chapter mainly reviews and discusses the recent advances in the field of redox-responsive nanomedicines, particularly the GSH-sensitive nanoDDS and ROS-sensitive nanoDDS, and provides an outlook of their future design.

4.2 GSH-Sensitive Drug Delivery Systems

The majority of the initial work on GSH-sensitive systems focused on introducing disulfide bonds into conventional drug delivery systems (Fig. 4.2). The disulfide bonds were stable under the normal condition in the extracellular compartment but underwent reversible conversion to free thiols in the presence of reducing agents like GSH in the intracellular compartment [4]. A summary of the different types of redox-sensitive nanoDDS is shown in Table 4.1. Generally, disulfide groups were incorporated in the backbone and side chain or as cross-linker in the nanocarriers. In addition, the disulfide groups could be used as the linker to synthesize the GSH-responsive drug conjugates. In this part, we will provide an introduction to different types of GSH-responsive drug delivery systems and discuss representative examples from recent studies.

Table 4.1 Representative GSH-sensitive nanoDDS

Components	Payloads	Key results	Ref.
Nanomicelles			
TAT-PEG-Asp_8-Dox	DOX	Redox-responsive drug release Increased cellular uptake Bypassing P-glycoprotein-mediated drug efflux	[25]
mPEG-b-P(Des-alt-Cys)	CPT	Dual redox-responsive drug release High antitumor efficacy both in vitro and in vivo	[26]

(*Continued*)

Components	Payloads	Key results	Ref.
FA-Pluronic	DOX	FA-mediated endocytosis Dual stimuli (pH and redox) triggered drug release Inhibition of multidrug resistance (MDR)	[27]
PC5MA-ss-PEO	DOX	Tumor targeting	[28]
mPEG-ss-PTX/TPGS	PTX	Inhibition of MDR Enhanced anticancer effect	[29]
TPGS-ss-PTX	PTX	Redox-responsive rapid micelle disassociation and drug release High cytotoxicity in PTX-resistant cancer cells	[30]
PEG-PAsp(MEA)-PEI	siRNA	Reduction and pH dual-sensitive tumor-targeted siRNA delivery	[31]
Liposomes			
Bioreducible lipid-like materials	siRNA	GSH-responsive siRNA delivery GFP downregulation	[32]
Nanogels			
P(OEGMAs-ss-AA)	DOX	Increased DOX release High proliferation inhibition in GSH-OEt-pretreated Bel7402 cells	[33]
PVA	DOX	Fast and complete release of DOX Facilitated cell internalization	[34]
Inorganic nanoparticles			
Gold nanoparticles	FITC	Differential response to extra and intracellular glutathione in cancer cells	[35]
Gold nanoparticles	CPT	Inducing higher oxidative stress in lung cancer cells than in normal cells Intensified synergistic therapeutic effects	[21]
MMSSN	DOX	Efficient pH and redox responsiveness	[36]
Other biodegradable nanoparticles			
BSA	Tetracycline	pH and redox dual responsive drug release	[22]
HSA	Ce6	Increased singlet oxygen production Increased cell death upon radiation	[37]
Type B gelatin conjugated by 2-iminothiolane	plasmid DNA and GEM	Improved in vivo targeting efficiency Increased transfection efficiency and tumor growth suppression	[38]
Carboxymethyl chitosan	Methotrexate (MTX)	pH and redox dual-responsive drug release Remarkable cytotoxicity against cancer cells	[39]

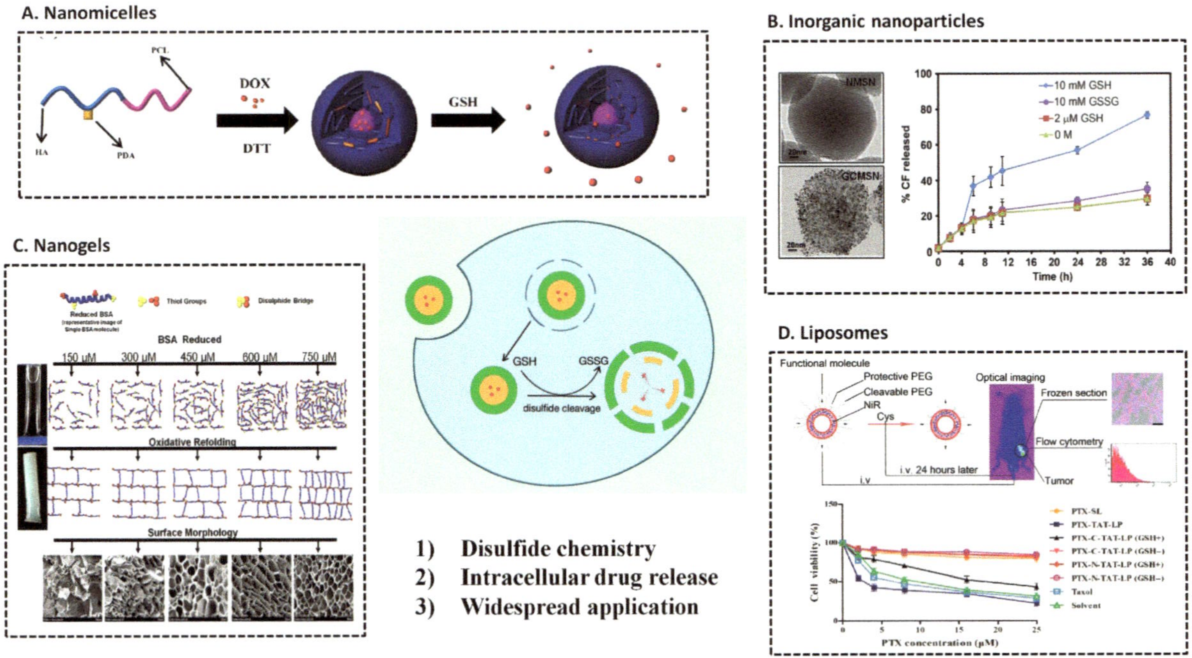

Figure 4.2 Different types of GSH-responsive nanomedicines. A, the delivery mechanism of the GSH-responsive cross-linked nanomicelles for delivery of DOX [20]; B, the GSH-responsive gold nanoparticle-modified mesoporous silica nanoparticles (GCMSN). TEM micrographs and drug release profile of the CF-loaded GCMSN [21]; C, hypothetical representation of the reduced BSA hydrogel undergoing oxidative refolding in non-native form and transforms into a gel and SEM of BSA gel for the selected concentrations studied [22]; D, Cleavable PEG modified liposomal delivery system and cytotoxicity of PTX-loaded GSH-responsive liposomes [23, 24]. Figures adapted with permission from [20–24]. Copyright (2011, 2015) American Chemical Society.

4.2.1 Nanomicelles

As mentioned before, nanoDDS have been widely utilized for anticancer therapy, among which nanomicelles as a new type of nanocarriers with an appropriate size (10–100 nm) are promising [24, 25]. Due to their unique structure formed by amphiphiles, nanomicelles could not only solubilize poorly water soluble drugs but also protect hydrolytically sensitive drugs from degradation by the enzymes and chemicals in the aqueous environment [40, 41]. However, slow degradation or dissociation of micelles may delay the drug release and compromise the therapeutic effect. Many studies on GSH-responsive micelles suggested that the disulfide linker had the potential to improve the drug delivery efficiency of the nanoparticles. GSH-responsive nanomicelles are usually assembled by amphiphilic polymers containing a disulfide linkage that was introduced through a disulfide functionalized initiator between their hydrophobic and hydrophilic blocks, thereby creating a "shell-sheddable" micellar structure. This disulfide linkage between the blocks is cleaved upon exposure to glutathione, destabilizing the micelles and releasing the therapeutic cargo. Other nanomicelles take advantage of the thiol-reducible linkers that are used to cross-link the micelle core or shell or employ disulfide linkages to attach drug molecules to one of the blocks. These reduction-triggered systems show a rapid burst release of the payloads in response to the high redox potential by the destruction of the nanocarrier [42].

The nanomicelles developed by Li et al. were one of the typical "shell-sheddable" micelles [43]. In this study, the backbone of the micelles was the conjugate of hyaluronic acid (HA) and deoxycholic acid (DOCA) linked by a disulfide bond (HA-ss-DOCA). Paclitaxel (PTX) was loaded into the micelles with high loading efficiency (34.1%). The desirable redox sensitivity of micelles has been verified via the size distribution change and in vitro drug release. HA-ss-DOCA micelles were essentially unchanged at 10 μM GSH but increased significantly at 10 and 20 mM GSH. However, both 10 and 20 mM GSH seemed to make no difference to the size distribution of HA-DOCA. In addition, in vitro drug release experiments showed that in the presence of 10 mM GSH, the release of PTX from insensitive nanomicelles was slow, while in the same condition, the release of PTX from HA-ss-DOCA micelles was dramatically accelerated: about 28.1% of PTX was released in 4 h and near 52.1% was released in 24 h. Furthermore, with a higher GSH concentration of 20 mM (close to intracellular GSH levels in tumor cells), more significant PTX release was observed: about 55.2% of PTX was released in 4 h and more than 90% was released in 24 h. Similarly, Nguyen et al. prepared shell-detachable micelles comprised a polymethacrylate bearing cholesterol and a polyethylene glycol block linked by disulfide bonds (PC5MA-ss-PEO) [28]. The PC5MA-ss-PEO nanomicelles displayed a rapid release of DOX in response to the reductive condition (10 mM DL-Dithiothreitol) and led to higher cytotoxicity to lung cancer cells compared with the non-reducible nanomicelles.

Han et al. prepared bioreducible nanoparticles constructed by an amphiphilic hyaluronic acid and polycaprolactone copolymer cross-linked via disulfide linkages, in which DOX was loaded into the nanoparticles as the model anticancer drug (DOX-HA-ss-NPs) [20]. In vitro release studies showed that in the presence of 10 mM GSH, more than 80% of the drug from the DOX-HA-ss-NPs was released within 48 h, whereas only nearly 30% DOX was released without GSH. On the other hand, there was no significant difference in drug release from the non-cross-linked nanoparticles in the presence or absence of GSH. Li et al. developed the core-cross-linked nanomicelles self-assembled by the thiolated linear-dendritic polymers containing a hydrophilic polyethylene glycol (PEG) and a dendritic cluster of cholic acids [44]. The thiol groups in the core of micelles were oxidized to disulfide bonds to form the disulfide cross-linked micelles (DCMs). The DCMs were stable in the solution with an extracellular GSH level (~2 μM). In the presence of intracellular GSH level (10 mM), the particle size of DCMs decreased suddenly after 30 min indicating that a critical number of disulfide bonds were reduced to dissociate the micelles. In vitro release studies showed that drug (PTX) release was accelerated when GSH (10 mM) was added. GSH-responsive nanomicelles could also be established by conjugating a polymer or a tumor-specific ligand to chemotherapy drugs via a cleavable bond, like disulfide bonds. Zhao and coworkers synthesized mPEG-ss-PTX by conjugating hydrophilic mPEG with PTX via a redox-sensitive disulfide bond [29]. The disulfide bond in the mPEG-ss-PTX conjugate could be broken in the reductive environment (in tumor cells) and PTX would be released to kill cancer cells. The redox-sensitive property was confirmed by the in vitro PTX release and in vivo pharmacokinetic data. Moreover, remarkably higher inhibition on MCF-7 cells exerted by mPEG-ss-PTX micelles was observed compared with that of free PTX.

Due to the heterogeneities of the cellular redox status in different tumor cells, the nanoDDS in response to more than one stimulus could serve as a strategy for effective drug release and anticancer treatment. For example, Chiang et al. developed a dual redox-responsive copolymer, mPEG-b-p(Des-alt-Cys), by co-polymerizing the ROS-sensitive structure, diethyl sulfide (Des), and GSH-sensitive structure, disulfide-containing cystamine (Cys), on the end of methoxy poly(ethylene glycol) (mPEG) [26]. The anticancer drug camptothecin (CPT) was incorporated into the micelles during self-assembly of the copolymers. The CPT-loaded dual redox-responsive micelles showed good stability at low levels of ROS and GSH in the blood circulation while high redox sensitivities in cancer cells with high levels of ROS or GSH. The ROS-sensitive nanomedicine will be discussed in detail in a later section.

4.2.2 Liposomes

Liposomes are one of the most widely studied drug delivery systems capable of carrying either hydrophilic or hydrophobic drugs in their inner core and bilayer membrane [43]. GSH-sensitive strategy has been used to improve the drug delivery efficiency of the liposome-based nanocarriers. Disulfide bonds can be used as targeting moities or utilized to prepare redox-sensitive lipids. Such GSH-responsive liposomes exhibit stability until reaching a reducing environment where the disulfide bonds can be cleaved, resulting in the disruption of the liposomal membrane and release of the liposomal contents [46]. In a study conducted by Wang et al., the bioreducible lipid-like materials (termed "lipidoid") containing disulfide bonds were synthesized. The lipidoid could load siRNA via electrostatic interaction. The gel electrophoresis and intracellular siRNA trafficking studies suggested that the bioreducible lipidoid in a strongly reductive intracellular environment boosted siRNA release, resulting from its degradation. Besides, green fluorescence protein (GFP) expression in MDA-MB-231 cells was suppressed by the lipidoid/siRNA complexes, while the naked siRNA had no effect on the GFP expression [32]. More recently, the liposomes decorated with trans-activating transcriptional activator (TAT), a cell-penetrating peptide, and PEG were reported by Fu et al. [24]. These liposomes were incorporated with a detachable disulfide-linked PEG coating, whereby it had the long blood circulation, improved tumor penetration, and redox-responsive intracellular PTX release. In the presence of GSH, the PTX loaded liposomes strongly inhibited the proliferation of murine melanoma B16F1 cells in vitro and in vivo (with a tumor inhibition rate of 69.4%). In addition, the serum aspartate transaminase level, alanine transaminase level, and creatine kinase level were in the normal range, indicating the low toxicity of these PTX-loaded liposomes.

4.2.3 Nanogels

Nanogels are biocompatible three-dimensional materials with high water content and sizes ranging from tens of nanometers to submicrons [47, 48]. Nanogels can be applied for encapsulation and delivery of various agents including anticancer drugs, proteins, plasmid DNA and imaging probes [48]. Matyjaszewski and coworkers described the preparation of well-defined reduction-sensitive functional nanogels using the inverse mini-emulsion atom transfer radical polymerization (ATRP) and disulfide–thiol exchange reaction [49, 50]. These nanogels could be used to load anticancer drugs, carbohydrates, and proteins [51, 52]. Hao et al. constructed the redox-sensitive hydrophilicity/hydrophobicity-reversable poly(oligo(ethylene glycol) methacrylates-ss-acrylic acid (p(OEGMAs-ss-AA)) nanogels [33]. DOX was loaded into the nanogels with a loading efficiency of 80%. Once internalized, DOX was released from the p(OEGMAs-ss-AA) nanogels in response to the intracellular GSH and entered nucleus to exert cytotoxicity. Chen et al. reported the synthesis of charge-conversional and reduction-sensitive poly(vinyl alcohol) (PVA) nanogels [34]. These PVA nanogels were prepared in a straightforward manner by inverse nanoprecipitation via the "click" reaction with an average diameter of 118 nm. Both MTT assay and real time cell analysis (RTCA) showed that these DOX-loaded charge-conversional and reducible PVA nanogels had much higher cytotoxicity than DOX-loaded non-charge-conversional or reduction-insensitive PVA nanogels. These results might result from the enhanced cellular uptake and efficient intracellular release of anticancer drugs in response to cytoplasmic glutathione as well as tumoral pH.

4.2.4 Inorganic Nanoparticles

Nanoparticles made of gold, semiconductor nanocrystals, super-paramagnetic, silicon and silica-based materials have been extensively studied as promising nanocarriers for drug delivery. Many redox-sensitive inorganic nanosystems have been reported for biomedical applications [53, 54]. Lu et al. developed inorganic material-based redox-sensitive systems [21]. This system was built to enable the synergistic effects of the amino-functionalized mesoporous silica nanoparticles (NMSNs) with high payload capacity for chemotherapeutic agents and gold nanoparticles (GNPs) as pore blocker and ROS elicitor to eradicate human non-small cell lung carcinoma (NSCLC). In this work, GNPs (5 nm in diameter) were attached onto the NMSNs via a relatively weak gold-nitrogen bond. A cyanine-based fluorescent probe (CF) was loaded in GCMSNs as a model guest molecule to investigate its GSH sensitivity and controlled release property. Results suggested that the release of CF reached 77% in 36 h after the addition of 10 mM GSH (known as intracellular level of GSH), while the control group (without GSH added) exhibited only 29% premature release of CF, which confirmed the GSH-responsive intracellular release of the payload from the designed GCMSNs. Furthermore, the authors demonstrated that the competitive binding of GNPs to intracellular GSH might induce oxidative stress to break down the defense of cancer cells, making them more susceptible to chemotherapy. Also, GCMSN exerted oxidative stress-induced autophagy in A549 lung cancer cells through the conversion of LC3-I to LC3-II, activation of Atg 5 expression, and disruption of bioenergetics, to synergistically eradicate cancer cells with chemotherapeutics. With the synergistic outcome, the designed CPT-loaded GCMSNs might be a unique and superior nanocarrier system for highly effective cancer therapy. Luo et al. reported another redox-sensitive silica and gold nanoparticle based nanoDDS [35]. In this work, a disulfide-linked silica (ReSi) was used to coat the gold nanoparticles. Furthermore, the cancer targeting ligand, transferrin, and fluorescein isothiocyanate (FITC) were conjugated to the nanoparticles. In vitro evaluation showed that with increasing concentration of GSH, DL-dithiothreitol (DTT), or beta-mercaptoethanol, the FITC release was obviously accelerated owing to the disulfide reduction and degradation of silica nanoshell. Nanographene is a class of two-dimensional inorganic carbon nanomaterials which attracted tremendous attentions in various fields in the past decade [55], including biosensing, imaging, and therapy, due to its unique physical and chemical properties. It was reported that with all atoms exposed on its surface, the nanographene possesses ultra-high surface area for efficient binding/loading of various biomolecules, and has been widely used as multifunctional nanocarriers for drug and gene delivery [56]. In a work by Shi et al., nanographene oxide (NGO) was modified by a PEG shell via a disulfide linkage (NGO-ss-mPEG nanoconjugates). NGO-ss-mPEG loaded with DOX via π–π stacking could be internalized by cancer cells via endocytosis. In the presence of intracellular GSH, the disulfide bond of NGO-ss-mPEG was rapidly cleaved to release the loaded DOX at an elevated rate for chemotherapy [57]. In another work, nanoGO-Ag nanocomposites were prepared. DOX was linked directly to the NGO-Ag nanocomposites via disulfide bonds, which could be cleaved by the intracellular GSH to trigger the DOX release [58]. In addition, Li et al. found that the biodegradation behaviors of GO could be regulated by its redox-sensitive coating [59]. The GO without surface coating could be gradually degraded through enzyme-induced oxidization by horseradish peroxidase (HRP) and became toxic to macrophages, while GO coated with biocompatible macromolecules such as PEG or bovine serum albumin (BSA) exhibited no obvious toxicity to cells because it could hardly be degraded by enzymes. In order to obtain biocompatible functionalized GO that can still undergo enzymatic degradation, PEG was conjugated to GO via a cleavable disulfide bond to obtain the GO-ss-PEG with negligible toxicity and considerable degradability.

4.2.5 Other Biodegradable Nanoparticles

As the most abundant protein in human plasma, human serum albumin (HSA) has been used as a delivery system for transporting many metabolic compounds and drugs [60–62]. Molina et al. prepared the photosensitizer (PS) chlorine6 (Ce6)-decorated human serum albumin nanoparticles (HSA NPs) with a diameter of 295 ± 5 nm by a nanoprecipitation method [37]. The NPs were stabilized by a redox-sensitive cross-linker to confer a smart drug delivery that was activated only under the reducing conditions. In vitro experiments showed that the redox-sensitive nanoparticles could exhibit significant cytotoxicity after internalization by cancer cells while remained non-toxic prior to irradiation. It was also reported that HSA could be used to prepare the hybrid nanoparticles with calcium phosphate (CaP) for cisplatin delivery. These nanoparticles were stable in the extracellular condition and released the loaded platinum drugs upon reaching the hypoxic microenvironment [63].

Amiji group reported the long-circulating redox-responsive thiolated type B gelatin (SH-Gel-PEG) that showed tremendous potential as stimuli-responsive gene delivery vehicles [64–68]. It was shown that the thiol cross-linkers of the nanoparticles could be disrupted in the intracellular environment, resulting in payload release and transgene expression. In addition, they have successfully formulated a GSH-responsive gelatin nanoparticle-based system that could efficiently deliver site-specific wt-p53 expressing plasmid DNA as well as gemcitabine (GEM) [38]. The gene/drug combination treatment showed synergistic anti-tumor activity compared to the gene or drug treatment alone indicating these GSH-responsive gelatin nanoparticles had great potential as nanocarriers for delivery of gene and chemodrugs.

4.3 ROS-Responsive Drug Delivery Systems

As another important redox-sensitive nanoDDS, ROS-responsive nanosystems are expected to stabilize the drug at the physiological condition and rapidly release the drug in response to ROS (Fig. 4.3). In order to achieve such goals, several ROS-responsive drug delivery strategies have been investigated. One approach is to introduce the functional groups, such as PPS, to nanomaterials as the "solubility switch." These groups undergo phase transition from hydrophobicity to hydrophilicity in the oxidative environment. Another approach is to use the "ROS-induced degradation" to control drug release. This can be done either by covalent attachment of drug molecules to the existing nanocarriers or by constructing new ROS-responsive nanocarriers for physical drug loading. In this part, both strategies will be discussed (Table 4.2).

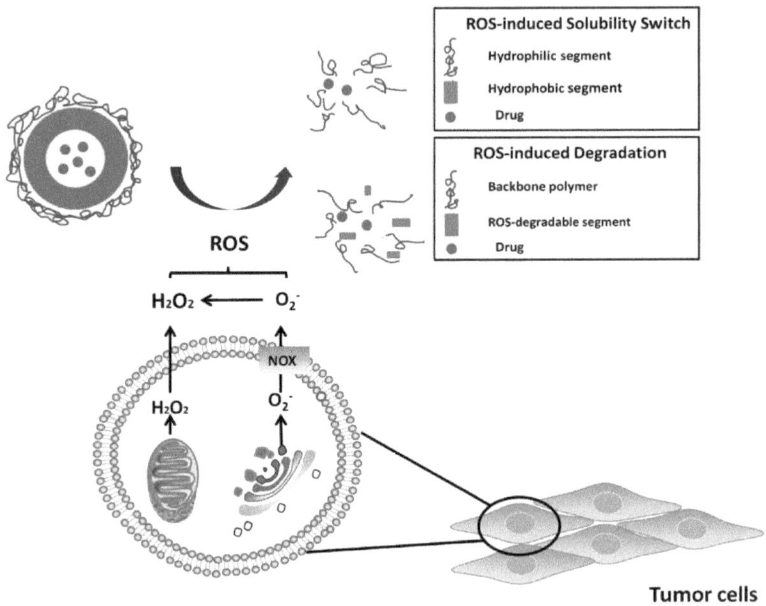

Figure 4.3 Schematic of ROS-responsive drug release via the solubility switch or degradation.

4.3.1 ROS-Responsive "Solubility Switch" Nanomedicines

4.3.1.1 Poly(propylene sulfide) containing nanomaterials

Poly(propylene sulfide) is susceptible to ROS oxidation. This organic sulfide undergoes phase transition from the hydrophobic sulfide to hydrophilic sulfoxide or sulfone in oxidative environment [69].

Table 4.2 Representative ROS-responsive polymeric materials and their mechanisms

ROS-responsive materials	Mechanisms	Ref.
ROS-induced solubility		
Poly(propylene sulfide)	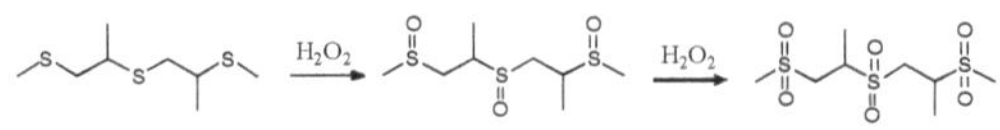	[70, 71]
Selenium	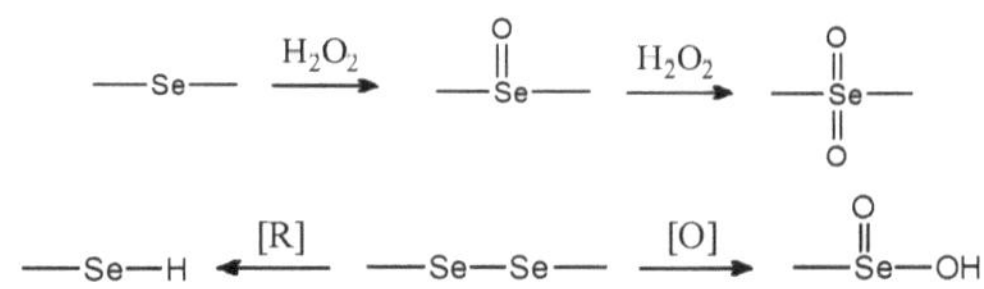	[72, 73]
Tellurium	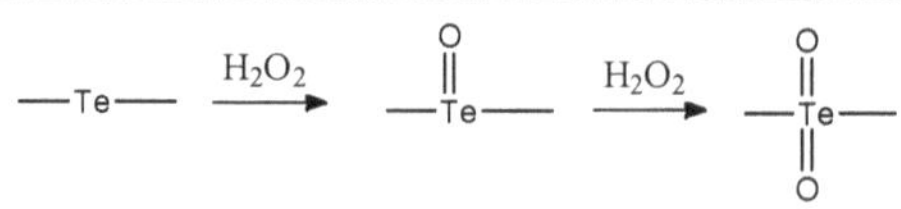	[74, 75]
ROS-induced degradation		
Phenylboronic acid	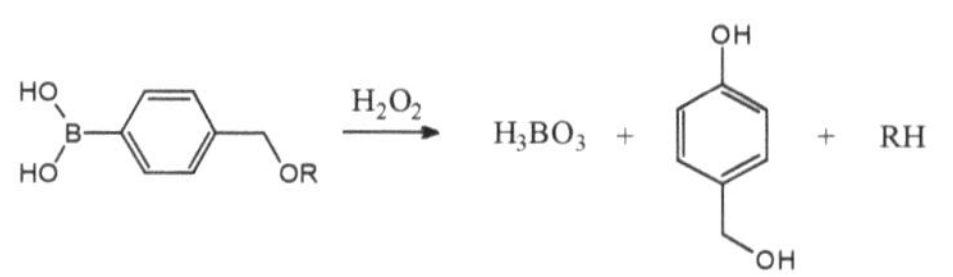	[76, 77]
Proline oligomers	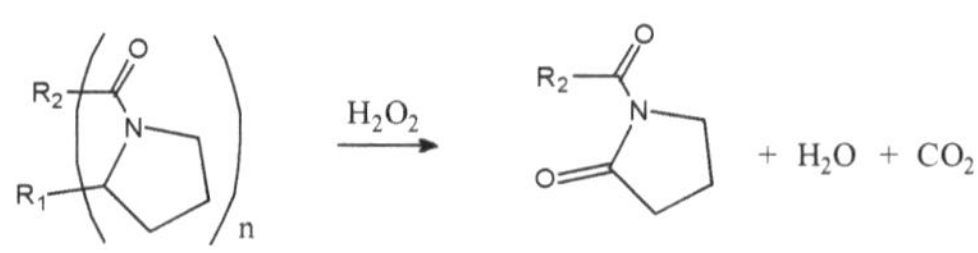	[78]
Polythioketal	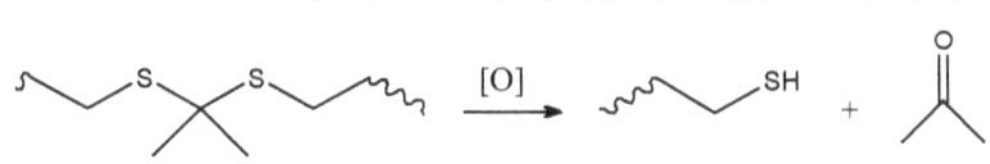	[79, 80]

This unique property makes it a promising candidate for the preparation of ROS-sensitive DDS [13]. In an early study, Hubbell et al. developed the oxidation-responsive polymeric vesicles based on the ABA block polymers of PEG and PPS [81]. The copolymer self-assembled into stable vesicles in aqueous solution at room temperature; however, after treating with H_2O_2, the most prevalent ROS in biological systems, the assembled vesicles underwent destabilization.

In a similar work, Dai and coworkers developed the ROS-sensitive polymeric micelles with an average diameter of 80 nm based on an amphiphilic polymer of poly(propylene sulfide)-polyethylene glycol-serine-folic acid (PPS-mPEG-Ser-FA) [71], for co-loading of the photosensitizer, zinc phthalocyanine (ZNPC), and anticancer drug DOX. The endocytosed ZNPC could generate high concentration of ROS when exposed to laser irradiation to compensate the relatively low intracellular ROS concentration (50–100 mM) and disassemble the micelles (DOX release) via the ROS-induced hydrophobic to hydrophilic transition of the PPS core. Furthermore, high ROS concentration produced by ZNPC could efficiently kill tumor cells. Zhu laboratory conjugated the anticancer drug 7-ethyl-10-hydroxy-camptothecin (SN38) with hyperbranched polyglycerol (HPG) via a H_2O_2 responsive thioether bond to form HPG-2S-SN38 which could self-assemble into stable nanomicelles for encapsulating the cinnamaldehyde (CA). In this system, the thioether was easily oxidized in response to the oxidative condition, leading to the H_2O_2 dependent release of CA and SN38. Moreover, ROS production induced by the released CA not only facilitated the tumor cells apoptosis but also promoted the breakdown of micellar structure, thereby synergistically exerting an anti-tumor effect with SN38 [82].

4.3.1.2 Selenium containing nanomaterials

Another commonly used ROS responsive DDS is based on selenium containing compounds which are known for their oxidation and reduction dual sensitivity [72, 73]. Similar to the use of sulfide groups in the PPS to achieve the hydrophobic to hydrophilic transition, water-insoluble selenides are able to be oxidized to hydrophilic selenoxides and selenones. In addition, selenium-containing materials have a relatively higher sensitivity to oxidation due to weaker energy of the C-Se bond compared to that of the C–S bond [83].

A series of ROS-sensitive selenium-containing nanomedicines developed by Zhang's group could serve as good examples [84, 85]. In their works, the selenium-containing amphiphilic triblock copolymer (PEG-PUSe-PEG) was synthesized [84]. The copolymer could self-assemble into nanomicelles for loading of

DOX and be oxidized by ROS. 72% of DOX was released from the PEG-PUSe-PEG micelles in 10 h when exposed to 0.1% H_2O_2, while 41% of DOX was released from the PEG-PU micelles (nonsensitive micelles) in the same condition. In addition, a selenium-containing surfactant, phenylselenide-1-undecyl triethylammonium bromide (SeQTA) was prepared for fabrication of the polymeric superamphiphile via the electrostatic interaction between the surfactant and the hydrophilic poly(ethylene glycol)-*block*-acrylic acid (PEG-*b*-PAA) copolymer [86]. Owing to the oxidation sensitivity of SeQTA, the micelles underwent the disassembly and released the cargos under mild oxidation (0.1% H_2O_2).

Additionally, Se–Se bonds could either be oxidized to seleninic acid in the presence of oxidants (0.01% ROS) or reduced to selenol in the reducing environment (0.01% GSH). Inspired by the unique dual responsive nature, Zhang et al. developed a diselenide-containing triblock copolymer (PEG-PUSeSe-PEG) containing PEG as a hydrophilic block and polyurethane-diselenide as a hydrophobic block. Collapse of the micelles was observed in the presence of either H_2O_2 or GSH, leading to the efficient cargo release [85].

4.3.1.3 Tellurium-containing nanomaterials

Although compared with aforementioned ROS responsive nanomaterials, tellurium-containing materials have been infrequently studied, they have gained increasing attention recently [87–90]. As an element of chalcogens, tellurium resembles selenium in terms of its oxidation-responsive properties. What is more, because of its lower electronegativity, telluride showed a higher ROS sensitivity than selenide. As shown in Table 4.2, divalent tellurium compounds are easy to be oxidized to the tetravalent and hexavalent states. Therefore, tellurium-containing nanomaterials are good choice for preparing the ROS-responsive drug delivery systems. Xu and coworkers synthesized an amphiphilic hyperbranched polymer containing tellurium and PEG [91]. In 1.0×10^{-4} M H_2O_2, the micelles formed by this polymer swelled resulting from the oxidation of tellurium to the tetravalent states, and transformed to irregular aggregates. The

ROS-responsiveness of the micelles could be adjusted by changing the cross-linking degree of the hyperbranched polymer. The ROS and radiation sensitivities made the nanomaterials a potential platform for chemo-radiation combination therapy. In another study, in order to improve the biocompatibility of tellurium-containing nanomaterials, Xu et al. developed ROS-responsive coassemblies by adding tellurium-containing molecules to biodegradable phospholipids (DPPE) which are the main component of the plasma membrane [92]. Owing to the reversible redox property of tellurium-containing molecules, the coassemblies could be oxidized and reduced by 2 equiv of H_2O_2 and ascorbic acid (Vc), respectively. The redox-sensitive properties were demonstrated by the fact that these coassemblies could respond to 100 μM H_2O_2, which is a biologically relevant ROS concentration.

4.3.2 Nanomedicines in Response to ROS-Induced Degradation

4.3.2.1 Boronic ester-containing nanomaterials

Boronic esters have been shown to undergo oxidation-induced degradation (Table 4.2) for decades. They can be oxidized under oxidizing condition then hydrolyzed in water. Recently, boronic esters have been used as ROS-degradable protecting groups for various applications [93]. The anticancer drugs, imaging agents and matrix metalloproteinase (MMP) inhibitors have been conjugated to boronic esters to realize the site-specific activation in the ROS-rich environment [94, 95]. Several boronic ester derivatives have been studied, among which aryl boronic esters with either ester or ether linkages showing superior degradation kinetics were used most frequently. As shown in Fig. 4.4, in the work of Lux et al., a ROS-sensitive boronic ester-containing polymer was synthesized (polymer 1) [77]. In order to improve the hydrolytic stability and cleavage kinetics, polymer 2 was developed through introducing an ether linkage between the aryl boronic ester groups and the backbone. The Nile red was used as a probe to evaluate the release of payloads from the polymer-assembled nanoparticles in oxidative environment.

Both nanoparticles had H_2O_2-dependant release kinetics, while polymer 2 nanoparticles were much more sensitive to H_2O_2 compared with polymer 1 nanoparticles. The release profiles of both nanoparticles in the absence H_2O_2 were comparable. When exposed to 1 mM H_2O_2, polymer 1 nanoparticles released only 50% of the dye in 26 h, while polymer 2 nanoparticles did so in 10 h. In addition, transmission electron microscopy (TEM) revealed that more polymer 2 nanoparticles ripped and collapsed than polymer 1 nanoparticles when incubated at low concentrations of H_2O_2 (100 and 50 μM), while similar morphological changes of both nanoparticles were observed at relatively high concentrations of H_2O_2 (100 and 250 mM).

Moreover, Chen and coworkers developed a series of ROS-activated aromatic nitrogen mustard prodrugs in order to selectively kill chronic lymphocytic leukemia (CLL) cells [96]. These agents were composed of two separate domains joined by a linker: one is H_2O_2 responsive moiety (boronic eaters); the other is cell-damaging group (nitrogen mustard). These prodrugs showed strong DNA cross-linking abilities in the presence of H_2O_2, whereas little DNA cross-linking was detected without H_2O_2. The cytotoxicity data indicated that the prodrugs induced 40–80% apoptosis in CLL patient-derived leukemic lymphocytes but less than 25% apoptosis in normal lymphocytes obtained from healthy donors. More recently, Lee and coworkers fabricated a H_2O_2-activatable antioxidant prodrug (BRAP) [97] in response to high concentration of H_2O_2 in the ischemia/reperfusion (I/R) injury site. BRAP was synthesized through the simple conjugation of (4-(hydroxymethyl)phenyl)boronic acid to 2-(hydroxymethyl)-2-methylpropane-1,3-diol at room temperature. In the prodrug, the p-hydroxybenzyl alcohol (HBA) served as an antioxidant moiety. BRAP containing boronic esters was anticipated to be oxidized by H_2O_2 to generate HBA. In addition, BRAP was added to H_2O_2-containing D_2O and the signal changes of HBA were monitored by LC-MS/MS. The results showed that BRAP was oxidized and generated HBA in a H_2O_2-responsive manner. In addition, the Amplex Red assay indicated that BRAP could scavenge H_2O_2 in a concentration-dependent manner.

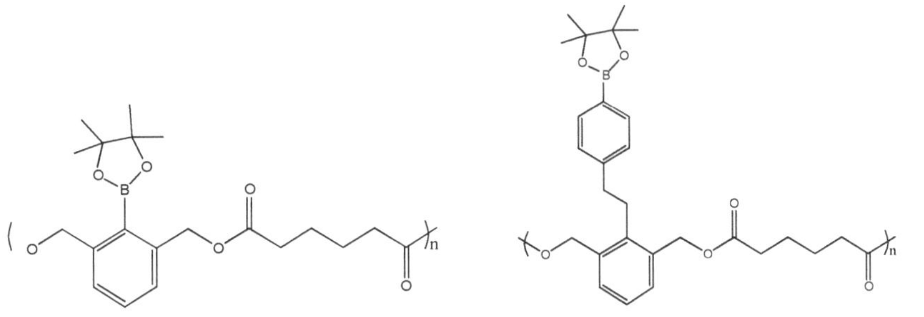

Figure 4.4 Structures of Polymer 1 and Polymer 2.

4.3.2.2 Proline oligomer-containing nanomaterials

The proteins or peptides containing aspartic acid, glutamic acid and proline residues are prone to be cleaved by ROS, causing protein fragmentation [98, 99]. Among the three residues, proline is mostly studied (Table 4.2), in developing the ROS-responsive nanoDDS [100]. Due to the superior biocompatibility to other ROS-responsive materials such as selenium, proline oligomers are more favorable for tissue engineering. Different from other ROS-responsive nanomaterials that usually show rapid ROS-induced degradation in a matter of hours or days, proline oligomers are degraded for few weeks. Thus, this type of nanomaterials is perfect for tissue engineering and controlled drug release in the chronic oxidative stress conditions, such as the inflammatory response to implants and atherosclerotic lesions. In a work by Sung group, poly(e-caprolactone) (PCL) and poly(ethylene glycol) (PEG) based polymeric scaffolds were fabricated using proline oligomers as cross-linkers [79]. Oligoproline peptides of varying lengths were conjugated to PEG and the proline oligomers were tested for degradability in 5 mM H_2O_2 and 50 μM Cu(II) at 37°C. After 6 days of incubation under this metal-catalyzed oxidation condition, all proline residues were degraded completely while PEG molecules were retained. In order to verify that the polyproline peptide as a cross-linker was responsible for the ROS degradability of the polymeric scaffolds, the authors synthesized PEG-P_7-PEG-cross-linked scaffolds ("7" refers to the length of the oligo(proline) peptide) and PEG-dihydrazide-cross-linked scaffolds (as the control). The ROS-dependent degradation of the PEG-P_7-PEG scaffolds was observed when exposed to the typical ROS generator SIN-1 which could produce peroxynitrite and hydroxyl radicals in situ. In addition, the activated murine macrophages were used to mimic an oxidative stress environment in vitro. Under this condition, the proline oligomer-containing scaffolds were degraded more effectively than the control scaffolds cross-linked with PEG-dihydrazide.

4.3.2.3 Polythioketal-containing nanomaterials

ROS can destabilize the thioketal bond in polymers, leading to chain scission and polymer breakdown with acetone as a byproduct (Table 4.2). In order to protect siRNA from degradation in the harsh environment of gastrointestinal tract and deliver siRNA to the inflamed intestinal tissue, Wilson et al. developed the thioketal-containing nanoparticles (TKNs) containing poly-(1,4-phenyleneacetone dimethylene thioketal) [19]. In in vitro studies, incubation of the CMFDA (dye)-loaded TKNs with the activated macrophages containing high level of ROS resulted in the intracelluar dye release seven-fold higher than that of incubation with the non-activated macrophages. In addition, the dye release could be mitigated by treating the activated macrophages with ROS scavenger, indicating that the dye release from the TKNs was triggered by ROS. The data was further confirmed by the in vivo biodistrubution data that the TKNs were accumulated in the inflamed intestinal tissues featured by a high level of ROS. Wang and co-workers used the ROS-sensitive thioketal (TK) as a linker between photodynamic nanoparticles and DOX. Singlet oxygen ($^1O_2$), generated by photodynamic therapy (PDT) under the excitation of NIR light, was able to not only trigger the release of DOX but also damage cancer cells to compensate for the detrimental effects of chemotherapy [80].

In order to accelerate wound healing, Tang et al. tried to deliver the stromal cell-derived factor-1-alpha by the polythioketal-containing nanomaterial named (poly-(1,4-phenyleneacetone dimethylene thioketal (SDF-1α-PPADT) [101]. Increased drug release was observed in the presence of ROS (60% of drugs released in 8 h) compared with that in PBS (23% of drugs released in 48 h). Upon administration of SDF-1α-PPADT in the mice with the full-thickness skin defects, the SDF-1α level increased significantly and stayed stable for 24 h in the wounds.

4.3.2.4 Silicon nanoparticles

It was reported that the inner pores of silicon (Si) particles could be the shelter for various drugs [102, 103]. It was also reported that the nanostructured porous silicon (PSi) could be degraded by ROS in the diseased environment [104]. However, simple adsorption of drug molecules to the Si surface usually results in the uncontrolled, rapid burst release of loaded drugs, compromising drug efficacy. To address this issue, Sailor group conjugated the fluorescent dye Alexa Fluor 488 or anticancer drug DOX with the Si particles through Si-C bonds [105]. Before the Alexa Fluor 488 or DOX attachment, the authors modified the surface of the Si particles to activate the carboxylic groups. The Si-C bonds were stable in aqueous media, but could be cleaved in ROS-rich environments. By this way, drug release could be controlled by the cleavage of the covalent bonds of Si matrix.

4.4 Conclusions and Perspectives

In this chapter, we review the design and mechanism of redox-sensitive nanosystems. Current studies have shown that redox-responsive nanoDDS inspire a new way to resolve the stability dilemma of drug carriers. The redox-responsive targeted/intracellular drug delivery approach can significantly enhance drug efficacy, overcome multi-drug resistance (MDR), and reduce drug and carrier-associated side effects.

Although the redox-responsive nanoparticle-based platforms have proven to be a powerful and effective strategy for drug delivery, several challenges remain. For a stimuli-responsive/sensitive system, the sensitivity is the key. There have been numerous reports on redox-sensitive systems, but a majority of them utilized single stimulus, which may show the limited sensitivity. However, the human body is a complex collection of different microenvironments containing various potential "stimuli." Considering the complexity, the design of DDS which can sense different changes and respond to multiple stimuli, including redox, in a predictable manner may be desired [105, 106]. Ideally, these systems would be able to harmonize their responses to each stimulus to precisely control drug delivery and release. For example, the reduction sensitivity can be combined with the pH responsiveness for intracellular drug delivery because certain pathological tissues (e.g., tumor tissues) are commonly characterized by both the reducing conditions and low pH [106]. In addition, these internal stimuli may be combined with the external stimuli (such as temperature and light) or other targeting ligands (such as the monoclonal antibody) for contribution of multifunctional drug delivery systems. The external stimuli or ligands are responsible for drug targeting and the internal stimuli are responsible for drug release.

It should be also noted that the exact intracellular fate of redox-sensitive nanovehicles remains unclear. It was reported that the cell surface thiols affected the cell entry of disulfide-conjugated peptides [107]. Despite controversy, some studies also indicated that disulfide bonds were cleaved in endosomal and lysosomal compartments [108, 109]. However, the direct evidence of the redox-responsive nanomedicine's fate is little. It is hard to conclude if the reduction-sensitive nanovehicles are finally destructed in the cytosol and/or cell nucleus. The mechanism study is needed because the successful redox-responsive drug delivery requires better understanding of the intracellular trafficking and fate of nanovehicles.

In addition, many of the reported systems were not constructed by the biodegradable and/or biocompatible materials, increasing the risk of undesirable toxicity. In the future, more efforts shall be directed to development of novel redox-sensitive biodegradable nanomaterials to meet the clinical safety requirement.

References

1. Torchilin VP. *Nanoparticulates as Drug Carriers: Imperial College Press*; 2006.
2. Senior JH. Nanoparticulate Drug-Delivery Systems. *Drug Development & Industrial Pharmacy*, 2007; 34: 116.
3. Ferrari M. Cancer nanotechnology: opportunities and challenges. *Nature Reviews Cancer*, 2005; 5: 161.
4. Mura S, Nicolas J, Couvreur P. Stimuli-responsive nanocarriers for drug delivery. *Nature Materials*, 2013; 12: 991–1003.
5. Xin W, Hai C. Novel insights into redox system and the mechanism of redox regulation. *Molecular Biology Reports*, 2016; 43: 607–628.

6. Estrela JM, Ortega A, Obrador E. Glutathione in cancer biology and therapy. *Critical Reviews in Clinical Laboratory Sciences*, 2006; 43: 143–181.
7. Schafer FQ, Buettner GR. Redox environment of the cell as viewed through the redox state of the glutathione disulfide/glutathione couple. *Free Radical Biology & Medicine*, 2001; 30: 1191–1212.
8. Wu G, Fang YZ, Yang S, Lupton JR, Turner ND. Glutathione metabolism and its implications for health. *Journal of Nutrition*, 2004; 134: 489–492.
9. Go YM, Jones DP. Redox compartmentalization in eukaryotic cells. *Biochimica Et Biophysica Acta General Subjects*, 2008; 1780: 1273–1290.
10. Arunachalam B, Phan UT, Geuze HJ, Cresswell P. Enzymatic Reduction of Disulfide Bonds in Lysosomes: Characterization of a Gamma-Interferon-Inducible Lysosomal Thiol Reductase (GILT). *Proceedings of the National Academy of Sciences of the United States of America*, 2000; 97: 745–750.
11. Kurz T, Eaton JW, Brunk UT. Redox activity within the lysosomal compartment: implications for aging and apoptosis. *Antioxidants & Redox Signaling*, 2010; 13: 511–523.
12. Ong W, Yang Y, Cruciano AC, Mccarley RL. Redox-Triggered Contents Release from Liposomes. *Journal of the American Chemical Society*, 2008; 130: 14739–14744.
13. Napoli A, Valentini M, Tirelli N, Müller M, Hubbell JA. Oxidation-responsive polymeric vesicles. *Nature Materials*, 2004; 3: 183–189.
14. Papa S, Skulachev VP. Reactive oxygen species, mitochondria, apoptosis and aging. *Molecular & Cellular Biochemistry* 1997; 174: 305–319.
15. Winterbourn CC. Reconciling the chemistry and biology of reactive oxygen species. *Nature Chemical Biology*, 2008; 4: 278.
16. Trachootham D, Alexandre J, Huang P. Targeting cancer cells by ROS-mediated mechanisms: a radical therapeutic approach? *Nature Reviews Drug Discovery*, 2009; 8: 579.
17. Szatrowski TP, Nathan CF. Production of Large Amounts of Hydrogen Peroxide by Human Tumor Cells. *Cancer Research*, 1991; 51: 794–798.
18. Johansson A, Jesaitis AJ, Lundqvist H, Magnusson KE, Sjölin C, Karlsson A, et al. Different subcellular localization of cytochrome b and the dormant NADPH-oxidase in neutrophils and macrophages:

effect on the production of reactive oxygen species during phagocytosis. *Cellular Immunology*, 1995; 161: 61–71.

19. Wilson DS, Dalmasso G, Wang L, Sitaraman SV, Merlin D, Murthy N. Orally delivered thioketal nanoparticles loaded with TNF-α-siRNA target inflammation and inhibit gene expression in the intestines. *Gastroenterology*, 2010; 9: 923.
20. Han HS, Thambi T, Choi KY, Son S, Ko H, Lee MC, et al. Bioreducible shell-cross-linked hyaluronic acid nanoparticles for tumor-targeted drug delivery. *Biomacromolecules*, 2015; 16: 447–456.
21. Lu HY, Chang YJ, Fan NC, Wang LS, Lai NC, Yang CM, et al. Synergism through combination of chemotherapy and oxidative stress-induced autophagy in A549 lung cancer cells using redox-responsive nanohybrids: a new strategy for cancer therapy. *Biomaterials*, 2015; 42: 30–41.
22. Kumara RST, Thiruselvi T, Baran MA, Gnanamani A. pH and redox sensitive albumin hydrogel: A self-derived biomaterial. *Scientific Reports*, 2015; 5: 15977.
23. Kuai R, Yuan W, Li W, Qin Y, Tang J, Yuan M, et al. Targeted delivery of cargoes into a murine solid tumor by a cell-penetrating peptide and cleavable poly (ethylene glycol) comodified liposomal delivery system via systemic administration. *Molecular pharmaceutics*, 2011; 8: 2151–2161.
24. Fu H, Shi K, Hu G, Yang Y, Kuang Q, Lu L, et al. Tumor-targeted paclitaxel delivery and enhanced penetration using TAT-decorated liposomes comprising redox-responsive poly (ethylene glycol). *Journal of Pharmaceutical Sciences*, 2015; 104: 1160–1173.
25. Lee SC, Huh KM, Lee J, Cho YW, Galinsky RE, Park K. Hydrotropic polymeric micelles for enhanced paclitaxel solubility: in vitro and in vivo characterization. *Biomacromolecules*, 2007; 8: 202–208.
26. Kim SC, Yoon HJ, Lee JW, Yu J, Park ES, Chi SC. Investigation of the release behavior of DEHP from infusion sets by paclitaxel-loaded polymeric micelles. *International Journal of Pharmaceutics*, 2005; 293: 303–310.
27. Cheng R, Feng F, Meng F, Deng C, Feijen J, Zhong Z. Glutathione-responsive nano-vehicles as a promising platform for targeted intracellular drug and gene delivery. *Journal of Controlled Release*, 2011; 152: 2–12.
28. Li J, Huo M, Wang J, Zhou J, Mohammad JM, Zhang Y, et al. Redox-sensitive micelles self-assembled from amphiphilic hyaluronic acid-

deoxycholic acid conjugates for targeted intracellular delivery of paclitaxel. *Biomaterials*, 2012; 33: 2310–2320.

29. Chi TN, Tran TH, Amiji M, Lu X, Kasi RM. Redox-sensitive nanoparticles from amphiphilic cholesterol-based block copolymers for enhanced tumor intracellular release of doxorubicin. *Nanomedicine Nanotechnology Biology & Medicine*, 2015; 11: 2071–2082.
30. Li Y, Xiao K, Luo J, Xiao W, Lee JS, Gonik AM, et al. Well-defined, reversible disulfide cross-linked micelles for on-demand paclitaxel delivery. *Biomaterials*, 2011; 32: 6633–6645.
31. Zhao D, Zhang H, Yang S, He W, Luan Y. Redox-sensitive mPEG-SS-PTX/TPGS mixed micelles: An efficient drug delivery system for overcoming multidrug resistance. *International Journal of Pharmaceutics*, 2016; 515: 281–292.
32. Chiang YT, Yen YW, Lo CL. Reactive oxygen species and glutathione dual redox-responsive micelles for selective cytotoxicity of cancer. *Biomaterials*, 2015; 61: 150–161.
33. Pan ZZ, Wang HY, Zhang M, Lin TT, Zhang WY, Zhao PF, et al. Nuclear-targeting TAT-PEG-Asp 8-doxorubicin polymeric nanoassembly to overcome drug-resistant colon cancer. *Acta Pharmacologica Sinica*, 2016; 37: 1110.
34. Nguyen DH, Lee JS, Bae JW, Choi JH, Lee Y, Son JY, et al. Targeted doxorubicin nanotherapy strongly suppressing growth of multidrug resistant tumor in mice. *International Journal of Pharmaceutics*, 2015; 495: 329–335.
35. Bao Y, Guo Y, Zhuang X, Li D, Cheng B, Tan S, et al. D-α-tocopherol polyethylene glycol succinate-based redox-sensitive paclitaxel prodrug for overcoming multidrug resistance in cancer cells. *Molecular Pharmaceutics*, 2014; 11: 3196–3209.
36. Li J, Yu X, Wang Y, Yuan Y, Xiao H, Cheng D, et al. A Reduction and pH Dual-Sensitive Polymeric Vector for Long-Circulating and Tumor-Targeted siRNA Delivery. *Advanced Materials*, 2014; 26: 8217–8224.
37. Wang M, Alberti K, Varone A, Pouli D, Georgakoudi I, Xu Q. Enhanced Intracellular siRNA Delivery using Bioreducible Lipid-Like Nanoparticles. *Advanced Healthcare Materials*, 2014; 3: 1398–1403.
38. Hao Y, Qin W, Wei C, Zhao Y, Yong T, Lu G, et al. Hydrophilicity/Hydrophobicity Reversable and Redox-Sensitive Nanogels for Anticancer Drug Delivery. *Molecular Pharmaceutics*, 2015; 12: 1636–1647.

39. Chen W, Achazi K, Schade B, Haag R. Charge-conversional and reduction-sensitive poly(vinyl alcohol) nanogels for enhanced cell uptake and efficient intracellular doxorubicin release. *Journal of Controlled Release*, 2015; 205: 15–24.
40. Luo YL, Huang XC, Tu WM, Hsu HY. Differential in situ sensing of extra- and intracellular glutathione by a novel redox-responsive silica matrix-Au nanoprobe. *Analytica Chimica Acta*, 2015; 902: 196–204.
41. Li X, Zhao W, Liu X, Chen K, Zhu S, Shi P, et al. Mesoporous Manganese Silicate Coated Silica Nanoparticles as Multi-Stimuli-Responsive T1-MRI Contrast Agents and Drug Delivery Carriers. *Acta Biomaterialia*, 2016; 30: 378–387.
42. Molina AM, Morales-Cruz M, Benítez M, Berríos K, Figueroa CM, Griebenow K. Redox-sensitive cross-linking enhances albumin nanoparticle function as delivery system for photodynamic cancer therapy. *Journal of Nanomedicine & Nanotechnology*, 2016; 6: 3.
43. Xu J, Singh A, Amiji MM. Redox-responsive targeted gelatin nanoparticles for delivery of combination wt-p53 expressing plasmid DNA and gemcitabine in the treatment of pancreatic cancer. *Bmc Cancer*, 2014; 14: 75.
44. Gao C, Liu T, Dang Y, Yu Z, Wang W, Guo J, et al. pH/redox responsive core cross-linked nanoparticles from thiolatedcarboxymethyl chitosan for in vitro release study of methotrexate. *Carbohydrate Polymers*, 2014; 111: 964–970.
45. Langner M, Kral TE. Liposome-based drug delivery systems. *Polish Journal of Pharmacology*, 1999; 51: 211–222.
46. Maeda T, Fujimoto K. A reduction-triggered delivery by a liposomal carrier possessing membrane-permeable ligands and a detachable coating. *Colloids & Surfaces B Biointerfaces*, 2006; 49: 15–21.
47. Oh JK, Drumright R, Siegwart DJ, Matyjaszewski K. The development of microgels/nanogels for drug delivery applications. *Progress in Polymer Science*, 2008; 33: 448–477.
48. Kabanov AV, Vinogradov SV. Nanogels as Pharmaceutical Carriers: Finite Networks of Infinite Capabilities. *AngewandteChemie*, 2009; 40: 5418–5429.
49. Oh JK, Tang C, Gao H, Tsarevsky NV, Matyjaszewski K. Inverse miniemulsion ATRP: a new method for synthesis and functionalization of well-defined water-soluble/cross-linked polymeric particles. Journal of the *American Chemical Society*, 2006; 128: 5578–5584.
50. Oh JK, Bencherif SA, Matyjaszewski K. Atom transfer radical polymerization in inverse miniemulsion: A versatile route toward

preparation and functionalization of microgels/nanogels for targeted drug delivery applications. *Polymer*, 2009; 50: 4407–4423.

51. Oh JK, Siegwart DJ, Matyjaszewski K. Synthesis and Biodegradation of Nanogels as Delivery Carriers for Carbohydrate Drugs. *Biomacromolecules*, 2007; 8: 3326–3331.
52. Oh JK, Siegwart DJ, Lee HI, Sherwood G, Peteanu L, Hollinger JO, et al. Biodegradable Nanogels Prepared by Atom Transfer Radical Polymerization as Potential Drug Delivery Carriers: Synthesis, Biodegradation, in Vitro Release, and Bioconjugation. *Journal of the American Chemical Society*, 2007; 129: 5939–5945.
53. Liu R, Zhao X, Wu T, Feng P. Tunable redox-responsive hybrid nanogated ensembles. *Journal of the American Chemical Society*, 2008; 130: 14418–14419.
54. Liu R, Zhang Y, Feng P. Multiresponsive Supramolecular Nano-gatedEnsembles. *Journal of the American Chemical Society*, 2009; 131: 15128–15129.
55. Geim AK, Novoselov KS. The rise of graphene. *Nature Materials*, 2007; 6: 183–191.
56. Sun X, Liu Z, Welsher K, Robinson JT, Goodwin A, Zaric S, et al. Nano-Graphene Oxide for Cellular Imaging and Drug Delivery. *Nano Research*, 2008; 1: 203–212.
57. Wen H, Dong C, Dong H, Shen A, Xia W, Cai X, et al. Engineered Redox-Responsive PEG Detachment Mechanism in PEGylated Nano-Graphene Oxide for Intracellular Drug Delivery. *Small*, 2012; 8: 760–769.
58. Hui C, Zhuyuan W, Shenfei Z, Lei W, Peng C, Dan Z, et al. SERS-fluorescence monitored drug release of a redox-responsive nanocarrier based on graphene oxide in tumor cells. *Acs Applied Materials & Interfaces*, 2014; 6: 17526–17533.
59. Li Y, Feng L, Shi X, Wang X, Yang Y, Yang K, et al. Surface coating-dependent cytotoxicity and degradation of graphene derivatives: towards the design of non-toxic, degradable nano-graphene. *Small*, 2014; 10: 1544–1554.
60. Gallo J, Hung C, Perrier D. Analysis of albumin microsphere preparation. *International journal of pharmaceutics*, 1984; 22: 63–74.
61. Lin W, Coombes A, Davies M, Davis S, Illum L. Preparation of sub-100 nm human serum albumin nanospheres using a pH-coacervation method. *Journal of Drug Targeting*, 1993; 1: 237–243.

62. Müller BG, Leuenberger H, Kissel T. Albumin nanospheres as carriers for passive drug targeting: an optimized manufacturing technique. *Pharmaceutical Research*, 1996; 13: 32–37.

63. Shi H, Cheng Q, Yuan S, Ding X, Liu Y. Human Serum Albumin Conjugated Nanoparticles for pH and Redox-Responsive Delivery of a Prodrug of Cisplatin. *Chemistry*, 2015; 21: 16547–16554.

64. Kommareddy S, Amiji MM. Cell Transfection and Analysis Using DNA-Loaded Gelatin Nanoparticles. *CSH protocols*, 2008; 2008: pdb-prot4887.

65. Kommareddy S, Amiji MM. Intracellular trafficking studies using gold-encapsulated gelatin nanoparticles. *CSH protocols*, 2008; 2008: pdb-prot4886.

66. Kommareddy S, Amiji MM. Preparation and loading of gelatin nanoparticles. *CSH protocols*, 2008; 2008: pdb-prot4885.

67. Kommareddy S, Amiji M. Biodistribution and pharmacokinetic analysis of long-circulating thiolated gelatin nanoparticles following systemic administration in breast cancer-bearing mice. *Journal of pharmaceutical sciences*, 2007; 96: 397–407.

68. Kommareddy S, Amiji M. Preparation and evaluation of thiol-modified gelatin nanoparticles for intracellular DNA delivery in response to glutathione. *Bioconjugate chemistry*, 2005; 16: 1423–1432.

69. Vo CD, Kilcher G, Tirelli N. Polymers and Sulfur: what are Organic Polysulfides Good For? Preparative Strategies and Biological Applications. *Macromolecular Rapid Communications*, 2010; 30: 299–315.

70. Napoli A, Tirelli N, Kilcher G, Hubbell A. New Synthetic Methodologies for Amphiphilic Multiblock Copolymers of Ethylene Glycol and Propylene Sulfide. *Macromolecules*, 2001; 34: 8913–8917.

71. Dai L, Yu Y, Luo Z, Li M, Chen W, Shen X, et al. Photosensitizer enhanced disassembly of amphiphilic micelle for ROS-response targeted tumor therapy in vivo. *Biomaterials*, 2016; 104: 1–17.

72. Poole KM, Nelson CE, Joshi RV, Martin JR, Gupta MK, Haws SC, et al. ROS-responsive microspheres for on demand antioxidant therapy in a model of diabetic peripheral arterial disease. *Biomaterials*, 2015; 41: 166–175.

73. Bergson G, Weidler AM, Mandell L, Kvande PC, Meisingseth E, Bergson G, et al. Proton-Mobility in the Indene Ring-System. *Acta Chemica Scandinavica*, 1963; 17: 862–864.

74. Ogawa A, Nishiyama Y, Kambe N, Murai S, Sonoda N. Selenium, carbon monoxide and water as a new reduction system: Reductive cleavage

of disulfides and diselenides to thiols and selenols. *Tetrahedron Letters*, 1987; 18: 3271–3274.

75. Isikgor FH, Becer CR. Lignocellulosic biomass: a sustainable platform for the production of bio-based chemicals and polymers. *Polymer Chemistry*, 2015; 6: 4497–4559.

76. Xu W, Wang H, Xie F, Chen J, Cao H, Xu J-B. Facile and environmentally friendly solution-processed aluminum oxide dielectric for low-temperature, high-performance oxide thin-film transistors. *ACS applied materials & interfaces*, 2015; 7: 5803–5810.

77. Lux CDG, Joshibarr S, Nguyen T, Mahmoud E, Schopf E, Fomina N, et al. Biocompatible polymeric nanoparticles degrade and release cargo in response to biologically relevant levels of hydrogen peroxide. *Journal of the American Chemical Society*, 2012; 134: 15758–15764.

78. Jäger E, Höcherl A, Janoušková O, Jäger A, Hrubý M, Konefał R, et al. Fluorescent boronate-based polymer nanoparticles with reactive oxygen species (ROS)-triggered cargo release for drug-delivery applications. *Nanoscale*, 2016; 8: 6958–6963.

79. Yu SS, Koblin RL, Zachman AL, Perrien DS, Hofmeister LH, Giorgio TD, et al. Physiologically Relevant Oxidative Degradation of Oligo(proline) Cross-Linked Polymeric Scaffolds. *Biomacromolecules*, 2011; 12: 4357–4366.

80. Wang X, Meng G, Song Z, Liu X. A Reactive1O2- Responsive Combined Treatment System of Photodynamic and Chemotherapy for Cancer. *Scientific Reports*, 2016; 6: 29911.

81. Yue C, Zhang C, Alfranca G, Yao Y, Jiang X, Yang Y, et al. Near-Infrared Light Triggered ROS-activated Theranostic Platform based on Ce6-CPT-UCNPs for Simultaneous Fluorescence Imaging and Chemo-Photodynamic Combined Therapy. *Theranostics*, 2016; 6: 456–469.

82. Liu B, Wang D, Liu Y, Zhang Q, Meng L, Chi H, et al. Hydrogen peroxide-responsive anticancer hyperbranched polymer micelles for enhanced cell apoptosis. *Polymer Chemistry*, 2015; 6: 3460–3471.

83. Xu H, Cao W, Zhang X. Selenium-containing polymers: promising biomaterials for controlled release and enzyme mimics. Accounts of *Chemical Research*, 2013; 46: 1647–1658.

84. Ma N, Li Y, Ren H, Xu H, Li Z, Zhang X. Selenium-containing block copolymers and their oxidation-responsive aggregates. *Polymer Chemistry*, 2010; 1: 1609–1614.

85. Ma N, Li Y, Xu H, Wang Z, Zhang X. Dual Redox Responsive Assemblies Formed from Diselenide Block Copolymers. *Journal of the American Chemical Society*, 2010; 132: 442–443.

86. Han P, Ma N, Ren H, Xu H, Li Z, Wang Z, et al. Oxidation-Responsive Micelles Based on a Selenium-Containing Polymeric Superamphiphile. *Langmuir the Acs Journal of Surfaces & Colloids*, 2010; 26: 14414–14418.

87. Yu F, Li P, Wang B, Han K. Reversible near-infrared fluorescent probe introducing tellurium to mimetic glutathione peroxidase for monitoring the redox cycles between peroxynitrite and glutathione in vivo. *Journal of the American Chemical Society*, 2013; 135: 7674–7680.

88. Mao S, Dong Z, Liu J, Li X, Liu X, Luo G, et al. Semisynthetic tellurosubtilisin with glutathione peroxidase activity. *Journal of the American Chemical Society*, 2005; 127: 11588–11589.

89. Thomas J, Dong Z, Dehaen W, Smet M. Selenium/Tellurium-Containing Hyperbranched Polymers: Effect of Molecular Weight and Degree of Branching on Glutathione Peroxidase-Like Activity. *Macromolecular rapid communications*, 2012; 33: 2127–2132.

90. Liu X, Silks LA, Liu C, Ollivault-Shiflett M, Huang X, Li J, et al. Incorporation of tellurocysteine into glutathione transferase generates high glutathione peroxidase efficiency. *Angewandte Chemie International Edition*, 2009; 48: 2020–2023.

91. Fang R, Xu H, Cao W, Yang L, Zhang X. Reactive oxygen species (ROS)-responsive tellurium-containing hyperbranched polymer. *Polymer Chemistry*, 2015; 15: 2817–2821.

92. Wang L, Fan F, Cao W, Xu H. Ultra-sensitive ROS-responsive Coassemblies of Tellurium-Containing Molecules and Phospholipids. *Acs Applied Materials & Interfaces*, 2015; 7: 16054–16060.

93. Webb KS, Levy D. A facile oxidation of boronic acids and boronic esters. *Tetrahedron Letters*, 1995; 36: 5117–5118.

94. Chung C, Srikun D, Chang SL, Chang CJ, Cho BR. A two-photon fluorescent probe for ratiometric imaging of hydrogen peroxide in live tissue. *Chemical Communications*, 2011; 47: 9618–9620.

95. Kuang Y, Balakrishnan K, Gandhi V, Peng X. Hydrogen peroxide inducible DNA cross-linking agents: targeted anticancer prodrugs. *Journal of the American Chemical Society*, 2011; 133: 19278–19281.

96. Chen W, Balakrishnan K, Kuang Y, Han Y, Fu M, Gandhi V, et al. Reactive Oxygen Species (ROS) Inducible DNA Cross-Linking Agents and Their Effect on Cancer Cells and Normal Lymphocytes. *Journal of Medicinal Chemistry*, 2014; 57: 4498–4510.

97. Lee D, Park S, Bae S, Jeong D, Park M, Kang C, et al. Hydrogen peroxide-activatable antioxidant prodrug as a targeted therapeutic

agent for ischemia-reperfusion injury. *Scientific Reports*, 2015; 5: 16592.

98. Stadtman ER, Levine RL. Free radical-mediated oxidation of free amino acids and amino acid residues in proteins. *Amino Acids*, 2003; 25: 207–218.

99. Garrison WM. ChemInform Abstract: Reaction Mechanisms in the Radiolysis of Peptides, Polypeptides and Proteins. *Cheminform*, 1985; 87: 381–398.

100. Amici A, Levine RL, Tsai L, Stadtman ER. Conversion of amino acid residues in proteins and amino acid homopolymers to carbonyl derivatives by metal-catalyzed oxidation reactions. *Journal of Biological Chemistry*, 1989; 264: 3341–3346.

101. Tao T, Hao J, Yuan Y, Fang H, Ji SZ, Liu YY, et al. A new method of wound treatment: targeted therapy of skin wounds with reactive oxygen species-responsive nanoparticles containing SDF-1α. *International Journal of Nanomedicine*, 2015; 10: 6571–6585.

102. Anglin EJ, Schwartz MP, Ng VP, Perelman LA, Sailor MJ. Engineering the chemistry and nanostructure of porous silicon Fabry-Pérot films for loading and release of a steroid. *Langmuir*, 2004; 20: 11264–11269.

103. Salonen J, Laitinen L, Kaukonen AM, Tuura J, Björkqvist M, Heikkilä T, et al. Mesoporous silicon microparticles for oral drug delivery: loading and release of five model drugs. *Journal of controlled release*, 2005; 108: 362–374.

104. Tzurbalter A, Shatsberg Z, Beckerman M, Segal E, Artzi N. Mechanism of erosion of nanostructured porous silicon drug carriers in neoplastic tissues. *Nature Communications*, 2015; 6: 6208.

105. Choi GH, Min KH, Sang CL. Block copolymer-templated mineralization for pH-responsive robust nanocarriers of 5-fluorouracil. *Macromolecular Research*, 2014; 22: 329–336.

106. Cheng R, Meng F, Ma S, Xu H, Liu H, Jing X, et al. Reduction and temperature dual-responsive crosslinked polymersomes for targeted intracellular protein delivery. *Journal of Materials Chemistry*, 2011; 21: 19013–19020.

107. Aubry S, Burlina F, Dupont E, Delaroche D, Joliot A, Lavielle S, et al. Cell-surface thiols affect cell entry of disulfide-conjugated peptides. *The FASEB Journal*, 2009; 23: 2956–2967.

108. Yang J, Chen H, Vlahov IR, Cheng JX, Low PS. Evaluation of disulfide reduction during receptor-mediated endocytosis by using FRET

imaging. *Proceedings of the National Academy of Sciences of the United States of America*, 2006; 103: 13872–13877.

109. Collins DS, Unanue ER, Harding CV. Reduction of disulfide bonds within lysosomes is a key step in antigen processing. *Journal of Immunology*, 1991; 147: 4054–4059.

Chapter 5

Hypoxia-Responsive Nanomedicines

Federico Perche[a] **and Kanjiro Miyata**[b]

[a]*Centre for Molecular Biophysics,*
CNRS UPR4301, 45071 Orléans, France
[b]*Graduate School of Medicine, The University of Tokyo,*
7-3-1 Hongo, Bunkyo, Tokyo, 113-0033, Japan

federico.perche@cnrs-orleans.fr

Hypoxia, low oxygen tension, is a critical regulator of metabolism in both normal and diseased tissues. In normal conditions, oxygen gradients are essential for the maintenance of adult stem cells. In diseased conditions, ectopic activation of hypoxia-regulated pathways is associated with chronic inflammation and supporting cancer cell proliferation. Accordingly, hypoxia-responsive nanomedicines have been developed to assess hypoxic status and to counter effects of aberrant hypoxia. This chapter provides a current view of hypoxia-responsive nanomedicines and possible perspectives.

Stimuli-Responsive Nanomedicine
Edited by Lin Zhu

ISBN 978-981-4800-70-9 (Hardcover), 978-0-429-29529-4 (eBook)
www.jennystanford.com

5.1 Introduction

Most drug treatments and diagnosis probes rely on molecules repeatedly administered to monitor, stop, or revert disease progression. These pharmaceutics are rapidly cleared from the circulation by host defense mechanisms such as the mononuclear phagocyte system (MPS) after injection [1, 2]. To avoid MPS clearance and increase their therapeutic index, drugs are integrated into nanoparticles (NP), a dosage form reducing elimination by biological defense mechanisms and achieving increased safety and survival over free drugs, with several approved NP formulations [1, 3, 4]. After integration into NP, drugs and probes are amenable to diseased tissue targeting by conjugation of specific ligands to the surface of NP such as antibodies against proteins enriched at the cell surface of cancer cells for tumor targeting [1, 5]. However, the presence of ligands at the NP surface, that is a coat of exogenous antigens, activates biological defense mechanisms among which the MPS and the complement system, limiting the benefits of targeting [5, 6]. Another challenge for the disease-specific ligand conjugation strategy involves the identification of consensus motifs, shared by patients affected by most frequent pathologies. This drawback is especially visible for cancer targeting, where the need to select ligands for the 200 different types of cancer, is coupled with inter- and intra-patient heterogeneity of the tumor proteome [7, 8].

One of the solutions proposed to address these challenges consists of a minimal library of stimuli-responsive NP administered as inactive precursors with confined activation at the disease site by either physical stimuli (ultrasound, light, temperature) or broad pathophysiological traits [9–15]. Such physiological stimuli (redox imbalance, hypoxia, acidosis, inflammation, and shared alterations of biochemical pathways) represent the visible tip of an iceberg of distinct molecular pathways affected by different pathologies [5, 10]. Whereas hypoxia is chiefly considered as a tumor trait, decreased oxygen tension is a characteristic shared by both normal and pathological processes making hypoxia an important clinical factor both for tissue engineering and disease remediation [16]. In this chapter, we will first describe physiological and pathophysiological processes with existing hypoxia. Then, we will present hypoxia-responsive NP where endocytosis (NP with a therapeutic payload) or signal emission (NP with a probe) are triggered either directly by hypoxia or indirectly by stimuli associated with hypoxia. Finally, we will propose possible avenues for further development of hypoxia-responsive NP.

5.2 Physiological Roles of Hypoxia

Although oxygen concentration in the atmosphere is 21%, the oxygen tension in tissues is significantly lower due to limited oxygen diffusion. Thus, tissues face different oxygen tensions. Because of the importance of oxygen tension in tissue physiology, one should distinguish between normoxia, physioxia, and hypoxia [17]. Normoxia corresponds to the culture of cells (primary or immortalized) in incubators with 21% O_2, physioxia to the actual oxygen tension of a given tissue in the organism (usually 2–9%), and hypoxia (<1%) to an oxygen concentration significantly lower than other tissues [18]. Such a distinction is required for the interpretation of the results of in cellulo studies, which are most frequently performed without considering the actual oxygen tension of the corresponding normal and diseased tissue in human patients.

Hypoxia in normal tissues plays a critical role both during development and adult life, mainly based on the role of hypoxic niches in development and maintenance of stem cells [16, 18]. Indeed, oxygen gradients in the developing embryo and placenta are patterns of organogenesis [16]. In adults, the primary lymphoid organs (bone marrow and thymus) are hypoxic, an hypoxic environment crucial to the persistence of hematopoietic stem cells via control of stemness pathways such as Notch and Wnt by HIF-1α [16, 18, 19].

5.2.1 HIFs as Molecular Sensors of Hypoxia

Response to hypoxia is coordinated by the transcription factors Hypoxia Inducible Factors (HIFs), considered as the molecular sensors of hypoxia [20, 21]. HIFs are dimeric proteins with, a β unit (HIF-1β, HIF-2β), and a α subunit (HIF-1α, HIF-2α, HIF-3α) [22]. Response to oxygen tension is achieved by the control of HIF-α

stability in the presence of oxygen via hydroxylation of the prolines in the ODDD (Oxygen-Dependent Degradation Domain) by prolyl hydroxylases (PHDs), an hydroxylation resulting in rapid ubiquitin-mediated HIF-α proteasome processing under normoxia [23]. Under hypoxia, α subunits get stabilized by inhibition of PHDs by reactive oxygen species (ROS), a stabilization allowing HIF-α translocation to the nucleus where it dimerizes with HIF-β to control gene transcription of more than 200 genes containing hypoxia responsive elements (HRE) in their promoter [23]. Fine response to hypoxia gradients is achieved by different O_2-dependent stabilization thresholds for HIF-1α (0–2% O_2) and HIF-2α (2–5% O_2) [24]. In addition to induction of the transcription of genes containing HRE, hypoxia also controls cellular metabolism by oxygen-dependent expression of a set of miRNA defined as hypoxia regulated miRNA [25, 26].

5.2.2 Hypoxia-Associated Oxidative Stress

ATP generation by the respiratory chain in mitochondria requires oxygen as electron acceptor [27]. This oxidative phosphorylation is associated with the production of ROS such as $O^{\cdot-}_2$ and H_2O_2 by incomplete reduction of oxygen [28]. Due to the importance of oxygen for energy production, mitochondria have been proposed as oxygen sensors to adapt ATP production to local oxygen tension [29]. Hypoxia leads to a mitochondria-dependent increase in ROS production, with ROS-dependent stabilization of HIF-1α to shift ATP generation to anaerobic glycolysis [30].

5.2.3 Wound Healing

Hypoxia and HIF-1α are involved in several tissue repair mechanisms such as bone regeneration and wound healing. Disruption of blood vessels after injury, visualized by bleeding, results in acute local hypoxia. Not surprisingly, this local hypoxia-gradient is one of the regulators of wound healing with induction of several growth factors such as PDGF (Platelet-Derived Growth Factor) and FGF (Fibroblast Growth Factor) [31, 32]. As example, Clemens et al. demonstrated HIF-1α activation during bone repair and increased bone repair after pharmacological HIF-1α activation [33].

5.3 Pathophysiological Roles of Hypoxia in Cancer

Hypoxia is considered as a hallmark of solid tumors, supporting both cancer proliferation, metastasis, drug resistance, and associated with a poor prognosis in cancer patients [20, 21, 34–37]. Hypoxia occurs as cancer cells become distant from blood vessels. This separation from the circulatory system limits oxygen and nutrients delivery to cancer cells in hypoxic regions, pausing tumor growth. Hypoxia in tumors is illustrated in Fig. 5.1, by hyperspectral imaging of hemoglobin saturation, and shows a hypoxic gradient and a hypoxic core. This hypoxic core, far from vessels and with high expression of efflux drugs transporters represents a *terra incognita* for anticancer therapies, as demonstrated by the recurrence of colorectal cancer liver metastases after dramatic initial tumor reduction following treatment with styrene maleic acid/pirarubicin micelles, a recurrence due to residual hypoxic tumor cells [38].

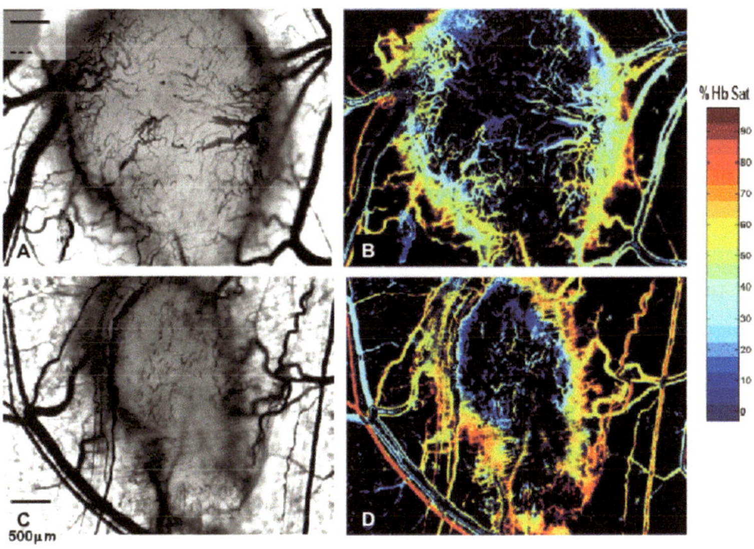

Figure 5.1 Hypoxia in eight day old 4T1 tumors. (A) dorsal skin window chamber viewed with light microscopy shows diffuse tumor micro-vascularity, (B) Corresponding hyperspectral imaging of the same tumors exhibits hemoglobin saturations #10% over a 70% of the tumor surfaces. Scale bar represents 500 μm. Reproduced from [39].

Correlation between HIF-1α and progression was also reported in patients, with HIF-1α overexpression in patient tumors of distinct types, correlated with decreased p53 activity and enhanced cell proliferation [40]. Moreover, the critical role of hypoxia in cancer progression is supported by two meta-analysis of clinical trials involving 7, 000 and 10,108 cancer patients treated with radiotherapy alone or in combination with hypoxia-remediation therapy [41, 42]. Both analyses reported significantly improved overall survival when radiotherapy was combined with high oxygen concentrations or a hypoxia-sensitive prodrug over control.

5.3.1 Cancer Drug Resistance

Hypoxia sustains resistance both to physical and chemical treatments. Decreased DNA damage after ionizing radiation of hypoxic tumors limits apoptosis induction and clinical efficacy of radiotherapy [21, 42]. This leads to increased radiation exposure to achieve a therapeutic effect, along with more off-target toxicity. Photodynamic therapy (PDT) of the generation of cytotoxic singlet oxygen ($^1O_2$) from molecular oxygen after irradiation of a photosensitizer [43]. However, such a dependence on molecular oxygen limits PDT's potential to treat hypoxic regions [44].

In addition to decreasing the activity of radiotherapy and PDT, hypoxia increases cancer cell resistance to anticancer drugs by upregulation of efflux pumps such as Pgp (P glycoprotein), inhibition of apoptosis activity to reduce the activity of remaining intracellular drugs, and impaired metabolism to support aberrant cellular proliferation (Fig. 5.2) [45]. This is complemented with inhibition of apoptotic effects of anticancer drugs by HIF-dependent upregulation of anti-apoptotic factors such as Bcl-2 and Bcl-xl [45–47]. Finally, HIF-dependent upregulation of self-renewal pathways such as OCT3/4 and Nanog leads to propose hypoxic tumor regions as niches for drug resistant tumor initiating and re-initiating cancer stem cells [48, 49].

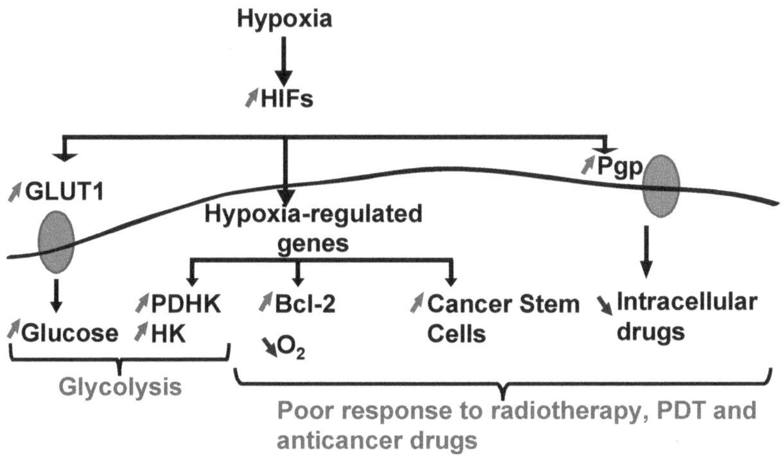

Figure 5.2 Principal modes of hypoxia-induced treatment resistance. Activation of HIFs in tumors, drives their nuclear translocation to control genes with HRE in their promoters. Glycolysis is induced for ATP generation at low oxygen tensions. HIFs also upregulate anti-apoptotic pathways and efflux pumps for tumor resistance. Moreover, decreased molecular oxygen levels reduce the activity of radiotherapy and photodynamic therapies (PDT). Finally, hypoxic microenvironments act as cancer stem niches important for tumor recurrence.

5.3.2 Promotion of Tumor Neoangionesis and Metastasis

HIFs also stimulate tumor neoangiogenesis to direct nutrients to cancer cells via activation of angiogenic genes such as VEGF (Vascular Endothelial Growth Factor) and PDGFB (Platelet Derived Growth Factor B) [50]. These factors are secreted by cancer cells to produce a gradient pattern for the synthesis of immature tumor vessels.

Promotion of angiogenesis is accompanied with that of epithelial-to-mesenchymal transition (EMT) to support metastasis [21, 51]. EMT consists of converting cells strongly attached to the basement membrane into individual cells able to digest the extracellular matrix to access lymph and/or blood and develop secondary metastases.

5.3.3 Metabolic Adaptation of Hypoxic Cancer Cells

Since hypoxia reduces ATP production by oxidative phosphorylation in the mitochondria, hypoxic tumor cells use anaerobic glycolysis as energy source [52]. HIF-1α regulates glycolysis at several steps from increased glucose endocytosis (upregulation of glucose transporter 1, GLUT1), glucose degradation (pyruvate dehydrogenase kinase, PDHK), to exocytosis of the by-product lactate (monocarboxylate transporter 4, MCT-4) [53, 54]. In addition to direct ATP generation, glycolysis further supports cancer progression as intermediate products are shared with synthesis pathways of lipids, proteins, and nucleic acids [52, 55]. Contribution of glycolysis to nucleotide synthesis is so important to support cancer cell proliferation that enhanced glycolysis is also evident in oxygenated tumor tissues, a phenomenon called aerobic glycolysis, the Warburg effect [52].

Integration of all these hypoxia-regulated pathways to drive tumor progression has been illustrated by Milane et al. [56]. They compared the time required for human breast cancer cells cultured under normoxia or hypoxia before their orthotopic implantation in mammary fat pads, to develop into 100 mm^3 tumors. Whereas it required 7–8 weeks for tumors developed from normoxia cultured cells, hypoxia preconditioning reduced this time to 3 weeks. Faster tumor growth was correlated with elevated levels of hypoxia and hypoxia-induced pathways markers such as HIF-1α, promotion of neoangiogenesis (CD31), metabolic adaptation to hypoxia (GLUT1), multidrug resistance (Pgp), all pathways sustaining tumor growth and metastasis.

5.3.4 Increased Tumor Hypoxia after Chemotherapy

Hypoxia's support to cancer progression is reinforced by the increased fraction of hypoxic cells after therapy, with increased HIF-1α expression by cancer cells after doxorubicin, gemcitabine and paclitaxel treatments [57, 58]. Incubation of several human breast cancer cell lines with IC50 doses of paclitaxel increased HIF-1α expression. Paclitaxel-induced HIF-1α expression enriched the percentage of cancer stem cells from 1% to 12% in cellulo, and promoted mammosphere formation, phenotypes correlated with metastasis and recurrence after treatment [59]. In orthotropic models, paclitaxel increased the fraction of cancer stem cells from 1% to 2.5%, a promotion of the stem cell phenotype which was abrogated by co-treatment with the HIF-1α inhibitor digoxin. Finally, HIF-1α transcriptional signature after chemotherapy was correlated with decreased survival of breast cancer patients.

5.3.5 Role of Hypoxia in Adipose Tissue

Lastly, evidence suggests the involvement of hypoxia and HIF-1 in the impaired metabolism of adipose tissue in diseases such as diabetes and obesity [60]. Hypoxia, and activation of the hypoxia sensor HIF-1 exert these effects through decreased lipolysis and induction of glucose transporters, resulting in increased lipid retention and glucose accumulation in adipocytes [61, 62].

Altogether, these data indicate targeted delivery of anticancer treatments to hypoxic regions as a critical parameter to counter cancer progression.

One should also consider that hypoxic tumor regions are frequently the most distant from blood vessels [63]. Accordingly, a primary concern for the design of tumor hypoxia-targeted formulations should be to reach the hypoxic niche. Recent work suggests that production of NP of less than 50 nm is critical for cargo distribution to poorly permeable tumors and lymph node metastases [64, 65].

5.4 Hypoxia-Activated Payload Delivery

Hypoxic status of inflamed and tumor tissues has been used as stimulus to induce content release from hypoxia-responsive nanoparticles (HR NP). HR NP encapsulate a hydrophobic drug in the core of the NP, with a hydrophilic shell to decrease non-specific endocytosis in normoxic tissues and improve biocompatibility. Tuning of morphological transition at corresponding hypoxic disease levels results in detachment of the shell, selectively triggering drug endocytosis by target cells. We will cover strategies based on hypoxia-dependent hydrophobic-to-hydrophilic transition, hypoxia-dependent cross-linker cleavage, hypoxia-dependent oxygen generation to induce micelle burst and hypoxia tropism of sickle red blood cells (Table 5.1). Moreover, a platform featuring reversible hypoxia-controlled insulin delivery will be presented.

Table 5.1 Characteristics of hypoxia-activated nanocarriers

Probe or cargo	Size	Targeting	Mechanism	Refs
Cisplatin and catalase	230 nm	H_2O_2-sensitive	Catalase-mediated O_2 generation to induce disruption, cisplatin release and decrease local hypoxia	[77]
Methylene blue and catalase	205 nm	H_2O_2-sensitive and $\alpha_v\beta_3$ conjugated	H_2O_2-conditioned FRET to induce O_2 generation and enhance photodynamic therapy	[76]
Porphyrin derivative	7 μm	Hypoxic cellular tropism	Selective accumulation of sickle Red Blood Cells in hypoxic tumor areas with programmed delayed hemolysis	[81]
Doxorubicin	66 nm	Hypoxia-sensitive	Hypoxia-mediated azobenzene linker cleavage to disrupt micelles and induce content release	[73]
siRNA	100 nm	Hypoxia-sensitive	Hypoxia-mediated azobenzene linker cleavage to induce positive charge exposure and promote cellular uptake	[74]
Doxorubicin	200 nm	Hypoxia-sensitive	Hydrophobic-to-hydrophilic transition of the 6-(2-nitroimidazole) linker resulting in micelle disruption and content release	[67]
Porphyrin derivative	50 nm	Hypoxia-sensitive	Oxygen-dependent phosphorescence, with light-harvesting polymer to increase energy transfer to the sensor in the core of the micelles	[103]
Iridium complex	50 nm	Hypoxia-sensitive	Oxygen-dependent phosphorescence, identification of hypoxic primary and secondary tumors 1h after intravenous injection in vivo	[100]
Ruthenium complex	50 nm	Hypoxia-sensitive	Oxygen-dependent phosphorescence, with hierarchical architecture and NIR light upconversion for increased sensitivity and selectivity	[107]

5.4.1 Nanoparticles with Nitroimidazole Derivatives

The potential of 2-nitroimidazole to treat hypoxic conditions was reported as early as 1953, with cumulative accumulation in anaerobic bacteria resulting in lethality through inhibition of protein synthesis [66]. These earlier findings directed synthesis of 2-nitroimidazole derivatives in the 1990s to map hypoxic regions in tissues, championed by the commercial hypoxia marker pimonidazole hydrochloride (Hydroxyprobe Inc.) [67].

However, HR NP incorporating 2-nitroimidazole derivatives were only introduced in 2014 [68]. HR NP take advantage of the hydrophobicity of nitroimidazole derivatives to use them as hypoxia-labile cross-linkers [20, 68–70]. These linkers are cleaved by nitroreductases by a two-step reduction, with the first step being reversed by molecular oxygen [69].

Thambi et al. substituted carboxymethyl dextran (hydrophilic) with 6-(2-nitroimidazole)hexylamine (hydrophobic), thereby forming an amphiphilic polymer capable of self-assembly into 200 nm HR NP and of encapsulating doxorubicin [68]. While biocompatible dextran allows stability in the circulation, the hydrophobic-to-hydrophilic transition of 6-(2-nitroimidazole)hexylamine under hypoxia drives content release by NP disassembly. Indeed, a dramatic increase in doxorubicin cellular uptake was achieved under hypoxia over normoxia.

Intravenous injection of HR NP to mice bearing subcutaneous SCC7 tumors led to 4-fold doxorubicin accumulation in tumors over liver and spleen, with confirmed distribution to hypoxic tumor regions. The therapeutic potential of this system is suggested by its enhanced therapeutic activity in vivo over free drug.

Azobenzene derivatives represent a large class of stimuli-sensitive nanocarriers; their hydrophobicity and extensive chemistry of azobenzene substitutions promoted the development of azobenzene-based compounds for hypoxia imaging and hypoxia-targeted delivery [71–73].

Liu et al. proposed to use hypoxia-induced cleavage of azobenzene instead of hydrophobic-to-hydrophilic transition for hypoxia-triggered doxorubicin release [74]. They synthesized poly(ethylene glycol)-hexanethiol-azobenzene-combrestatin A4 (PEG-C6-AZO-CA4) amphiphilic polymers with each group responsible for a specific task. PEG serves as hydrophilic stealth layer. Hydrophobic C6 promotes self-assembly. AZO acts as hypoxia-labile linker. CA4, an inhibitor of tubulin polymerization, acts as an antineoplastic agent. Note that the hydrophobicity of AZO and CA4 further promoted and stabilized self-assembly, and allowed doxorubicin solubilization in the hydrophobic core of 70 nm PEG-C6-AZO-CA4 micelles. They reported hypoxia-mediated doxorubicin release in vitro, which was due to micelles' disassembly as evidenced by light scattering and transmission electron microscopy, whereas minimal micelles' disassembly and drug release were detected after 4 days at 37°C. Accordingly, PEG-C6-AZO-CA4 micelles were capable of hypoxia-dependent cellular uptake and cytotoxicity.

Also using azobenzene as the hypoxia-labile linker, Torchilin and coworkers designed tumor hypoxia-activated siRNA NP (Fig. 5.3) [75]. They used poly(ethylene glycol)-azobenzene-polyethylenimine-dioleylphosphatidylethanolamine (PAPD) amphiphilic copolymers where hydrophilic PEG confers stability in physiological fluids, hydrophilic azobenzene (Azo) and di-oleyl-phosphatidyl-ethanolamine (DOPE) promote self-assembly into ~100 nm NP, and polyethylenimine (PEI) condenses siRNA and protects it from RNase degradation (Fig. 5.3).

In this system, oxygen control of cellular uptake is based on PEG shedding after azobenzene cleavage under hypoxia, PEG release resulting in the exposure of PEI's positive charges to promote interaction with negatively charged cellular membranes and internalization. In vitro incubation of PAPD under hypoxia with rat liver microsomes as source of nitroreductases led to hypoxia-induced charge conversion from neutral (0.1 ± 6.5 mV) to positive (13.2 ± 6.7 mV) (Fig. 5.4B). Culture of cancer cell spheroids with PAPD under hypoxia dramatically increased the penetration of PAPD in these 3D cell cultures compared to normoxia and compared to spheroids cultured with non-responsive NP (azobenzene-free PEG-PEI-DOPE) under hypoxia (Fig. 5.4C). Finally, after systemic administration, PAPD accumulated in tumor tissues and achieved downregulation of the reporter gene GFP (Green Fluorescent Protein) in ovarian tumor xenografts expressing GFP.

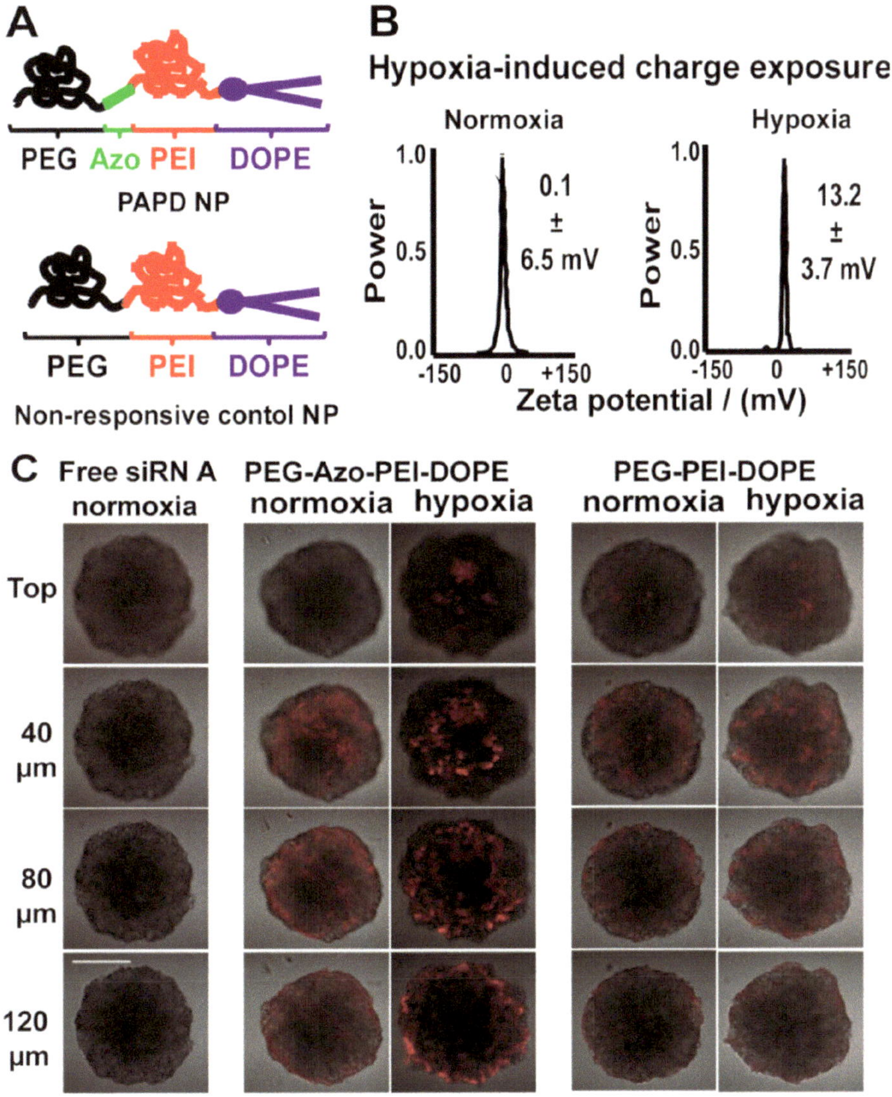

Figure 5.3 Hypoxia-induced siRNA cellular internalization using PAPD NP. (A) Structure of hypoxia-responsive PAPD and, of non-responsive control NP. (B) Hypoxia-induced charge exposure after incubation of PAPD under hypoxia. (C) Hypoxia-dependent penetration of cancer cell spheroids cultures under hypoxia. Scale bar represents 200 μm. Adapted from [75] with permission.

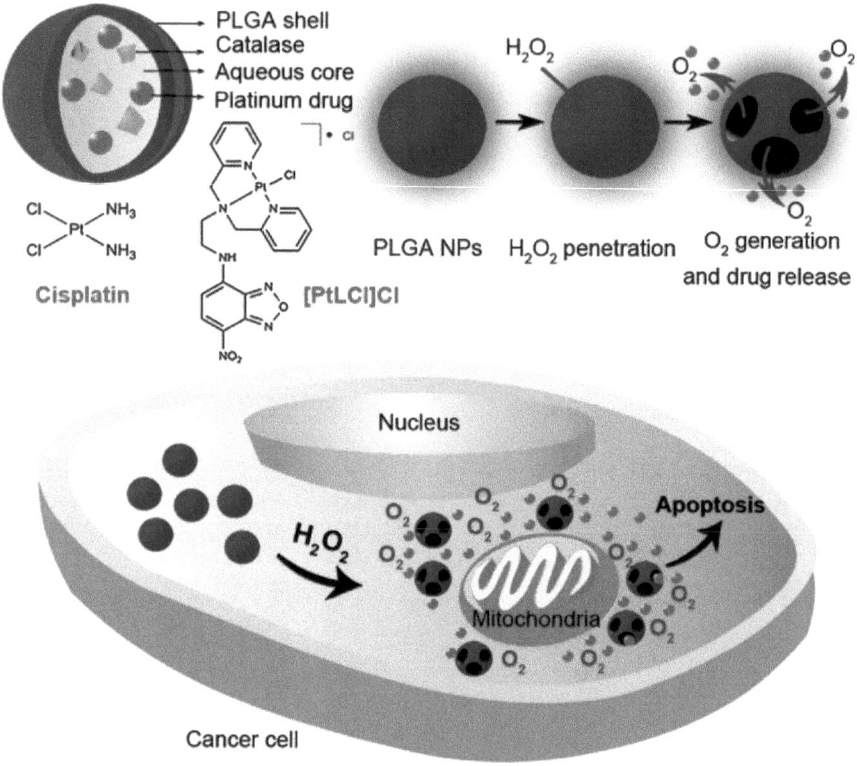

Figure 5.4 H_2O_2-responsive nanoparticles. Schematic illustration of the PGLA NP structure and of intracellular drug release after O_2 generation. Reproduced with permission from [78].

5.4.2 H_2O_2-Responsive Nanocarriers

Hydrogen peroxide (H_2O_2) is a hypoxia-induced ROS commonly elevated in cancer cells [27, 76]. Guo's group developed H_2O_2-responsive NP carrying catalase and either cisplatin for chemotherapy or methylene blue for PDT (Fig. 5.4) [77, 78].

Their strategy relies on H_2O_2-dependent activation of spherical PLGA (poly(D,L-lactic-*co*-glycolic acid)) NP with a size of ~200 nm encapsulating the oxygen generating enzyme catalase. In this design, NP encapsulating catalase and cargo (either cisplatin or methylene blue) are stable in the absence of H_2O_2 with virtually no content release in vitro over 24 h. However, incubation with a physiologically relevant H_2O_2 concentration (100 μM), induces

H_2O_2 penetration and the production of O_2 bubbles inside the NP; bubbles which lead to destabilization and content release. Because of this H_2O_2 activated content release by oxygen production, this platform was coined H_2O_2-activatable and O_2-evolving NP (HAO NP).

It is worth noting that in addition to control of content release, oxygen generation potentiated chemotherapy by cisplatin. Authors demonstrated that catalase-containing NP loaded with cisplatin were able to revert cisplatin resistance in SGC 7901/DDP cells (80% cytotoxicity), compared with minimal activity of non-sensitive NP without catalase (20% cytotoxicity), promising results to tackle drug resistance. Next, the same group elegantly addressed hypoxia-associated resistance to PDT. They used HAO NP for PDT (HAOP NP) encapsulating catalase, the photosensitizer methylene blue, and black hole quencher 3 (BHQ-3) to avoid $^1O_2$ production in the absence of irradiation. To increase selectivity towards tumor tissues, these PLGA NP harbor the tumor neovasculature ligand cyclo-Arg-Gly-Asp (cRGD) peptide [79]. They reported in cellulo H_2O_2-activated $^1O_2$ production and apoptosis, with no apoptosis in un-irradiated cells, achieving H_2O_2-dependent phototoxicity. Intravenous injection of HAOP NP inhibited the growth of tumor xenografts in mice, while no therapeutic effect was observed in un-irradiated HAOP NP and irradiated catalase-free NP groups. Improved therapeutic activity was correlated with decreased HIF-1α staining in tumors, demonstrating delivery to hypoxic areas, with a potentiation of PDT consistent with HIF-1α's role in drug resistance and cancer recurrence.

5.4.3 Sickle Erythrocytes as Trojan Horses to Hypoxic Niches

Sickle cell diseases are characterized by erythrocytes with impaired morphology due to mutations in hemoglobin [39, 80]. Sickle red blood cells (SSRBC) selectively accumulate in hypoxic regions over normal red blood cells (nRBC) because they harbor vascular zip codes recognizing adhesion molecules such as PCAM-1 (Platelet Cell Adhesion Molecule 1) and VCAM-1 (Vascular Cell Adhesion Molecule 1) expressed on hypoxic tumor vessels [39, 81]. Choe et al. proposed to take advantage of this property for SSRBC-mediated cargo delivery to hypoxic tumors [82]. SSRBC were isolated from transgenic sickle cell disease models. Their distribution to hypoxic 4T1 tumors was compared to that of nRBC after intravenous injection. 3-fold more SSRBC accumulated in tumors over nRBC during a 12–24 h period. Moreover, SSRBC were tumor selective with >4-fold accumulation in tumors over liver and spleen. Having confirmed selectivity, Authors assessed the potential of SSRBC for cargo delivery using calcein as a model drug. Calcein was loaded by hypotonic pre-swelling, leading to 85% calcein-positive cells. However, no calcein release was detected in tumors 24 h after injection. To address this, Authors co-loaded RBC with the photosensitizer protoporphyrin IX (ppIX) to promote delayed hemolysis after irradiation. The ppIX loading amount and irradiation time were tuned to achieve hemolysis 12–24 h after irradiation, a timeframe corresponding to their maximal tumor accumulation profile. Photo-irradiation of SSRBC just before their systemic administration enhanced intratumoral calcein release by more than 2-fold after 24 h. Conversely, no significant release was detected with nRBC. Importantly, ppIX loading and irradiation did not affect tumor specificity as no calcein release was detected in muscles 24 h after injection of SSRBC. Given the importance of hypoxic regions in cancer resistance, this strategy could be expanded to target unidentified tumors and intractable metastases. Finally, co-loading with a drug and an imaging agent would impart theranostic properties.

5.4.4 Hypoxia-Responsive Insulin Patch

Current response to diabetes involves monitoring of blood glucose levels and self-administration of insulin by the patient with associated risks of low patient compliance and improper glucose control. Existing solutions consist in dual devices with a glucose monitoring device dialoguing with an insulin reservoir [83]. A new closed-loop single device solution has been introduced as microneedle array patches containing hypoxia-sensitive insulin-loaded vesicles [84]. The device consists of conical microneedles of 10 μm tip radius and 600 μm height loaded with hypoxia-sensitive glucose-responsive vesicles (GRVs). GRVs are self-assembled from 2-nitroimidazole-conjugated hyaluronic acid.

2-nitroimidazole exhibits a hydrophobic to hydrophilic conversion under bioreductive hypoxic conditions, triggering hypoxia-induced release of the encapsulated insulin. Dynamic correlation between glucose monitoring and hypoxia-induced release of insulin was achieved by co-loading of glucose oxidase. The smart insulin patch functions as follows: high glucose concentrations lead to oxygen consumption by glucose oxidase, generated hypoxia induces insulin release, insulin release regulates blood glucose levels. In vitro, these GRVs displayed a significantly increased insulin release at hyperglycemic glucose levels (400 mg/dL) over normoglycemic levels (100 mg/dL) whereas no insulin release was detected in the absence of glucose. Integration of GRVs in a microneedle transcutaneous patch regulated blood glucose in a mouse model of type 1 diabetes for prolonged periods. Finally, the formulation is amenable to personalization as insulin release kinetics can be adjusted by varying glucose oxidase dose.

5.5 Hypoxia Imaging

Although several molecular tracers for hypoxia imaging have been developed, most of these strategies identify zones of lower oxygen tension rather than quantitate oxygen content [85]. On the other hand, encapsulation of tracers in NP which selectively accumulate and/or activate in hypoxic environments allows increased brightness by dye concentration leading to increased sensitivity [86]. Optical methods (phosphorescence or luminescence) are the most frequently used for NP-based oxygen sensing.

Establishment of the commercially available 2-nitroimidazole pimonidazole hydrochloride for hypoxia detection influenced research towards similar derivatives [69]. Although this approach successfully imaged hypoxia in the whole range of hypoxia and physioxia levels [72, 87, 88], their dependence on reduction for signal conversion /activation could represent a limitation for tumor hypoxia imaging. Indeed, the redox imbalance of tumor tissues with 2- to 10-fold higher glutathione concentration in tumors over normal tissues [89], could be responsible for unspecific reduction of such sensors and result in inaccurate mapping of hypoxic zones or even false positives.

This shortcoming has been addressed by the change of focus towards hypoxia-sensitive transition metals with porphyrin, ruthenium, and iridium complexes [90]. These systems face quenching of triplet state phosphorescence after NIR irradiation by molecular oxygen, oxygen-dependent conversion from exited triplet state to singlet state [91]. To estimate oxygen concentration, signal conversion is measured in different environments to establish non-linear Stern–Volmer plots of phosphorescence or luminescence intensity ratio under oxygen and nitrogen atmospheres (I_{N2}/I_{O2}) (Fig. 5.5) [92, 93]. The decay time (τ) of dyes is also affected by molecular oxygen, is less sensitive to environmental pH and photobleaching, and does not require dye excitation [91, 93].

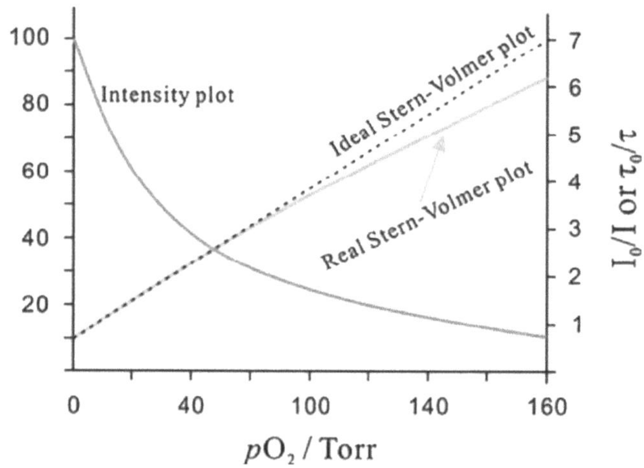

Figure 5.5 Luminescence quenching by oxygen, and respective Stern–Volmer plot. This figure and legend are reproduced from [93] under Creative Common Attribution License (CCBY).

Imaging shows dynamic evaluation of the hypoxic fraction in solid tumors, without the problems associated with the invasiveness and sampling heterogeneity of biochemical assessment of hypoxia. Details on oxygen sensing by physical techniques, materials, and probes can be found in comprehensive reviews [86, 90, 94].

We will limit ourselves to NP-based oxygen sensors. Indeed, encapsulation of sensors into NP is expected to reduce artifacts resulting from interactions with circulating biomolecules. Moreover, these undesired interactions of probes could affect their properties resulting in inaccurate O_2 quantitation. Finally, NP allow ratiometric imaging to consider heterogeneous distribution of the sensor and inhomogeneous irradiation. An oxygen-insensitive dye is co-encapsulated to increase diagnostic accuracy and facilitate inter-patient comparisons [90].

5.5.1 Hypoxia-Activated Signal

The polycyclic structure of transition metal complexes renders them particularly insoluble in water, and amenable to formulation as dendrimers [90]. Porphyrin-based dendrimers were among the first kind of oxygen-sensing NP, with commercial examples such as the second generation dendrimer Oxyphor R2 (Pd-meso-tetra(4-carboxyphenyl) porphyrin, Oxygen Enterprises) [90, 95]. Whereas conventional dendrimer-based oxygen sensors have limited cellular penetration, click-assembled oxygen-sensing nanoconjugates (CAOS) were introduced for depth-resolved oxygen imaging [96]. This platform was built by click chemistry of alkyne-functionalized dendron subunits on an azide-functionalized palladium tetra-carboxy-tetra-benzoporphyrin (PdTCTBP) core. CAOS nanoconjugates showed inverse dependence of phosphorescence intensity on pO_2 both in cancer cells monolayers and spheroids. Moreover, CAOS were able to distribute in cancer cell spheroids farther than surface layers whereas commercial Oxyphor G2 dendrimers failed to penetrate spheroids.

Another solution to improve solubility of metal complexes is the formation of micelles, a remarkable example of metal-based hypoxia nanosensor capable of primary tumor imaging and metastases detection will be described [97]. Zheng and coworkers prepared iridium-based hypoxia-activated co-micelles (Ir-CM) from iridium (III) conjugated poly(*N*-vinylpyrrolidone) (Ir-PVP) and poly(ε-caprolactone)-*b*-poly(*N*-vinylpyrrolidone) (PCL-PVP) (Fig. 5.6). Self-assembly is driven by mixing hydrophobic Ir and PCL with hydrophilic PVP, forming 50 nm spherical NP. Ir-CM are excited at 560 nm (a wavelength compatible with two-photon excitation for depth-resolved imaging) and emit in the near infrared (710 nm), also for deep tissue imaging. Ir-CM demonstrated sharp

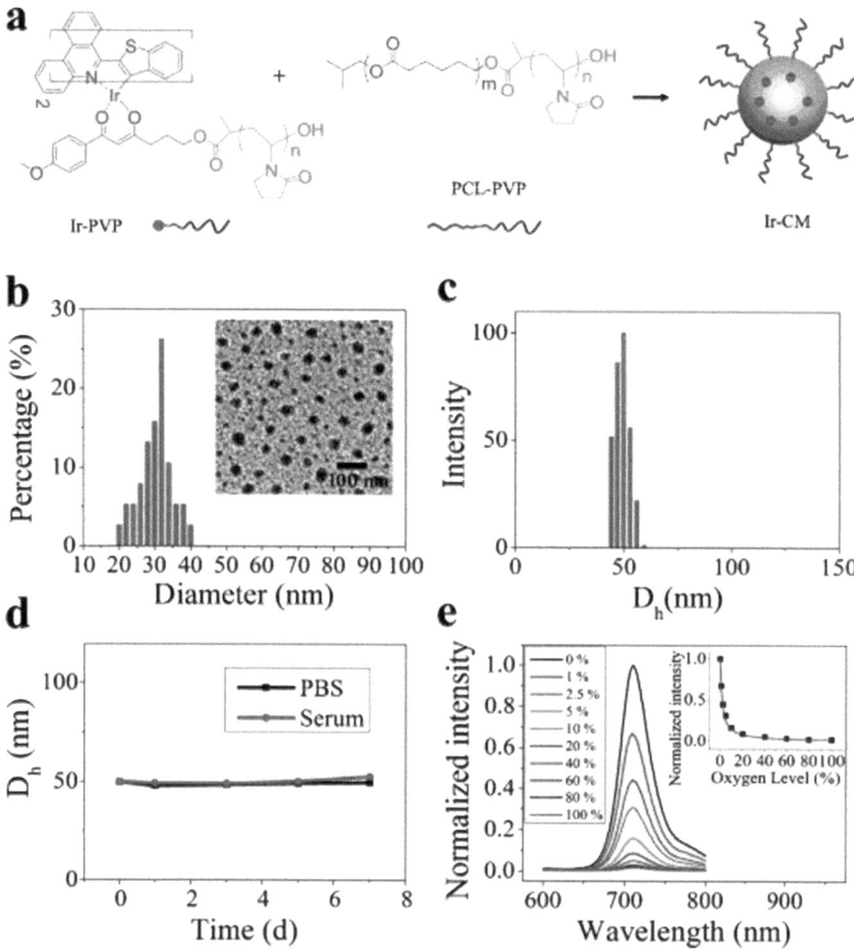

Figure 5.6 Iridium-based hypoxia-activated co-micelles (Ir-CM) to track cancer metastasis. (A) Synthesis of Ir-CM. The molar ratio of PCL-PVP and Ir-PVP is 4:1. (B) Phosphorescence emission spectra of Ir-CM in distilled water at different oxygen levels. Excitation: 560 nm. Insert: decay curve of peak intensities of Ir-CM with increasing oxygen levels. Reproduced with permission from [97].

oxygen-dependent phosphorescence emission with a dynamic range spanning physioxic and hypoxic oxygen concentrations (0–10% O_2), and a 60-fold increase in phosphorescence from saturated oxygen to 0% O_2 in vitro (Fig. 5.6B). Hypoxia response was demonstrated in cellulo both with monolayer and spheroid cultures, with a signal intensity proportional to HIF-1α staining. Authors then took advantage of the small size of Ir-CM (50 nm, monodisperse) to detect primary and secondary tumors using experimental and spontaneous metastasis models. One week after intravenous injection of cancer cells to establish experimental lung cancer metastases, systemic administration of Ir-CM readily identified lung metastases by whole body imaging as early as 1 h after injection, with a signal detected up to 48 h post-injection. Interestingly, signal was 10-fold higher in HIF-1α positive lung metastases over liver and spleen which are common non-specific accumulation sites for NP. To discern whether this was due to a cumulative effect (with an expected 10-fold higher accumulation of IrCM in lungs over spleen and liver) or to confined activation in the tumor, Authors compared ex vivo fluorescence profiles of major organs (activation) with the injected dose per g of tissue (ID/g, accumulation). At 48 h post-injection only 4% ID/g (10-fold phosphorescence) was measured in metastases-bearing lungs compared with 11% ID/g in the spleen (1-fold phosphorescence). Data demonstrate selective Ir-CM activation in hypoxic tumors rather than an accumulation effect. Selectivity was reinforced by the 60-fold stronger fluorescence in experimental metastases-bearing lungs over naïve lungs.

Next, Authors evaluated Ir-CM in the more stringent and physiologically relevant model of spontaneous lymph node metastasis from primary tumors. To do so, they administered Ir-CM to mice bearing subcutaneous tumors in the right paw. One hour after injection, two regions of strong phosphorescence were detected, one corresponding to the primary tumor and the second to HIF-1α positive lymph node metastases. Metastases were detected in the inguinal lymph node, the draining lymph node in their model, with a 30-fold stronger phosphorescence over inguinal lymph nodes from naïve mice. Altogether, Ir-CM appear as a significant advancement both for cancer detection and patient stratification.

Even with the development of NIR dyes and thinner optical fibers, signal attenuation in tissues [98, 99] may reduce the energy reaching hypoxic niches below the threshold required for signal conversion and result in inaccurate diagnosis.

5.5.2 Energy Transfer for Improved Signal Response

As only a fraction of illuminated light reaches the dye in the core of NP, light harvesting and energy transfer from donor polymer in the shell to the oxygen probe in the core has been proposed for enhanced sensitivity [100–102]. In these constructs, the donor polymer stabilizes NP and serves as a reference dye. Since energy transfer, either by fluorescence (FRET, Fluorescence Resonance Energy Transfer) and luminescence (LRET) is highly distance-dependent [103], this approach represents a solution both to decrease inaccuracy due to leakage of co-encapsulated reference dye and to discriminate between free and NP-conjugated probe. We will describe examples with a polymer as FRET donor and with silica NP as LRET donor.

Wu et al. formed NP by precipitation of a porphyrin derivative and polyfluorene donor polymers [100]. They used PDHF (poly(9,9-dihexylfluorene)) and PFO (poly(9,9-dioctylfluorene)) polymers to encapsulate Pt(II) octaethylporphyrin (PtOEP) to form spherical NP of 50 nm designated π-conjugated polymer NP (CPdots). In CPdots, the polyfluorenes are used as a FRET donor and PtOEP as a FRET acceptor. Both PDHF (shell) and PtOEP (core) absorb at 380 nm, with a relative contribution correlated with their access to illumination (80% and 20%, respectively). Excitation results in oxygen-insensitive fluorescence from polyfluorenes at 420 nm, and oxygen-dependent phosphorescence at 650 nm for PtOEP after FRET. Whereas donor fluorescence was similar under saturated oxygen and nitrogen, phosphorescence was dramatically increased under nitrogen, together with an increase in decay lifetime from 10 μs (saturated oxygen) to 37 μs (nitrogen) (Fig. 5.7). This NP assembly allowed several cycles of oxygen imaging in vitro.

To allow stronger signal amplification from low energy light, Shi's group used core-void-shell hollow mesoporous silica NP featuring lanthanide upconversion nanoparticles (UCNP) as light harvesting antennae to transfer energy to hypoxia-sensitive ruthenium complexes by LRET [104]. They introduced core-void-shell silica NP with UCNP in the core, ruthenium (II) dichloride/tris (4,7-diphenyl-1,10-phenanthroline) ($[Ru(dpp)_3]^{2+}Cl_2$) oxygen sensor in the void, and a mesoporous silica shell (Fig. 5.8). UCNP

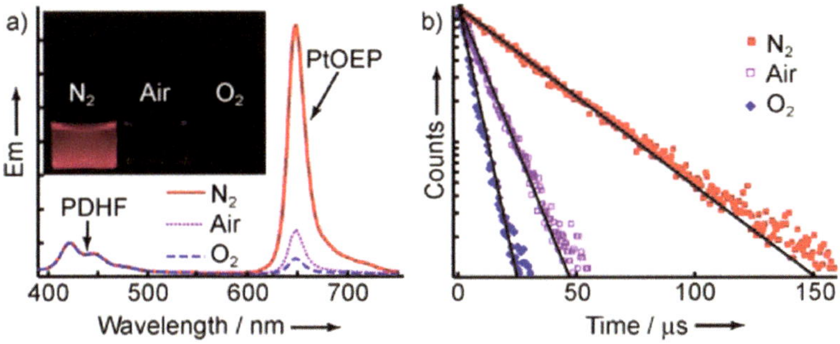

Figure 5.7 Oxygen sensor with light harvesting polymer. (a) Oxygen-dependent emission spectra of PtOEP-doped PDHF dots after 350 nm excitation. Inset shows CPdots in aqueous solutions saturated with nitrogen, air, and oxygen, respectively, under a UV lamp. (b) Phosphorescence decays of the 10% PtOEP-doped PDHF dots with different oxygen concentrations. Scattered symbols represent experimental data, and solid lines correspond to fits. Reproduced with permission from [100].

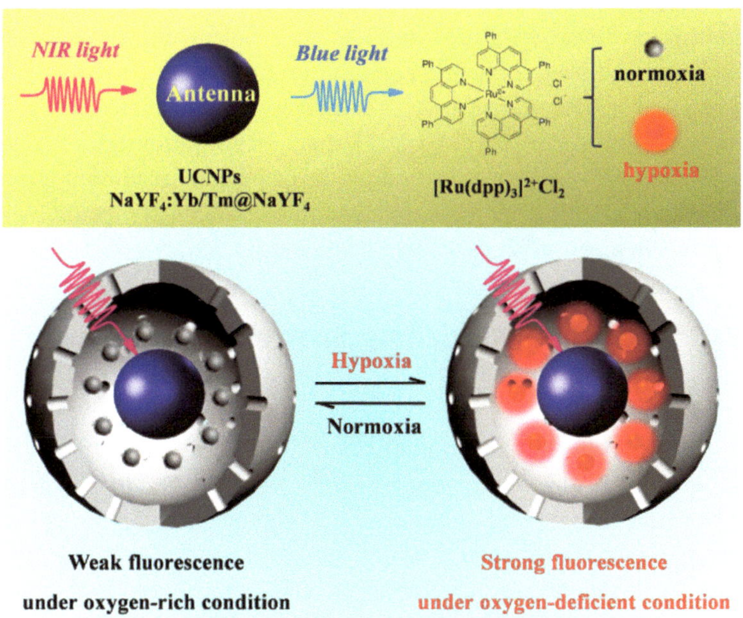

Figure 5.8 Illustration of UCNP/[Ru(dpp)$_3$]$^{2+}$Cl$_2$ hollow mesoporous core-void-shell silica nanoparticles and their oxygen sensing mechanism. NIR light is of too low energy to excite the sensor. However, it induces emission of blue photons by the light-harvesting core after up-conversion to allow highly sensitive oxygen-dependent phosphorescence. Reproduced with permission from [104].

are $NaYF_4$ nanocrystals which incorporate Yb^{3+} lanthanide sensitizer ions to induce emission of high energy (visible) photons after irradiation with low energy NIR light (980 nm). Such conversion is based on sequential absorption of 980 nm photons by a dopant (here Yb^{3+}) with long-lived energy states, resulting in emission of higher energy photons (450 and 475 nm) amenable to excite proximal dyes with absorbance in the visible but not in the NIR range [105]. The cavity/void of the hollow mesoporous NP is used to encapsulate $[Ru(dpp)_3]^{2+}Cl_2$ whose luminescence is quenched by oxygen [106]. Whereas $[Ru(dpp)_3]^{2+}Cl_2$ is not directly exited by NIR light, its absorption spectra overlaps that of UCNP upconverted NIR light, and results in oxygen-dependent emission at 613 nm after irradiation at 980 nm. The outermost optically transparent mesoporous shell stabilizes the assembly in physiological fluids and does not restrict oxygen diffusion to the cavity. Altogether, this hierarchical architecture confines interaction between donor and acceptor within the hollow cavity for increased selectivity and sensitivity with imaging at low oxygen pressures (<2% O_2). Of note, the oxygen-insensitive 477 nm emission band of $[Ru(dpp)_3]^{2+}Cl_2$ permitted ratiometric imaging (643 nm/477 nm) to refine mapping of oxygen gradients. Finally, $[Ru(dpp)_3]^{2+}Cl_2$/UCNP hollow mesoporous silica NP could detect hypoxia in a zebrafish cerebral anoxia model.

5.6 Hypoxia-Mimicking Scaffolds for Tissue Engineering

Hypoxia's role in development and bone repair also make it important for tissue engineering [107]. Using an original approach, Quinlan et al. reported enhanced bone repair using a porous glass/collagen-glycosaminoglycan hypoxia mimicking scaffold [108]. Authors used a porous collagen I/glycosaminoglycan scaffold doped with cobalt ions (Co^{2+})-conjugated bioactive glass. In this design, cobalt ions mimic hypoxia by stabilization of HIF-1α [109]. The sustained release of cobalt ions thereby achieved improved VEGF production by HUVEC (Human Umbilical Vein Endothelial Cells), and increased alkaline phosphate activity and calcium production by pre-osteoblastic cells; demonstrating improved tissue engineering by hypoxia mimicry.

A further advancement in the design of 3D tissue environments mimicking physiological oxygen levels for each tissue was reported with hypoxia inducible (HI) hydrogels [110]. HI hydrogels are made of gelatin and ferulic acid, with a sol-gel transition achieved by laccase-mediated formation of diferulic acid cross-links. As the laccase reaction consumes oxygen, oxygen levels in the matrix can be tuned based on laccase concentration, with a further control of O_2 tension based on hydrogel thickness and substitution degree. These hydrogels encapsulating endothelial colony forming cells were superior stimulators of blood vessel recruitment and neovascularization after in vivo implantation. Increased tissue repair was dependent on the activation of hypoxia inducible factors and MMP, contributing to the development of better tissue models and improved grafts solutions.

5.7 Challenges and Perspectives

We covered several classes of HR NP where formulation improves the distribution to hypoxic areas and enhances the responsiveness to oxygen gradients. However, as hypoxia-activated assemblies were recently introduced, several challenges remain before turning them into a popular and profitable technology. Due to their earliness in development, HR NP face the common issues of NP development without any approved examples to support investors, patients, and physicians adhesion [111, 112].

5.7.1 Limitations of the Models Used to Evaluate HR NP

In most studies introducing novel HR NP, authors demonstrate activation of HR NP under hypoxic conditions in cancer cell lines under hypoxia or in the best case scenario in HIF-1α positive orthotopic tumors. However, such results may need to be reexamined based on the actual hypoxic levels measured in patients. Indeed, demonstrating activation of a HR NP under acute hypoxia ($<2\%$ O_2), as promising results to treat pathologies with oxygenation status slightly inferior to matched tissues in control patients may lead to very high attrition rates of HR NP candidates at early stages of clinical trials due to irrelevant modeling of action mechanism [113]. One solution could be to implement hypoxia-mimicking scaffolds [108, 110] in the screening process for HR NP candidates.

5.7.2 Hypoxia Heterogeneity Challenges

In addition to disease-specific hypoxia levels, HR NP face inter-patient heterogeneity [114, 115]. The ever growing catalog of hypoxia-responsive nanoprobes could be used to stratify patients into groups (e.g., <1% O_2, <3% O_2, <5% O_2). To broaden the population of patients eligible to treatment, HR NP would be tuned to respond to distinct hypoxia-levels by creating different HR NP versions. Fine-tuning of oxygen sensitivity would be achieved by substitutions of the hypoxia-responsive moiety, an approach comparable with the library of hypoxia nanoprobes developed by the group of Pr. Nagano [72, 88].

5.7.3 Cost and Complexity Challenges

Although very promising, the increasing complexity of HR NP, a tendency already observed with other targeted nanomedicines, may limit their translation potential because of unfavorable cost versus benefit [116, 117]. Cost of current approaches represent a particular challenge to treat the predicted 9.6 billion World population projected for 2050, especially considering that 86% will inhabit the less developed countries [118]. Even in developed countries, this targeting cost may further disadvantage medical support towards existing solutions over HR NP, a critical risk [119].

5.7.4 Adipose Tissue Targeting

Evidence of hypoxia in diseased adipose tissues [61, 62] suggests evaluation of HR NP in obesity models. In addition to hypoxia responsiveness, HR NP would benefit from the accumulation of NP in fat tissue [120]. Hossen et al. reported a 35-fold higher accumulation in fat of dietary induced obesity mice over normal mice. Harashima and coworkers attributed these results to features of obese fat tissue: Adipocyte growth is sustained by a dense vascular network, clusters of neovascularization, and local inflammation resulting in increased vascular permeability of obese visceral fat. In light of these data, HR NP formulation may improve the treatment of the 42 million worldwide overweight children under the age of five. Children obesity requires safe treatments effective at low dosage, and could greatly benefit from increased drug delivery to fat tissue [121].

Development of HR NP involves biologists, chemists, physicists and pharmaceutical scientists. It is of no doubt that HR NP will benefit from advances in their respective fields and their dialogue to improve HR NP fitness to distinct conditions. As more stimuli-sensitive NP advance into clinical studies, feedback will refine HR NP design resulting in nanoparticles able to treat highly complex and rapidly evolving human diseases [15].

Acknowledgment

This work was financially supported by JSPS KAKENHI Grant Number 14F04352.

References

1. Chow, E. K.-H., and D. Ho (2013). Cancer nanomedicine: From drug delivery to imaging, *Science Translational Medicine*. **5**(216): pp. 216rv4–216rv4.
2. Davis, M. E., and D. M. Shin (2008). Nanoparticle therapeutics: An emerging treatment modality for cancer, *Nature Reviews Drug Discovery*. **7**(9): pp. 771–782.
3. Torchilin, V., and M. M. Amiji (2010). *Handbook of Materials for Nanomedicine*. vol. 1. Pan Stanford Publishing.
4. Mahato, R. I., and A. S. Narang (2011). *Pharmaceutical Dosage Forms and Drug Delivery*. CRC Press.
5. Bertrand, N., et al. (2014). Cancer nanotechnology: The impact of passive and active targeting in the era of modern cancer biology, *Advanced Drug Delivery Reviews*. **66**: pp. 2–25.
6. Bae, Y. H., R. J. Mrsny, and K. Park (2013). *Cancer Targeted Drug Delivery: An Elusive Dream*. Springer Science & Business Media.

7. Sawyers, C. L., et al. (2013). AACR Cancer progress report 2013, *Clinical Cancer Research.* **19**(20 Supplement): pp. S1–S98.
8. Yap, T. A., et al. (2012). Intratumor heterogeneity: Seeing the wood for the trees, *Science Translational Medicine.* **4**(127): pp. 127ps10–127ps10.
9. Ryu, J. H., et al. (2012). Tumor-targeting multi-functional nanoparticles for theragnosis: New paradigm for cancer therapy, *Advanced Drug Delivery Reviews.* **64**(13): pp. 1447–1458.
10. Chan, A., et al. (2013). Remote and local control of stimuli responsive materials for therapeutic applications, *Advanced Drug Delivery Reviews.* **65**(4): pp. 497–514.
11. Zhu, L., and V. P. Torchilin (2013). Stimulus-responsive nanopreparations for tumor targeting, *Integrative Biology.* **5**(1): pp. 96–107.
12. Perche, F., S. Biswas, and V. Torchilin (2015). Stimuli-sensitive polymeric nanomedicines for cancer imaging and therapy. In: Thakur, V. K., Thakur, M. K., eds. *Handbook of Polymers for Pharmaceutical Technologies, Processing and Applications,* vol. 2, p. 311.
13. Ge, Z., and S. Liu (2013). Functional block copolymer assemblies responsive to tumor and intracellular microenvironments for site-specific drug delivery and enhanced imaging performance, *Chemical Society Reviews.* **42**(17): pp. 7289–7325.
14. Perche, F., and V. P. Torchilin (2013). Recent trends in multifunctional liposomal nanocarriers for enhanced tumor targeting, *Journal of Drug Delivery.* **2013**: p. 32.
15. Mura, S., J. Nicolas, and P. Couvreur (2013). Stimuli-responsive nanocarriers for drug delivery, *Nature Materials.* **12**(11): pp. 991–1003.
16. Simon, M. C., and B. Keith (2008). The role of oxygen availability in embryonic development and stem cell function, *Nature Reviews Molecular Cell Biology.* **9**(4): pp. 285–296.
17. McKeown, S (2014). Defining normoxia, physoxia and hypoxia in tumours—implications for treatment response, *The British Journal of Radiology.* **87**(1035): p. 20130676.
18. Imanirad, P., and E. Dzierzak (2013). Hypoxia and HIFs in regulating the development of the hematopoietic system, *Blood Cells, Molecules and Diseases.* **51**(4): pp. 256–263.
19. Spencer, J. A., et al. (2014). Direct measurement of local oxygen concentration in the bone marrow of live animals, *Nature.* **508**(7495): pp. 269–273.

20. Wilson, W. R., and M. P. Hay (2011). Targeting hypoxia in cancer therapy, *Nature Reviews Cancer*. **11**(6): pp. 393–410.
21. Brown, J. M., and W. R. Wilson (2004). Exploiting tumour hypoxia in cancer treatment, *Nature Reviews Cancer*. **4**(6): pp. 437–447.
22. Brahimi-Horn, M. C., and J. Pouysségur (2009). HIF at a glance, *Journal of Cell Science*. **122**(8): pp. 1055–1057.
23. Kaluz, S., M. Kaluzová, and E. J. Stanbridge (2008). Regulation of gene expression by hypoxia: Integration of the HIF-transduced hypoxic signal at the hypoxia-responsive element, *Clinica Chimica Acta*. **395**(1): pp. 6–13.
24. Keith, B., R. S. Johnson, and M. C. Simon (2012). HIF1α and HIF2α: Sibling rivalry in hypoxic tumour growth and progression, *Nature Reviews Cancer*. **12**(1): pp. 9–22.
25. Kulshreshtha, R., et al. (2008). A microRNA component of the hypoxic response, *Cell Death & Differentiation*. **15**(4): pp. 667–671.
26. Kulshreshtha, R., et al. (2007). A microRNA signature of hypoxia, *Molecular and Cellular Biology*. **27**(5): pp. 1859–1867.
27. Murphy, M. (2009). How mitochondria produce reactive oxygen species, *Biochemical Journal*. **417**: pp. 1–13.
28. Hielscher, A., and S. Gerecht (2015). Hypoxia and free radicals: Role in tumor progression and the use of engineering-based platforms to address these relationships, *Free Radical Biology and Medicine*. **79**: pp. 281–291.
29. Guzy, R. D., and P. T. Schumacker (2006). Oxygen sensing by mitochondria at complex III: The paradox of increased reactive oxygen species during hypoxia, *Experimental Physiology*. **91**(5): pp. 807–819.
30. Chandel, N., et al. (1998). Mitochondrial reactive oxygen species trigger hypoxia-induced transcription, *Proceedings of the National Academy of Sciences*. **95**(20): pp. 11715–11720.
31. Ruthenborg, R. J., et al. (2014). Regulation of wound healing and fibrosis by hypoxia and hypoxia-inducible factor-1, *Molecules and Cells*. **37**(9): pp. 637–643.
32. Murdoch, C., M. Muthana, and C. E. Lewis (2005). Hypoxia regulates macrophage functions in inflammation, *Journal of Immunology*. **175**(10): pp. 6257–6263.
33. Wan, C., et al. (2008). Activation of the hypoxia-inducible factor-1alpha pathway accelerates bone regeneration, *Proceedings*

of the National Academy of Sciences of the United States of America. **105**(2): pp. 686–691.

34. Dhani, N., et al. (2015). *The Clinical Significance of Hypoxia in Human Cancers.* in *Seminars in Nuclear Medicine.* Elsevier.
35. Höckel, M., and P. Vaupel (2001). Tumor hypoxia: Definitions and current clinical, biologic, and molecular aspects, *Journal of the National Cancer Institute.* **93**(4): pp. 266–276.
36. Chang, Q., et al. (2011). Hypoxia predicts aggressive growth and spontaneous metastasis formation from orthotopically grown primary xenografts of human pancreatic cancer, *Cancer Research.* **71**(8): pp. 3110–3120.
37. Melillo, G. (2013). *Hypoxia and Cancer: Biological Implications and Therapeutic Opportunities.* Springer Science & Business Media.
38. Daruwalla, J., et al. (2015). Styrene maleic acid copolymer–pirarubicin induces tumor-selective oxidative stress and decreases tumor hypoxia as possible treatment of colorectal cancer liver metastases, *Surgery.* **158**(1): pp. 236–247.
39. Terman, D. S., et al. (2013). Sickle erythrocytes target cytotoxics to hypoxic tumor microvessels and potentiate a tumoricidal response, *PloS One.* **8**(1): e52543.
40. Zhong, H., et al. (1999). Overexpression of hypoxia-inducible factor 1α in common human cancers and their metastases, *Cancer Research.* **59**(22): pp. 5830–5835.
41. Overgaard, J. (2007). Hypoxic radiosensitization: Adored and ignored, *Journal of Clinical Oncology.* **25**(26): pp. 4066–4074.
42. Overgaard, J. (2011). Hypoxic modification of radiotherapy in squamous cell carcinoma of the head and neck–a systematic review and meta-analysis, *Radiotherapy and Oncology.* **100**(1): pp. 22–32.
43. Gross, S., et al. (2003). Monitoring photodynamic therapy of solid tumors online by BOLD-contrast MRI, *Nature Medicine.* **9**(10): pp. 1327–1331.
44. Fuchs, J., and J. Thiele (1998). The role of oxygen in cutaneous photodynamic therapy, *Free Radical Biology and Medicine.* **24**(5): pp. 835–847.
45. Rohwer, N., and T. Cramer (2011). Hypoxia-mediated drug resistance: Novel insights on the functional interaction of HIFs and cell death pathways, *Drug Resistance Updates.* **14**(3): pp. 191–201.

46. Nardinocchi, L., et al. (2009). Inhibition of HIF-1alpha activity by homeodomain-interacting protein kinase-2 correlates with sensitization of chemoresistant cells to undergo apoptosis, *Molecular Cancer.* **8**(1): p. 1.

47. Chen, N., et al. (2009). BCL-xL is a target gene regulated by hypoxia-inducible factor-1α, *Journal of Biological Chemistry.* **284**(15): pp. 10004–10012.

48. Heddleston, J., et al. (2010). Hypoxia inducible factors in cancer stem cells, *British Journal of Cancer.* **102**(5): pp. 789–795.

49. Koh, M. Y., et al. (2011). The hypoxia-associated factor switches cells from HIF-1α-to HIF-2α-dependent signaling promoting stem cell characteristics, aggressive tumor growth and invasion, *Cancer Research.* **71**(11): pp. 4015–4027.

50. Rey, S., and G. L. Semenza (2010). Hypoxia-inducible factor-1-dependent mechanisms of vascularization and vascular remodeling, *Cardiovascular Research*, **86**(2): 236–242. p. cvq045.

51. Nisticò, P., M. J. Bissell, and D. C. Radisky (2012). Epithelial-mesenchymal transition: General principles and pathological relevance with special emphasis on the role of matrix metalloproteinases, *Cold Spring Harbor Perspectives in Biology.* **4**(2): p. a011908.

52. Tong, X., F. Zhao, and C. B. Thompson (2009). The molecular determinants of de novo nucleotide biosynthesis in cancer cells, *Current Opinion in Genetics & Development.* **19**(1): pp. 32–37.

53. Denko, N. C. (2008). Hypoxia, HIF1 and glucose metabolism in the solid tumour, *Nature Reviews Cancer.* **8**(9): pp. 705–713.

54. Kim, J.-W., et al. (2006). HIF-1-mediated expression of pyruvate dehydrogenase kinase: A metabolic switch required for cellular adaptation to hypoxia, *Cell Metabolism.* **3**(3): pp. 177–185.

55. Chen, X., Y. Qian, and S. Wu (2015). The Warburg effect: Evolving interpretations of an established concept, *Free Radical Biology and Medicine.* **79**: pp. 253–263.

56. Milane, L., Z. Duan, and M. Amiji (2011). Role of hypoxia and glycolysis in the development of multi-drug resistance in human tumor cells and the establishment of an orthotopic multi-drug resistant tumor model in nude mice using hypoxic pre-conditioning, *Cancer Cell International.* **11**(3): pp. 1–16.

57. Samanta, D., et al. (2014). Hypoxia-inducible factors are required for chemotherapy resistance of breast cancer stem cells, *Proceedings of the National Academy of Sciences.* **111**(50): pp. E5429–E5438.

58. Cao, Y., et al. (2013). Tumor cells upregulate normoxic HIF-1α in response to doxorubicin, *Cancer Research.* **73**(20): pp. 6230–6242.
59. Li, X., et al. (2008). Intrinsic resistance of tumorigenic breast cancer cells to chemotherapy, *Journal of the National Cancer Institute.* **100**(9): pp. 672–679.
60. Netzer, N., et al. (2015). Hypoxia, oxidative stress and fat, *Biomolecules.* **5**(2): pp. 1143–1150.
61. Michailidou, Z., et al. (2015). Adipocyte pseudohypoxia suppresses lipolysis and facilitates benign adipose tissue expansion, *Diabetes.* **64**(3): pp. 733–745.
62. Wood, I. S., et al. (2007). Hypoxia increases expression of selective facilitative glucose transporters (GLUT) and 2-deoxy-D-glucose uptake in human adipocytes, *Biochemical and Biophysical Research Communications.* **361**(2): pp. 468–473.
63. Jain, R. K. (2001). Delivery of molecular and cellular medicine to solid tumors, *Advanced Drug Delivery Reviews.* **46**(1): pp. 149–168.
64. Cabral, H., et al. (2015). Systemic targeting of lymph node metastasis through the blood vascular system by using size-controlled nanocarriers, *ACS Nano,* **9**(5): 4957–4967.
65. Cabral, H., et al. (2011). Accumulation of sub-100 nm polymeric micelles in poorly permeable tumours depends on size, *Nature Nanotechnology.* **6**(12): pp. 815–823.
66. Maeda, K., T. Osato, and H. Umezawa (1953). A new antibiotic, azomycin, *The Journal of Antibiotics.* **6**(4): p. 182.
67. Varia, M. A., et al. (1998). Pimonidazole: A novel hypoxia marker for complementary study of tumor hypoxia and cell proliferation in cervical carcinoma, *Gynecologic Oncology.* **71**(2): pp. 270–277.
68. Thambi, T., et al. (2014). Hypoxia-responsive polymeric nanoparticles for tumor-targeted drug delivery, *Biomaterials.* **35**(5): pp. 1735–1743.
69. Kizaka-Kondoh, S., and H. Konse-Nagasawa (2009). Significance of nitroimidazole compounds and hypoxia-inducible factor-1 for imaging tumor hypoxia, *Cancer Science.* **100**(8): pp. 1366–1373.
70. Lv, P.-C., and H.-L. Zhu (2015). Nitroimidazole derivatives: A patent review of US 2014/0141084 A1, *Expert Opinion on Therapeutic Patents.* **25**(4): pp. 489–492.
71. Mason, R. P., and J. L. Holtzman (1975). The role of catalytic superoxide formation in the O_2 inhibition of nitroreductase, *Biochemical and Biophysical Research Communications.* **67**(4): pp. 1267–1274.

72. Piao, W., et al. (2013). Development of azo-based fluorescent probes to detect different levels of hypoxia, *Angewandte Chemie International Edition.* **52**(49): pp. 13028–13032.

73. Deloncle, R., and A.-M. Caminade (2010). Stimuli-responsive dendritic structures: The case of light-driven azobenzene-containing dendrimers and dendrons, *Journal of Photochemistry and Photobiology C: Photochemistry Reviews.* **11**(1): pp. 25–45.

74. Liu, H., et al. (2015). Development of hypoxia-triggered prodrug micelles as doxorubicin carriers for tumor therapy, *RSC Advances.* **5**(27): pp. 20848–20857.

75. Perche, F., et al. (2014). Hypoxia-targeted siRNA delivery, *Angewandte Chemie.* **126**(13): pp. 3430–3434.

76. Szatrowski, T. P., and C. F. Nathan (1991). Production of large amounts of hydrogen peroxide by human tumor cells, *Cancer Research.* **51**(3): pp. 794–798.

77. Chen, H., et al. (2015). H_2O_2-activatable and O_2-evolving nanoparticles for highly efficient and selective photodynamic therapy against hypoxic tumor cells, *Journal of the American Chemical Society.* **137**(4): pp. 1539–1547.

78. Chen, H., W. He, and Z. Guo (2014). An H_2O_2-responsive nanocarrier for dual-release of platinum anticancer drugs and O_2: Controlled release and enhanced cytotoxicity against cisplatin resistant cancer cells, *Chemical Communications.* **50**(68): pp. 9714–9717.

79. Arap, W., R. Pasqualini, and E. Ruoslahti (1998). Cancer treatment by targeted drug delivery to tumor vasculature in a mouse model, *Science.* **279**(5349): pp. 377–380.

80. Stuart, M. J., and R. L. Nagel (2004). Sickle-cell disease, *The Lancet.* **364**(9442): pp. 1343–1360.

81. Brown, M. D., T. M. Wick, and J. R. Eckman (2001). Activation of vascular endothelial cell adhesion molecule expression by sickle blood cells, *Fetal & Pediatric Pathology.* **20**(1): pp. 47–72.

82. Choe, S.-W., et al. (2013). Drug-loaded sickle cells programmed ex vivo for delayed hemolysis target hypoxic tumor microvessels and augment tumor drug delivery, *Journal of Controlled Release.* **171**(2): pp. 184–192.

83. Bratlie, K. M., et al. (2012). Materials for diabetes therapeutics, *Advanced Healthcare Materials.* **1**(3): pp. 267–284.

84. Yu, J., et al. (2015). Microneedle-array patches loaded with hypoxia-sensitive vesicles provide fast glucose-responsive insulin delivery,

Proceedings of the National Academy of Sciences of the United States of America. **112**(27): pp. 8260–8265.

85. Carreau, A., et al. (2011). Why is the partial oxygen pressure of human tissues a crucial parameter? Small molecules and hypoxia, *Journal of Cellular and Molecular Medicine*. **15**(6): pp. 1239–1253.
86. Dmitriev, R. I., and D. B. Papkovsky (2015). Intracellular probes for imaging oxygen concentration: How good are they?, *Methods and Applications in Fluorescence*. **3**(3): p. 034001.
87. Takahashi, S., et al. (2012). Reversible off–on fluorescence probe for hypoxia and imaging of hypoxia–normoxia cycles in live cells, *Journal of the American Chemical Society*. **134**(48): pp. 19588–19591.
88. Kiyose, K., et al. (2010). Hypoxia-sensitive fluorescent probes for in vivo real-time fluorescence imaging of acute ischemia, *Journal of the American Chemical Society*. **132**(45): pp. 15846–15848.
89. Jorgenson, T. C., W. Zhong, and T. D. Oberley (2013). Redox imbalance and biochemical changes in cancer, *Cancer Research*. **73**(20): pp. 6118–6123.
90. Roussakis, E., et al. (2015). Oxygen-sensing methods in biomedicine from the macroscale to the microscale, *Angewandte Chemie International Edition*. **54**: pp. 8340–8362.
91. Vanderkooi, J. M., et al. (1987). An optical method for measurement of dioxygen concentration based upon quenching of phosphorescence, *Journal of Biological Chemistry*. **262**(12): pp. 5476–5482.
92. Vinogradov, S. A., and D. F. Wilson (2010). Porphyrin-dendrimers as biological oxygen sensors, *Designing Dendrimers*. **1**: pp. 463–503.
93. Wolfbeis, O. S. (2015). Luminescent sensing and imaging of oxygen: Fierce competition to the Clark electrode, *BioEssays*. **37**(8): 921–928.
94. Do, Q. N., et al. (2014). Redox-and hypoxia-responsive MRI contrast agents, *ChemMedChem*. **9**(6): pp. 1116–1129.
95. Lo, L.-W., et al. (1997). A new, water soluble, phosphor for oxygen measurements in vivo, in *Oxygen Transport to Tissue XIX*., Springer. pp. 651–656.
96. Nichols, A. J., et al. (2014). Click-assembled, oxygen-sensing nanoconjugates for depth-resolved, near-infrared imaging in a 3D cancer model, *Angewandte Chemie International Edition*. **53**(14): pp. 3671–3674.

97. Zheng, X., et al. (2015). Tracking cancer metastasis in vivo by using an iridium-based hypoxia-activated optical oxygen nanosensor, *Angewandte Chemie*. **127**(28): pp. 8212–8217.

98. Gonçalves, M. S. T. (2008). Fluorescent labeling of biomolecules with organic probes, *Chemical Reviews*. **109**(1): pp. 190–212.

99. Liu, Y., Y.-C. Tseng, and L. Huang (2012). Biodistribution studies of nanoparticles using fluorescence imaging: A qualitative or quantitative method?, *Pharmaceutical Research*. **29**(12): pp. 3273–3277.

100. Wu, C., et al. (2009). Ratiometric single-nanoparticle oxygen sensors for biological imaging, *Angewandte Chemie International Edition*. **48**(15): pp. 2741–2745.

101. Roussakis, E., et al. (2014). Two-photon antenna-core oxygen probe with enhanced performance, *Analytical Chemistry*. **86**(12): pp. 5937–5945.

102. Grimland, J. L., et al. (2011). Photosensitizer-doped conjugated polymer nanoparticles with high cross-sections for one-and two-photon excitation, *Nanoscale*. **3**(4): pp. 1451–1455.

103. Sekar, R. B., and A. Periasamy (2003). Fluorescence resonance energy transfer (FRET) microscopy imaging of live cell protein localizations, *The Journal of Cell Biology*. **160**(5): pp. 629–633.

104. Liu, J., et al. (2014). Ultrasensitive nanosensors based on upconversion nanoparticles for selective hypoxia imaging in vivo upon near-infrared excitation, *Journal of the American Chemical Society*. **136**(27): pp. 9701–9709.

105. Haase, M., and H. Schäfer (2011). Upconverting nanoparticles, *Angewandte Chemie International Edition*. **50**(26): pp. 5808–5829.

106. Papkovsky, D. B., and R. I. Dmitriev (2013). Biological detection by optical oxygen sensing, *Chemical Society Reviews*. **42**(22): pp. 8700–8732.

107. Hadjipanayi, E., and A. F. Schilling (2013). Hypoxia-based strategies for angiogenic induction: The dawn of a new era for ischemia therapy and tissue regeneration, *Organogenesis*. **9**(4): pp. 261–272.

108. Quinlan, E., et al. (2015). Hypoxia-mimicking bioactive glass/collagen glycosaminoglycan composite scaffolds to enhance angiogenesis and bone repair, *Biomaterials*. **52**: pp. 358–366.

109. Buttyan, R., et al. (2003). Acute intravesical infusion of a cobalt solution stimulates a hypoxia response, growth and angiogenesis in the rat bladder, *The Journal of Urology*. **169**(6): pp. 2402–2406.

110. Park, K. M., and S. Gerecht (2014). Hypoxia-inducible hydrogels, *Nature Communications*. **5**: p. 4075.
111. Kaur, I. P., et al. (2014). Issues and concerns in nanotech product development and its commercialization, *Journal of Controlled Release*. **193**: pp. 51–62.
112. Ten Have, H. (2007). *Nanotechnologies, Ethics and Politics*. United Nations Educational, 244 p.
113. Kummar, S., et al. (2007). Compressing drug development timelines in oncology using phase'0'trials, *Nature Reviews Cancer*. **7**(2): pp. 131–139.
114. Coleman, C. N. (1988). Hypoxia in tumors: A paradigm for the approach to biochemical and physiologic heterogeneity, *Journal of the National Cancer Institute*. **80**(5): pp. 310–317.
115. Heinonen, I. H., et al. (2010). Regulation of human skeletal muscle perfusion and its heterogeneity during exercise in moderate hypoxia, *American Journal of Physiology-Regulatory, Integrative and Comparative Physiology*. **299**(1): pp. R72–R79.
116. Maeda, H., H. Nakamura, and J. Fang (2013). The EPR effect for macromolecular drug delivery to solid tumors: Improvement of tumor uptake, lowering of systemic toxicity, and distinct tumor imaging in vivo, *Advanced Drug Delivery Reviews*. **65**(1): pp. 71–79.
117. Cheng, Z., et al. (2012). Multifunctional nanoparticles: Cost versus benefit of adding targeting and imaging capabilities, *Science*. **338**(6109): pp. 903–910.
118. DeSA, U. (2013). World population prospects: The 2012 revision, *Population Division of the Department of Economic and Social Affairs of the United Nations Secretariat*, New York.
119. Eaton, M. A., L. Levy, and O. M. Fontaine (2015). Delivering nano-medicines to patients: A practical guide, *Nanomedicine: Nanotechnology, Biology and Medicine*. **11**(4): pp. 983–992.
120. Hossen, M. N., et al. (2012). Vascular-targeted nanotherapy for obesity: Unexpected passive targeting mechanism to obese fat for the enhancement of active drug delivery, *Journal of Controlled Release*. **163**(2): pp. 101–110.
121. World Health Organization (2012). *Population-Based Approaches to Childhood Obesity Prevention*. ISBN 9789241504782.

Chapter 6

Thermosensitive Nanomedicine

Stefan M. Cooper, Jr.,[a] Duc Ha,[b] Candace Snow-Davis,[b] Novia Watson,[b] Raymond E. Samuel,[a] Kesete Ghebreyessus,[c] Chengan Du,[b] Yongzhuo Huang,[d] and Feng Li[e]

[a]*School of Engineering, Hampton University, 100 E Queen St, Hampton, Virginia 23668, USA*
[b]*School of Pharmacy, Hampton University, 100 E Queen St, Hampton, Virginia 23668, USA*
[c]*School of Science, Department of Chemistry and Biochemistry, Hampton University, 100 E Queen St, Hampton, Virginia 23668, USA*
[d]*Shanghai Institute of Materia Medica, Chinese Academy of Sciences, 501 Hai-ke Rd, Shanghai 201203, China*
[e]*Harrison School of Pharmacy Auburn University, 702 S. Donahue Dr., Auburn, Alabama 36849, USA*

Fzl0023@auburn.edu

6.1 Introduction

Nanocarriers have received a great deal of attention because of their ability to enhance drug delivery. Due to their unique structures and physicochemical properties, nanocarriers such as micelles, liposomes, dendrimers, polymersomes, and nanogels have been

Stimuli-Responsive Nanomedicine
Edited by Lin Zhu

ISBN 978-981-4800-70-9 (Hardcover), 978-0-429-29529-4 (eBook)
www.jennystanford.com

successfully used to deliver a variety of drugs. In particular, they have demonstrated the ability to selectively deliver drugs into tumors. To further enhance drug delivery performance, stimuli-responsive nanocarriers have been developed as the next generation of delivery systems. Drug release from these systems can be triggered by stimuli such as pH, redox potential, light, ultrasound, temperature, and by various combinations of the above stimuli. Among these stimuli-responsive systems, temperature-responsive nanocarriers, or thermosensitive nanocarriers, have demonstrated great potential for the treatment of cancers and other diseases. In particular, thermosensitive nanomedicine, when used congruently with local hyperthermia, is a very promising strategy for treating solid tumors.

Thermosensitive materials experience a dramatic change in physical properties, such as conformation, solubility, and hydrophobicity, in response to a change in the environmental temperature. A variety of strategies have been utilized to trigger drug release from different thermosensitive nanocarriers [1]. There are three commonly used strategies which are summarized as follows:

(i) Many thermosensitive polymers exist as one phase in solution below their lower critical solution temperature (LCST) with their polymeric infrastructures adopting an expanded coil conformation. When temperature is above LCST, these polymers collapse to form compact globules and result in phase separation. The phase separation is often associated with a change in turbidity. The cloud point can be used to identify phase separations. The cloud point is defined as the temperature at the onset of an increase in turbidity. Thermosensitive polymers are highly hydrated and solubilized below the LCST, primarily through hydrogen bonds formed with water molecules. When temperature is above LCST, hydrogen bonding is no longer sufficient to keep the polymers soluble in water. Dehydration of polymers leads to a rapid phase separation. Conveniently, the LCST can be tuned by modulating the hydrophilic and hydrophobic monomer content of desired polymers [2]. Additionally, a sharp phase transition at a polymer's LCST is associated with a significant change in formulation volume. Strategically, this can be used to control drug release. Phase transition also changes the hydrophobicity of materials, leading to either formation or dissociation of thermosensitive nanocarriers (e.g., micelles, polymersome). Conversely, there are polymers which undergo a phase transition in the opposite way and experience an upper critical solution temperature (UCST). These polymers exhibit a single phase behavior in solution above their UCST. These materials are not commonly used for drug delivery.

(ii) Temperature-induced disruption of membranes is a strategy commonly used to trigger drug release from thermosensitive liposomes. Liposome membranes transition from a gel state to a crystalline-phase in solution at a specific temperature (T_c). This leads to the increase of membrane permeability and release of drugs.

(iii) Some polymer (e.g., Pluronic) solutions undergo a temperature-induced transition and can form micelles. The micellization process is mainly driven by hydrophobic effects and is dependent on the critical micelle temperature (CMT) and critical micelle concentration (CMC) of polymers.

Various types of materials have been used to prepare thermosensitive nanocarriers. Many of these materials are polymers. These polymeric materials can be divided into two general categories: (1) polymers based on amphiphilic balance (e.g., Pluronic); and (2) polymers based on LCST [3]. Pluronic is the commercial name of a series of amphiphilic triblock copolymers of poly(ethylene oxide)–poly(propylene oxide)–poly(ethylene oxide) (PEO–PPO–PEO) (Fig. 6.1A). It has a temperature-responsive micellization and complementary phase separation. Poly(*N*-substituted acrylamide) are thermosensitive polymers with a phase transition at the LCST. Among acrylamide-based polymers, poly(*N*-isopropylacrylamide) (PNIPAAm) is the most extensively studied polymer, with a reported LCST of 32°C, which is close to the physiological body temperature (Fig. 6.1B). Although most of the thermosensitive materials are synthetic polymers, natural materials or bio-inspired materials have also been used to prepare thermosensitive nanomedicines [4–7].

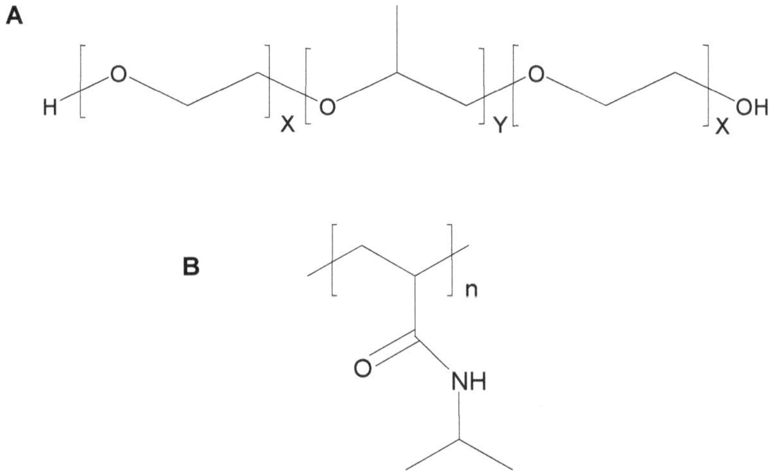

Figure 6.1 Examples of thermosensitive polymers. (A) poly(ethylene oxide)–poly(propylene oxide)–poly(ethylene oxide) (PEO–PPO–PEO); (B) Poly(*N*- isopropylacrylamide) (PNIPAAm).

In this chapter, we introduce different types of thermosensitive nanocarriers, including micelles, liposomes, dendrimers, polymersomes, and nanogels. Their design principles, properties, and applications will be discussed. The hybrid delivery systems which combine the advantages of multiple systems and multiple stimuli-responsive nanocarriers whose drug release is triggered by temperature and other stimuli such as pH, light, ultrasound, and redox are also presented.

6.2 Micelles

Polymeric micelles are prepared from amphiphilic polymers which form micelles through self-assembly when the polymer concentration is above the CMC. The CMC value is determined by the molecular weight and composition of polymers. The particle size of polymeric micelles ranges from 10 to 200 nm, which is dependent on the properties of polymers. The hydrophobic core of micelles has the capacity to load drugs or diagnostic agents, while the hydrophilic shell can stabilize the micelle and prolong the circulation in the blood. Thermosensitive micelles have been extensively studied due to their potential biomedical applications [8, 9]. Here, we will introduce two categories of thermosensitive micelles: conventional thermosensitive micelles and functionalized thermosensitive micelles.

6.2.1 Conventional Thermosensitive Micelles

pNIPAAm-based polymers are extensively used for fabricating thermosensitive polymeric micelles. They can be used either as the core forming block or corona forming entity, generating two possible variant thermosensitive micelles.

When pNIPAAm is copolymerized with a hydrophobic block, the resulting diblock copolymer forms micelles below the LCST of pNIPAAm. These micelles have a hydrophilic pNIPAAm corona and a hydrophobic core. When temperature is above LCST, the hydrophilic corona becomes increasingly more hydrophobic due to the loss of hydrogen-bonds and dehydration of the pNIPAAm block. This change disrupts the delicate hydrophilic/hydrophobic balance and facilitates destabilization and aggregation of micelles. Several of these types of polymers were synthesized which have pNIPAAm as the thermosensitive block and have several different hydrophobic blocks including poly(butylmethacrylate) [10], poly(caprolactone) [11], poly(DL-lactide) [12], and poly(methyl methacrylate) [13]. Due to the low LCST of pNIPAAm (around 32°C), micelles with pNIPAAm corona aggressively form aggregates upon injection at body temperature. Therefore, in vivo application of these micelles as intravenously injectable drug carriers is limited. To overcome this limitation, much work has been devoted to synthesizing polymers with a thermosensitive block composed of a mixture of NIPAAm and other monomers. The introduction of other monomers can modulate the LCST to nearly 37°C. Poly(*N*-isopropylacrylamide-co-*N*,*N*-dimethylacrylamide) (p(NIPAAm-co-DMAAm)) is a good example of one such copolymer [14, 15]. Thermosensitive micelles prepared from these polymers, with their higher LCST, are stable after injection at body temperature. Drug release can be triggered by local hypothermia. Appreciably enhanced anticancer activities were observed with the drugs delivered by these micelles [16].

pNIPAAm-polyethylene glycol (pNIPAAm-PEG) is another type of polymer that can be used to prepare thermosensitive micelles. These polymers are water soluble below the LCST, and form micelles above the LCST. In this case, pNIPAAm functions as the hydrophobic core, because of its increased hydrophobicity above the LCST. Due to this unique feature, micelles can be prepared by simply heating the polymer solution to above the LCST [17–21]. Drug release from these micelles can be triggered with local hypothermia.

Although pNIPAAm is the most commonly used material, many other polymers have also been used to prepare thermosensitive micelles [9]. A good example is poly(ethylene glycol)-*b*-poly(propylene oxide)-*b*-poly(ethylene glycol) (PEG-PPO-PEG or Pluronic). Pluronic polymers are commercially available with various molecular weights and PPO/PEO ratios. These polymers form micelles when the temperature is above the CMT. The dehydration of pluronic polymers at high temperatures may also cause phase separation. Pluronic micelles are composed of a PPO hydrophobic core and a PEG hydrophilic corona [22, 23]. A successful application of Pluronic micelles is the development of SP1049C. SP1049C is Pluronic L61 and F127 mixed micelles loaded with doxorubicin. This formulation is currently under phase III clinical trials for treating patients with multi-drug resistant and metastatic cancers [24].

6.2.2 Functionalized Thermosensitive Micelles

6.2.2.1 Degradable thermosensitive micelles

These types of micelles are prepared from copolymers with NIPAAm and other degradable monomers. Hennink's group synthesized PEG-poly(NIPAAm-co-(*N*-(2-hydroxypropyl)methacrylamide-dilactaten)) PEG-poly(NIPAAm-co-(HPMAm-Lacn)) [25]. Hydrolysis of side chains leads to an increased LCST. Once the LCST is increased above the environment temperature (e.g., body temperature for in vivo application), the hydrophobic block of the polymer changes to a hydrophilic state, leading to the dissociation of micelles and release of encapsulated drugs [26].

6.2.2.2 Cross-linked thermosensitive micelles

The stability of micelles is critical for their in vivo application. After intravenous injection, micelles are destabilized due to the dilution in the blood and/or the interaction with plasma proteins. The early disintegration of micelles in the blood may lead to premature drug release, which undermines targeting ability. Stability is a particular concern for thermosensitive micelles containing a hydrophobic block composed of pNIPAAm based polymers. When the environment temperature is below the LCST, the hydrophobic core of micelles quickly undergoes a hydrophobic to hydrophilic transition, leading to the destabilization of micelles. The situation becomes even worse when the LCST of the polymer is higher than the physiological body temperature. A useful strategy to stabilize thermosensitive micelles is covalently cross-linking micelles at the micelle core, corona, or interface.

For example, PEG-*b*-poly(DMAAmco-*N*-acryloxysuccinimide)-*b*-PNIPAAm (PEG-*b*-P(DMAAm-co-NAS)-*b*-PNIPAAm) was synthesized by McCormick et al. [27, 28]. This polymer dissolves in solution at 20°C and forms micelles when the temperature is above the LCST. Because of the presence of *N*-acryloxysuccinimide (NAS) groups, the micelle core can be cross-linked by cystamine. In addition, the disulfide bond within cystamine can be cleaved in a reductive environment, which facilitates the quick dissociation of micelles inside cells. Results showed that cross-linked micelles were more stable than these without cross-linking. Cross-linking also significantly delayed the leakage of a model drug, dipyridamole, from micelles under normal conditions. The release rate of drugs from cross-linked micelles can be increased by treating micelles with dithiothreitol (DTT). DTT is a reductive agent which can cleave the disulfide bonds and thus mimic intracellular environments [27, 28]. Similar cross-linking strategies were used to stabilize thermosensitive micelles prepared from other materials [29]. Linkers containing other cleavable bonds were also used. For example, adipic acid dihydrazide was used to cross-link micelles prepared from p(HPMAm)-*b*-p(AMPO)-*b*-p(HPMAm-Bz-co-HPMAm-Lac) by Hennink's group. Adipic acid dihydrazide forms hydrazone bonds with the ketone functional groups of poly(AMPO) block.

The hydrazone bond is a pH-sensitive bond which can be cleaved at an acidic pH. The cross-linked micelles showed good stability and retention of loaded paclitaxel drug at pH 7.4. Drug release was triggered at pH 5.0, which mimics the pH of late endosomes and lysosomes. The acidic pH-induced cleavage of hydrazone bonds led to the disintegration of micelles and triggered the release of loaded drugs [30]. The same research group also demonstrated the feasibility of using UV light to cross-link micelles. This was achieved by introducing photo-reactive monomers into copolymers used for preparing micelles [31].

6.3 Liposomes

Liposomes, similar to cell membranes, are vesicles consisting of phospholipids. Liposomes are capable of carrying both hydrophobic drugs at the phospholipid bilayer and hydrophilic drugs in the aqueous inner core. Liposome formulation can protect encapsulated drugs, improve their absorption, and increase half-life and bioavailability. In addition, liposomes are promising nanocarriers for site-specific drug delivery. They have been used successfully as drug formulations for cancer chemotherapy. Liposomes are generally considered as safe for biomedical use because they are primarily prepared from biocompatible and biodegradable lipids and other safe materials. Furthermore, the properties of liposomes (such as size, morphology, hydrophobicity, stability, specificity, and releasing profile) can be modified with flexibility through the selection of different lipids and other excipients in their formulations and through the use of different preparation methods. The development and application of thermosensitive liposomes have also been summarized in several previous reviews [1, 32, 33]. Typically, thermosensitive liposomes can be divided into three categories: (1) traditional thermosensitive liposomes (TTSL); (2) lysolipid-containing thermosensitive liposomes (LTSL); and (3) polymer-modified thermosensitive liposomes (PTSL) (Fig. 6.2). Here, we will briefly introduce basic concepts and typical examples of these thermosensitive liposomes.

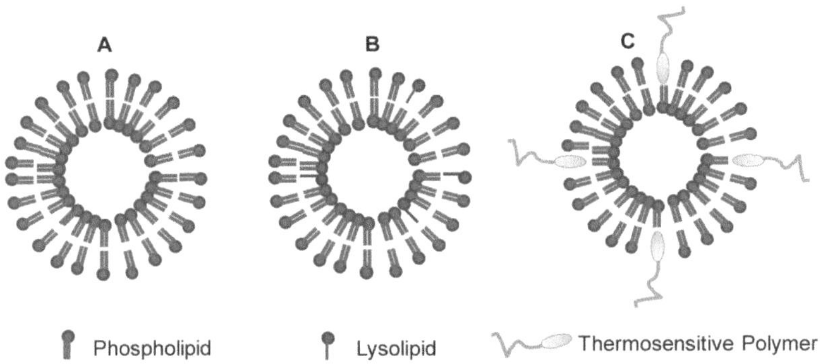

Figure 6.2 Different thermosensitive liposomes. (A) Traditional thermosensitivie liposomes. (B) Lysolipid-containing thermosensitive liposomes. (C) Polymer-modified thermosensitive liposomes.

6.3.1 Traditional Thermosensitive Liposomes

The first thermosensitive liposome was described by Yatvin et al. in 1978 [34]. This liposome was formulated from dipalmitoyl phosphatidylcholine (DPPC) and distearoyl phosphatidylcholine (DSPC). The transition temperature (T_c) is 41°C for DPPC and 54°C for DSPC, respectively. Modulation of the T_c was possible by varying the phospholipids proportions. A T_c could be set that ranged anywhere from 41 to 54°C. Neomycin was the drug of choice in their study. Liposomes with a 3:1 ratio of DPPC/DSPC experienced a release temperature from 42.5 to 45°C. Drug release was triggered by heat, which induced a gel-to-crystalline phase transition in liposome membranes and enhanced its permeability. To optimize therapeutic effects, liposomes with a T_c above body temperature but below 42°C are needed. This can be achieved by properly selecting lipid combinations in liposome formulations. The use of multiple different types of lipids with incompatible packing can enhance the permeability of liposomes. However, the heterogeneousness of lipid components may have a negative impact by increasing the width of the transition [35]. The major setback of TTSL is their poor or incomplete drug release. Although drug release from TTSL can be enhanced at mild hyperthermia, the quantity and rate of the release is insufficient to have desirable therapeutic benefits. Therefore, significant efforts have been made to improve the performance of thermosensitive liposomes through two major approaches: (1) incorporation of lysolipids, and (2) modification of liposomes with polymers.

6.3.2 Lysolipid-Containing Thermosensitive Liposomes

Lysolipids are lipids with a single acyl chain. They have relatively small single hydrophobic tails and thus a small hydrophobic/hydrophilic ratio. Lysolipids can favorably form micelles, which create defects in liposome membranes. Lysolipids incorporated into liposome formulations can also lower phase transition temperature and promote thermally triggered burst drug release [36, 37]. For example, thermosensitive liposomes containing lysolipids, distearoylphosphatidylethanolamine (DSPE) or monostearoyl phosphatidylcholine (MPSE) showed improved and rapid drug release and accumulation at tumor sites [36]. For instance, the inclusion of 10% of the lysolipid, monopalmitoyl-phosphatidylcholine (MPPC), in liposomes leads to a reduction of T_c from 43 to 39–40°C. The drug release rate was also increased by 50% at 42°C [36, 38]. LSTL formulation in combination with hyperthermia treatment resulted in increased cellular uptake of drugs by the tumor and showed complementary effective inhibition of tumor growth [39]. LTSL can significantly improve the release profile and are extremely effective for clinical applications [40]. This technology has been developed into a commercial product, ThermoDOX, which is currently under clinical trials sponsored by Celsion [41]. However, some studies indicated that lysolipids disassociate from liposomes upon dilution in serum containing medium [42]. This may result in premature drug release in the blood and compromise the performance of LTSL. Therefore, the timing of hyperthermia application and LTSL injection should be optimized to minimize premature drug release and improve clinical performance.

6.3.3 Polymer-Modified Thermosensitive Liposomes

Polymers are also used to modify liposomes to improve their thermo-sensitivities. Usually, these polymers showed improved membrane disruption abilities at an increased temperature. Both thermosensitive liposomes and/or conventional non-thermal responsive liposomes can be modified with appropriate polymers to create novel thermosensitive liposomes. The LCST of polymers is easily tuned by varying the polymeric infrastructure to trigger drug release at a desirable temperature [43, 44]. A variety of thermosensitive polymers have been used to modify liposomes, including poly(*N*-substituted acrylamides), poly(*N*-vinylethers), and others.

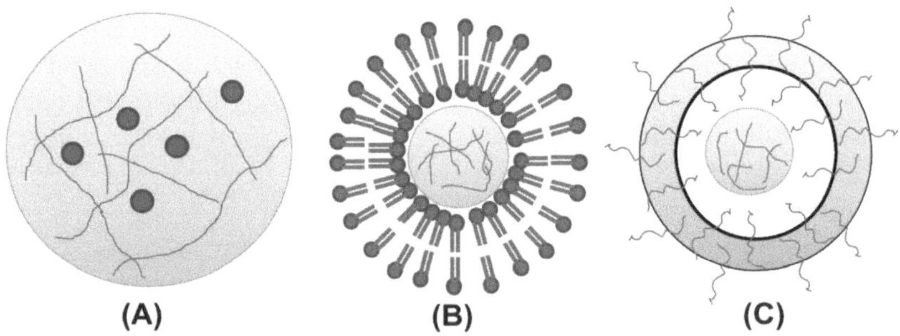

Figure 6.3 Hybrid thermosensitive nanocarriers. (A) Nanocarriers (e.g., nanoparticles, liposomes) loaded into thermosensitive hydrogels; (B) Liposomes or (C) polymersomes containing thermosensitive hydrogels.

6.3.3.1 Poly(*N*-substituted acrylamides)-modified liposomes

Pioneering studies of polymer-modified thermosensitive liposomes were performed by Kono and coworkers. They copolymerized NIPAAm with monomers containing hydrophobic lipid groups. The lipid groups can facilitate the interaction of thermosensitive polymers with liposome membranes [44, 45]. Liposomes modified with p(NIPAAm) copolymers showed increased drug release above the LCST of polymers, while slow drug release was observed below LCST. However, these initial proof-of-concept studies have limited clinical applications because the LCST of these polymers used was below physiological temperature. To adjust the LCST to nearby physiological temperature, other copolymers were synthesized by introducing various additional monomers together with NIPAAm and anchoring monomers. The introduction of hydrophilic acrylamide is a reported attempt which successfully raised the LCST of copolymers [46, 47]. Although liposomes modified with this polymer showed temperature triggered drug release, the release was still incomplete. Further studies were performed to understand governing factors that affect drug release profiles including polymer architecture and co-monomer composition. Liposomes modified with polymers bearing anchoring groups at their termini showed better performance than those modified with polymers bearing anchoring groups at multiple locations within the polymer backbone. Enhanced drug release observed in the former group might be due to greater flexibility of terminally anchored polymers [48–50]. The influence of different monomer structures on drug release was studied by Yoshino et al. In their study, three monomers with different structures (*N*-acryloylpyrrolidine, *N*,*N*-dimethylacrylamide, and *N*-isopropylmethacrylamide) were used to prepare copolymers with NIPAAm. Although these polymers have similar LCSTs (near 40°C), the copolymer composed of NIPAAm and *N*-isopropylmethacrylamide formed a hydrophobic domain above LCST. It showed a highly favorable interaction with liposome membranes, which led to enhanced membrane disruption and drug release [51].

6.3.3.2 Poly(*N*-vinylethers)-modified liposomes

Poly((*N*-vinylethers) has also been used for the modification of TSL. It has a similar thermosensitive mechanism as that of poly(NIPAAm). The LCST of desired polymers can be adjusted by introducing hydrophobic co-monomers, which may also serve as anchoring groups to facilitate the interaction of polymers with liposome membranes. Kono's group synthesized poly((2-ethoxy)ethoxyethyl vinyl ether-b-octadecyl vinyl ether) (p(EOEOVE-s-ODVE)) to generate a polymer-modified TSL. Enhanced drug release was shown in liposomes modified with polymers having a longer poly(EOEOVE) block. The longer poly(EOEOVE) block was more hydrophobic above LCST and had a stronger interaction with lipid membranes [52]. Thermosensitive liposomes modified with this type of polymer were successfully tested in vivo with animal tumor models as a promising delivery system for doxorubicin [53, 54].

In summary, TSLs have been widely used for various biomedical applications including cancer treatment [55, 56] and theranostics [57, 58]. However, the main concerns still remaining are specificity, rate of drug release, and stability of liposomes. Liposomes are versatile and can be modified to overcome these limitations.

6.4 Dendrimers

Dendrimers are macromolecules with well-defined structures and molecular weights. They have abundant peripheral groups which provide synthetic flexibility for functionalization. Nanocarriers prepared from dendrimers have a uniform structure and precisely controlled sizes. Many studies have explored the preparation and biomedical applications of thermosensitive dendrimers. Three major strategies are utilized to synthesize thermosensitive dendrimers: (1) modification of dendrimers with thermosensitive polymers; (2) design of dendrimers containing thermosensitive moieties; (3) design of collagen-mimic thermosensitive dendrimers.

6.4.1 Modification of Dendrimers with Thermosensitive Polymers

The first method of creating thermosensitive dendrimers is to modify the dendrimers with thermosensitive polymers (e.g., PNIPAM). The modification can occur either at the core or on the surface of dendrimers.

6.4.1.1 Modification of dendrimer surface

The first thermosensitive dendrimer was reported by Kimura et al. In their study, a functionalized poly(propylene imine) (PPI) dendrimer with peripheral thiol groups was used to initiate polymerization with NIPAM monomers. Poly(NIPAM) introduced to the surface of PPI made the modified dendrimer thermosensitive [59]. Several similar studies also demonstrated the feasibility of using pNIPAAm to modify the surface of other dendrimers [60, 61]. Usually, the modified dendrimers had different drug release profiles at different temperatures [61].

Besides the modification with pure poly(NIPAM), other co-monomers were also incorporated to fine-tune the thermosensitive proprieties. For example, dendrimer (Bolton H40) was modified with a copolymer poly(*N*-isopropylacrylamide)-

poly(2-(dimethylamino)ethyl methacrylate (PNIPAM-PDMA). Since the PNIPAM and PDMA homopolymers have different LCST, nanocarriers prepared from the above modified dendrimers showed two-stage thermo-induced collapse upon heating [62]. In addition, hydrophilic polymers (e.g., PEG) can be used to tune thermosensitive properties by adjusting the hydrophilic lipophilic balance (HLB) of prepared materials [63].

6.4.1.2 Modification of dendrimer core

Several other studies demonstrated the feasibility of preparing thermosensitive dendrimers by introducing thermosensitive polymers onto the dendrimer core. NIPAM monomers were polymerized onto the core of poly(benzyl ether) to synthesize thermosensitive dendritic–linear diblock copolymers [64, 65]. The modified dendrimers assemble into micelles with pNIPAM as the hydrophilic corona when temperature is below the LCST. Upon heating, the morphology of the system changes from spherical assembly into large entangled nanotubes [64]. In another study, Lee et al. synthesized poly(2-(2′-methoxyethoxy)ethyl methacrylate)-polyester dendron-based, double-hydrophilic linear–dendritic block copolymer. This polymer undergoes a thermally induced transition between double hydrophilic and amphiphilic structures. It forms micelles at 50°C and solutions at 25°C [66]. Kim et al. conjugated poly(*N*-isopropylacrylamide) PNIPAAM grafted poly(L-lactic acid) (PLLA) onto the core of poly(L-lysine) (PLL) dendrimer. The synthesized polymer contains a hydrophilic block (PLL), a hydrophobic block (PLLA), and a thermo-responsive (PNIPAAM) block [67].

The modification of dendrimers with thermosensitive polymers is a straightforward method to synthesize thermosensitive dendrimers. However, the modified dendrimers are shaped similarly to "star" polymers. The wide distribution in molecular weight of the incorporated polymers may lead to heterogeneousness in the molecular weight of the modified dendrimers. Several unique features of dendrimers, such as well-defined molecular weight, structure, and size, could be lost. Therefore, other approaches are being explored to preserve these unique features while introducing thermosensitive properties. Design of dendrimers containing thermosensitive moieties is one approach.

6.4.2 Design of Dendrimers Containing Thermosensitive Moieties

Instead of adding a block of thermosensitive polymer, the surface of dendrimers may be modified with thermosensitive moieties (or functional groups) to generate thermosensitive dendrimers. Isobutylamide (IBAM) is an exemplary thermosensitive polymeric side group. It was used to modify the surface of PAMAM and PPI dendrimers to generate thermosensitive dendrimers [68]. Similarly, Kono's group synthesized a PAMAM dendrimer with conjugated lipid cores and IBAM surface modifications. The introduction of an IBAM group permits macromolecular assembly with desired thermosensitivity [69]. Other thermosensitive moieties were also used to produce analogous thermosensitive dendrimers through modification of the dendrimer surface. Modification of PAMAM with phenylalanine can generate thermosensitive materials; on the contrary, PAMAM modified with leucine or isoleucine are not thermosensitive. The cloud point is dependent on the binding ratios of phenylalanine to the dendrimer [70]. Other additional thermosensitive moieties can be used in combination with IBAM to adjust thermosensitivity of dendrimers. For example, the hydrophobic amide groups (e.g., cyclopropanecarboxylic acid amide (CPCAM)) could reduce cloud point. With the modification of the same amount of IBAM and CPCAM, materials with a cloud point around body temperature could be synthesized [71]. These types of thermosensitive dendrimers demonstrated different thermosensitive properties when compared to linear polymers with similar thermosensitive moieties [72]. The dendrimer could have a sharper transition through dehydration of the surface functional groups, while maintaining the overall conformation of the whole molecule.

Oligo(ethylene glycol) (OEG) conjugated dendrimers were also studied as thermosensitive materials. The conjugation of hydrophilic OEG can adjust the hydrophilic/hydrophobic balance and thus change the LCST [73]. Kono's group conjugated hyperbranchedpoly(glycidol) with a variety of OEG. The cloud point of polymers is dependent on the composition of OEG [74]. Phase transition of these polymers can be induced at decreased pH value and at increased temperature, indicating the possibility of generating pH and thermo dual sensitive polymers [75]. Several other studies also demonstrated the use of dendritic OEG as thermosensitive materials. Kono's group modified the terminal of PAMAM with various alkoxy diethylene glycols. The LCST depends on the generation number of dendrimer and type of terminal alkoxyl groups [76]. Guo et al. prepared co-dendrimer PAMAM-co-OEG (PAG) with an "attach to" method. PAG demonstrated a sharp and fast phase transition. This carrier has been used for the delivery of methotrexate, which was encapsulated within the PAMAM core [77]. In vitro cell-based tests indicated that PAMAM dendrimers modified with OEG showed enhanced cellular uptake when incubated above the LCST. In addition, the modified PAMAM was less toxic than unmodified PAMAM [76].

Low generation PAMAM dendrimers are generally less toxic. However, it is more difficult to generate thermosensitive dendrimers with low generation numbers because of the small number of surface terminal groups available for conjugation. Zhao et al. modified low-generation PAMAM with hydroxylmethylamino-methane to increase the number of functional hydroxyl groups on the surface of the dendrimer. Two types of hydrophobic moieties were conjugated to generate two dendrimers (PAMAM-tris-IPAOBA and PAMAM-tris-DEAOBA). Thermosensitivity is dependent on the hydrophobicity of the dendrimer surface and the ionization state of the PAMAM core [78]. This type of dendrimer was used to deliver a model drug, vitamin E acetate. It was found that several factors may reduce cloud point, including higher drug loading, higher generation number of dendrimer, and elevated pH value. A decrease in pH value from 7 to 5 resulted in the transition of the dendrimer surface from a dehydrated status to a hydrated status, thus facilitating drug release at pH 5 [79].

6.4.3 Design of Collagen-Mimic Thermosensitive Dendrimers

Collagen is a natural protein present in animals and has natural thermosensitive properties. Collagen forms a triple helix under physiological conditions. Several research groups used elastin peptides conjugated dendrimers as artificial collagen materials. Unlike collagen, these peptides cannot form a triple helix by themselves. Once these peptides are conjugated with dendrimers,

dendrimers will assist in the formation of peptide triple helix by knotting strands. Several different elastin like peptide sequences, such as Gly-Pro-Nleu-(*N*-isobutylglycine), $(\text{Pro-Pro-Gly})_5$, and VPGVG, demonstrated their abilities to form a thermally reversible triple helix when conjugated to the surface of dendrimers [5–7]. Kono's group conjugated $(\text{Pro-Pro-Gly})_5$ peptide with dendrimers. The conjugated peptides formed a collagen like triple helix [6]. Kojima et al. also attached VPGVG peptide to a polyamidoamine (PAMAM) dendrimer. The phase transition of this dendrimer not only depends on temperature but is also affected by NaCl concentration and pH value. A model drug, rose Bengal, was used to study the dendrimer's ability to encapsulate and deliver drugs [7]. Both the folding status of the conjugated peptides and the ionization state of the dendrimer core may affect the LCST of VPGVG conjugated PAMAM in a pH-dependent manner [80]. The conformation of the main chain may also have a significant impact on the thermo proprieties of elastin like peptide conjugated dendrimers. Kojima et al. compared VPGVG conjugated linear poly-lysine and poly-lysine dendrimer. The results indicated that VPGVG peptide grafted linear poly-lysine had a lower phase transition temperature than VPGVG peptide grafted poly-lysine dendrimer [81].

6.5 Polymersomes

Polymersomes are a class of artificial polymeric vesicles. They can be conveniently prepared from various amphiphilic macromolecules. Polymersomes are capable of delivering both hydrophilic and hydrophobic drugs. In comparison to liposomes, polymersomes have advantages including good stability and tunable membrane property [82–84].

Typically, thermosensitive polymersomes are prepared from copolymers containing a thermosensitive block (e.g., PNIPAm). Below the LCST, the thermosensitive block of the copolymer becomes fully soluble in aqueous solution. Once the temperature is raised above LCST, the copolymer forms polymersomes because the thermosensitive block changes physicochemical properties and separates from water [85]. A similar phase transition governs micelle formation; however, the composition and properties of polymers regulate morphology. Temperature and polymer composition are critical factors that determine polymersomes' formation and properties. Ma et al. studied the formation of polymersomes by *ABABA* block copolymers with a thermosensitive hydrophilic *A* block. Spherical micelles were formed at high temperatures, while polymersomes were formed at lower temperatures. The length of the hydrophobic block in combination with temperature determines polymersome stability. The encapsulation efficiency of hydrophilic and/or hydrophobic drugs is dependent on the size of the hydrophilic inner cavity and the thickness of hydrophobic membranes of polymersomes, respectively. These properties, in turn, are determined by the length of both the hydrophilic and hydrophobic polymer blocks [86]. The effect of polymer composition on vesicle formation was demonstrated in another study. Liu et al. synthesized thermosensitive poly(*N*-vinylcaprolactam)$_n$-poly(dimethylsiloxane)$_{65}$-poly(*N*-vinylcaprolactam)$_n$ (PVCL$_n$-PDMS$_{65}$-PVCL$_n$) copolymers. This polymer formed polymersome at room temperature with a PVCL block ratio of 36–52% of the total polymer mass. The polymersome size directly correlated with the number of repeating units (*n* number) of PVCL block. PVCL length also determined the permeability of polymersomes [87]. Polypeptide-based block copolymers, poly[2-(dimethylamino)ethyl methacrylate]-*b*-poly(glutamic acid) (PDMAEMA-*b*-PGA), were also synthesized and used to prepare polymersomes [88]. Polymersomes were formed through electrostatic interactions between the two charged polymeric blocks near the isoelectric point. In basic conditions, the thermosensitive PDMAEMA block was uncharged and had a LCST of 40°C. This polymer dissolved in water as a single phase solution below its LCST, and self-assembled into micelles or polymersomes above its LCST. The morphology was dependent upon the length of PGA block. Qiao et al. synthesized copolymer poly(ethylene oxide)$_{45}$-*b*-poly(trans-*N*-(2-ethoxy-1,3-dioxan-5-yl) acrylamide)$_n$, PEO$_{45}$-*b*-PtNEA$_n$, with various PtNEA$_n$ block lengths (n = 22–172). The morphology of assembly was dependent on polymer concentration and length of the PtNEAn blocks. Stable polymersomes were formed from polymers with a longer PtNEAn block. This polymersome could load both hydrophobic doxorubicin and hydrophilic fluorescein isothiocyanate-lysozyme. Polymersomes prepared with PEO$_{45}$-*b*-

$PtNEA_{172}$were also more stable than spherical nanoparticles formed from PEO_{45}-*b*-$PtNEA_{44}$ [89].

Temperature-dependent assembly and disassembly of polymersomes can be utilized to trigger drug release. Drugs can be encapsulated in polymersomes when temperature is above the LCST. Once temperature is below the LCST, polymersomes disintegrate and release encapsulated drugs. For example, Qin et al. prepared polymersomes from Poly(ethylene oxide)-*b*-poly(*N*-isopropylacrylamide) (PEO-PNIPAm). This polymer formed polymersomes at body temperature and disassembled at lower temperatures, which makes it possible to induce drug release with local hypothermia [90]. Another study demonstrated that polymersomes formed by ($PVCL_n$-$PDMS_{65}$-$PVCL_n$) copolymers shrink gradually and form beadlike aggregates above their LCST, leading to sustained drug release [87].

Cross-linked polymersomes were also explored to further improve their stability and control drug release. For example, Chen et al. synthesized poly(2-cinnamoylethyl methacrylate)-*b*-poly(*N*-isopropylacrylamide) (PCEMA-*b*-PNIPAM). This copolymer formed polymersome above its LCST. The stability of polymersomes is further improved through UV cross-linking of PCEMA corona. Due to the presence of a PNIPAM thermosensitive block, temperature-dependent release of a model drug, 4-aminopyridine, was observed [91]. In another study, Cheng et al. synthesized two PEG-PAA-PNIPAM copolymers which formed polymersomes at 40°C. The polymersome structure can be maintained at a reduced temperature (37°C) through cross-linking its PAA segment with cystamine. This polymersome showed high loading efficiency (60–100%) for various protein drugs and has only minimal drug release in phosphate buffered saline. Drug release can be triggered with DTT treatment which mimics an intracellular reductive environment [85].

Hybrid polymersomes containing gold nanoparticles were developed to enhance thermo responsiveness. Using this strategy, thermosensitive polymersomes were prepared with a mixture of two copolymers: poly(*N*-isopropylacrylamide)-*b*-poly(lactic-co-glycolic acid) (PNIPAM-PLGA) and PEG-PLA. PNIPAM-PLGA was included to confer thermosensitivity to polymersome membranes. Dodecylthiol-stabilized gold nanoparticles, which can generate heat in response to photo-treatment, were incorporated to efficiently induce thermal transitions [92].

6.6 Nanogels

Thermosensitive hydrogels have been widely used for the delivery of drugs, proteins, nucleic acids, and cells. Hydrogels with a sol-gel transition temperature around body temperate have been extensively used in formulations for controlled drug release. These hydrogels form in situ gels upon injection [93]. Although nano-sized thermosensitive hydrogels are relatively under-explored compared to regular hydrogels, nanogels have several advantages including small particle size, prolonged circulation in the blood, and tumor targeting ability. In addition, temperature-induced collapse or expansion of nanogels can be used to control drug release and even achieve therapeutic effects in some cases [94].

Poly(NIPAM) or similarly derived copolymers are commonly used to prepare thermosensitive nanogels. Typically, nanogels are prepared when requisite components are allowed to polymerize while dispersed in aqueous solution containing surfactants, such as sodium dodecyl sulfate (SDS). Nanogels demonstrate temperature-dependent volume changes (i.e., swollen below LCST, while shrunken above LCST). Nanogel size is affected by the surfactant concentrations. Nanogels with small particle sizes can be produced with increased surfactant concentration during polymerization [95]. Kuckling et al. prepared photo-cross-linked nanogels with copolymerization of *N*-isopropylacrylamide (NIPAAm) and 2-dimethylmaleinimido ethylacrylamide (DMIAAm). Their work revealed that the size of nanogels is influenced by polymer solution concentration, the percentage of DMIAA in the polymer chain, and SDS concentration [96]. Several other groups studied poly(*N*-isopropylacrylamide-co-acrylamide) (PNIP/AAm) nanogels. These nanogels have been successfully used to deliver 5-fluorouracil, docetaxel, and near-infrared dye-12. In vivo studies showed that nanogels can selectively accumulate into tissues with higher temperatures. Docetaxel-loaded nanogels in combination with local hyperthermia showed improved inhibition of tumor growth in vivo [97–99]. Since many cross-linkers are toxic, a recent study developed a cross-linker free method to synthesize (*N*-isopropylacrylamidevinylpyrrolidone-acrylamide) (NIPAAm-VP-AAm) terpolymers nanogels with lower toxicity [100]. Other materials are used to prepare thermosensitive nanogels. For example, Jing et al. synthesized nanogels from the thermosensitive polymer poly(diethyleneglycolmethacrylate co-oligoethyleneglycolmethacrylate) grated hyaluronic acid [101]. This nanogel has been used to deliver hydrophobic anticancer drugs for treating ovarian cancers.

Another interesting application of nanogels is to exploit their thermo-induced volume expansion to directly kill cancer cells without the use of drugs. Lee et al. prepared nanogels by cross-linking oligo(L-lactic acid)-poly(ethylene oxide)-poly(propylene oxide)-poly(ethylene oxide)-oligo(L-lactic acid) and poly(ethylene glycol) grafted poly(L-lysine) with light. The prepared nanogels demonstrated reversible volume transition between 37°C (0.15 μm) and 15°C (1.4 μm). When treated with a brief cold-shock, the expansion of nanogels inside cells can cause significant damage to sub-cellular components and cell membranes, leading to cell necrosis. This application is quite innovative and represents a unique way of treating cancers with drug free polymer therapeutics [94].

6.7 Hybrid Thermosensitive Nanocarriers

Hybrid thermosensitive nanocarriers are delivery systems composed of at least two carrier types with the appropriate thermo-sensitivity. The following two hybrid thermo-sensitive nanocarrier classes will be discussed: (1) nanocarriers in thermosensitive hydrogels and (2) nanocarriers containing thermosensitive hydrogels (Fig. 6.3).

6.7.1 Nanocarriers in Thermosensitive Hydrogels

6.7.1.1 Liposomes in thermosensitive hydrogels

Thermosensitive hydrogels are commonly used for controlled drug release. In many applications, hydrogel depots formed in situ can provide sustained release of incorporated drugs. However, burst drug release and short retention time are major challenges for developing sustained release hydrogel formulations for small molecule hydrophilic drugs. Liposome-loaded chitosan-β-glycerophosphate hydrogel was developed to deliver hydrophilic small molecule drugs with improved release profiles [102]. Drugs were encapsulated in liposomes, and then the drug-loaded liposomes were dispersed in hydrogels. This system significantly slowed down the release of a model drug, carboxyfluoresein. The release rate was influenced by liposome size, lipid composition, and the presence of phospholipases. The study demonstrated the feasibility of developing injectable formulations for the slow release of hydrophilic small molecule drugs with a hybrid liposome in a gel system. In a similar study, Chen et al. dispersed curcumin-loaded liposomal bubbles in chitosan-β-glycerophosphate hydrogels. The system showed temperature and ultrasound dual sensitivities. The release of curcumin can be trigger by ultrasound [103].

6.7.1.2 Nanoparticles in thermosensitive hydrogels

Hybrid nanoparticles in thermosensitive hydrogels are prepared by physically dispersing NPs into hydrogels or coating NPs with hydrogel polymers through in situ polymerization. The hybrid system can extend drug release and introduce other favorable properties to hydrogels by using various functional NPs.

(a) Extended Release Formulations

Burst release is a common problem for drug-loaded NPs because of their large surface area and the presence of drugs loosely associated to the surface of NPs. Several studies incorporated drug-loaded NPs into thermosensitive hydrogels and effectively reduced burst release and prolonged drug release. For example, Date et al. prepared raltegravir and efavirenz-loaded PLGA NPs. NPs were incorporated in Pluronic-based thermosensitive vaginal gel for HIV prophylaxis [104]. Additionally, paclitaxel-loaded (polyethylene glycol)-poly(D,L-lactide-co-glycolide)-poly(L-lysine)-cyclic peptide NPs were dispersed in thermosensitive hydrogels composed of Pluronic F-127, Pluronic F-68, HPMC, methyl cellulose, and sodium alginate. NPs can effectively overcome drug resistance in cancer therapy while hydrogels can enhance retention and prolong release time. Improved anticancer effects were observed when implementing the aforementioned strategy [105]. Yin et al. prepared lidocaine-loaded poly(caprolactone)–poly(ethylene glycol)–poly(caprolactone) NPs. Incorporation of these NPs in thermosensitive Pluronic F-127 hydrogels extended drug release and promoted infiltration of anesthesia, revealing a potential application of this formulation for local delivery of lidocaine [106].

(b) Magnetic Field–Induced Drug Delivery

Magnetic-responsive metallic NPs (such as iron oxide NPs) have magnetic hyperthermia properties and can also be used as magnetic resonance imaging (MRI) contrast agents. Several studies incorporated iron oxide NPs in thermosensitive hydrogels. Heat generated from iron oxide NPs in response to an external magnetic field increases the local temperature and triggers drug release from hydrogels. For example, Huang et al. incorporated Fe_3O_4 NPs and a hydrophobic drug (ethosuximide) into a mixture of Pluronic F127 and polyvinyl alcohol (PVA). A magnetic field caused an increase of local temperature which led to an expansion of PVA and shrinkage of Pluronic F127. Drug release profiles were modified by adjusting the ratio of PVA and Pluronic F127 [107]. In another study, $Mn_{0.2}Zn_{0.8}Fe_2O_4$ NPs were dispersed in thermosensitive hydrogels composed of NIPAAm and *N*-hydroxymethyl acrylamide (HMAAm). Magnetic stimulation increased local temperature to 42.9°C. By adjusting the NIPAAm to HMAAm ratio, the LCST of the hydrogel can be tuned to around 40.1°C for possible magnetothermal triggered drug release [108]. In addition to the physical dispersion of NPs in hydrogels, thermosensitive gel can also be synthesized in situ onto the surface of NPs. Echeverria et al. prepared hybrid Fe_3O_4 NPs encapsulated in poly(acrylamide-acrylic acid) gels. The gels were synthesized on the surface of NPs through an inverse emulsion polymerization method [109]. Similarly, NIPAAM and methacrylate functionalized magnetic NPs poly(NIPAAM@Fe_3O_4 MNPs/TMSPMC/DOX) were synthesized as a temperature- and pH-sensitive delivery system for doxorubicin [110]. Additionally, PNIPAM-coated silica/magnetic nanoparticles (PNIPAM/SiO_2/Fe_3O_4 nanoparticles) were prepared for the delivery of vitamin D3. The drug loading and release profile are dependent on the LCST of the hybrid hydrogel [111].

(c) Light-Responsive Drug Delivery

Light-responsive NPs (e.g., gold NPs) have been extensively explored for photodynamic therapy (PDT) and photothermal therapy (PTT). Many studies incorporated NIR light-responsive gold NPs into thermosensitive hydrogels because NIR light has good penetration into biological tissues. Kim et al. loaded gold nanorods (GNRs) and chlorin e6 (Ce60) into chitosan functionalized Pluronic F68 nanogels for synergistic PDT and PTT. In this study, GNRs were used as a PTT agent and chlorin e6 (Ce6) as a photosensitizer. This hydrogel formulation enhanced drug delivery into tumors in vivo [112]. In situ polymerization of thermosensitive materials on the surface of gold NPs has also been explored [113]. Covalently bonding thermosensitive moieties onto NP surfaces rather than physically coating the surface improves the stability of augmented NPs [113, 114]. Shang et al. prepared nanogels by first coating NIR light-responsive gold nanorods with thermosensitive materials, followed by supplementary cross-linking to efficaciously cohere materials onto NP surfaces. The NP surfaces were coated with a monolayer of PEG-methacrylate (PEGMA). Then, the coated surfaces were further treated by in situ polymerization and cross-linking with PEGMA and NIPAAm monomers. This delivery system encapsulated a model drug, ZnPC4. NIR-induced drug release was evaluated [113]. NIR light-responsive NPs prepared from other materials have also been incorporated into thermosensitive hydrogels. For example, Barhoumi et al. used thermosensitive polymers polyNIPAAm and polyacrylamide to coat the surfaces of silica core–gold shell NPs. Strategically, the surface coating masked a peptide ligand targeting integrin β1. This peptide facilitates cellular uptake of NPs but lacks specificity, since integrin β1 is also present in normal tissues. Impressively, when NIR light is applied, heat is generated from silica core–gold shell NPs which led to the collapse of the copolymer coating on the surface of the NPs. Peptide ligands were then unmasked and thus enhanced cellular binding and cellular uptake of NPs [115]. In another study, Mohapatra et al. prepared chitosan-modified chemically reduced graphene oxide, which was further coated with thermosensitive nanogels composed of poly(NIPAM) and PEG-diacrylate. The system showed NIR light-induced thermal effects. The release rate of encapsulated doxorubicin was temperature dependent; drug release at 42°C was faster than at 37°C [4].

6.7.2 Nanocarriers Containing Thermosensitive Hydrogels

Here, we will discuss a few examples where thermosensitive hydrogels are incorporated inside of nanocarriers, such as liposomes and polymersomes. Several studies have reported the preparation of liposomes containing thermosensitive hydrogels within their interior cores to improve drug delivery performance. These types of hybrid delivery systems are named gel-liposomes or lipobeads [116]. The incorporation of hydrogels into the interior core of liposomes can increase strength and enhance drug loading. Problems associated with the physical instability of liposomes can be minimized by this strategy. For example, Zhang et al. prepared liposomes by a thin-film method with a poloxamer solution included in the interior aqueous core [117]. The incorporation of hydrogels was confirmed by the sustained disintegration of hydrogels after removing lipid bilayers with detergents. The change of liposome structures caused by a heat-cool cycle also indicated the gelation present in the interior core of the liposomes. This hybrid system showed a similar sol-gel transition temperature to that of Poloxamer solutions. The gelation in the interior core enhanced liposome membrane strength, improved physical stability, and prolonged drug release.

Similarly, thermosensitive hydrogels were also incorporated in polymersomes to improve drug release profiles. Lee et al. prepared PEG-PLA polymersomes containing PNIPAAm hydrogels, coined "hydrosomes." They confirmed hydrogel incorporation by confocal laser scanning microscopy. Hydrosomes showed reduced initial burst release and prolonged release of a model drug, FITC-dextran [118].

6.8 Multi-Stimuli-Responsive Systems

Although thermosensitive nanocarriers are very promising systems for drug delivery, systems responsive to multiple stimuli have also been extensively explored. Here, we illustrate examples of "smart" nanocarriers which respond to multiple stimuli and can further improve drug delivery performance.

6.8.1 Thermal and pH Dual Responsive Systems

PH-sensitive nanocarriers are ideal candidates for tumor targeted drug delivery. These nanocarriers are stable and have minimal drug release in blood with a neutral pH. Upon entering tumor cells, encapsulated drugs are quickly released in response to the mildly acidic pH in tumor cells. Several different strategies have been utilized to prepare temperature and pH dual responsive systems.

The first approach is to incorporate pH-sensitive moieties into thermosensitive polymers to modulate the LCST. The LCSTs of these polymers are dependent on pH. For example, copolymers of NIPAAm and pH-sensitive monomers showing pH-dependent LCST were synthesized. At neutral pH, the LCST of this polymer was above body temperature; contrarily the LCST was below body temperature at acidic pH values [119, 120]. Clearly, the thermal phase transition of these polymers was influenced by environment pH. In another study, Chen et al. synthesized a copolymer with a PCL hydrophobic block and a hydrophilic block composed of poly(triethylene glycol) copolymerized with 6-aminocaproic acid functionalized monomers [121]. This polymer formed micelles at neutral pH and body temperature. At mildly acidic pH levels and body temperature, the polymer underwent a phase transition and triggered drug release from micelles. Doxorubicin-loaded micelles were prepared with this polymer. The micelles were stable in the blood and had minimal drug leakage. Drug release rate was increased in an acidic pH environment.

A different method of generating thermal and pH dual response systems was also used in many other studies, where polymers were synthesized to have two functional blocks which are responsive to temperature and pH independently. For example, a dual responsive copolymer was synthesized by grafting pH-sensitive hydrophilic poly(2-carboxyethyl-2-oxazoline) to the backbone of a thermosensitive poly(NIPAAm) [122]. This polymer formed micelles which can be further cross-linked to yield core–shell nanogels with temperature and pH-dependent swelling properties.

The change of temperature led to the alteration of poly(NIPAAm) core confirmation, while the change in pH resulted in the alteration of poly(2-carboxyethyl-2-oxazoline) corona thickness. Several other polymers with similar features were also synthesized. Kuckling et al. synthesized nanogels with a thermosensitive poly(NIPAAm) core and pH-sensitive poly(2-vinylpyridine) arms [123]. In another study, hybrid nanogels were prepared from a hydroxypropylcellulose and poly(methacrylic acid) complex. The poly(methacrylic acid) was pH-sensitive and led to pH-induced phase transition in the nanogel [124].

Chiang et al. prepared nanogels with a polymer composed of acrylic acid and 2-methacryloylethyl acrylate polymer backbone grafted with PNIPAAm and PEG chain. At pH 7.4, doxorubicin was effectively loaded within this nanogel through favorable electrostatic interactions with ionized acrylic acid and hydrogen bonding with PNIPAAm. Doxorubicin-loaded nanogels were very stable and had minimal drug release at 4°C and pH 7.4. Rapid release of doxorubicin was observed at 37°C and pH 5.0. Drug release is triggered by two possible mechanisms: (1) disruption of the acrylic acid/doxorubicin interaction, due to the reduced ionization of acrylic acid molecules at pH 5.0; and (2) shrinkage of nanogels at high temperature [125].

In other studies, thermosensitive properties were utilized to facilitate the preparation of delivery systems with high drug loading efficiency and good stability, while changes in pH were used to control drug release. For example, Ahmed et al. developed carbohydrate-based nanogels for intracellular delivery of DNA and enzymes. Temperature sensitivity of the nanogels facilitated the encapsulation of macromolecule drugs in nanogels. The degradation of materials at acidic pH led to the dissociation of nanogels and triggered the release of loaded drugs [126]. In another study, mPEG2000-isopropylideneglycerol polymer was used to prepare a micelle formulation for paclitaxel. Micelles underwent a sol-gel phase transition at body temperature, which improved the micelle stability and enhanced drug retention. PH-sensitive hydrophobic ketal groups within the polymer were cleaved at acidic pH, resulting in the dissociation of micelles and quick drug release [127].

6.8.2 Thermal and Light Dual Responsive Systems

Thermal and light dual responsive systems can be prepared through two different strategies. The first strategy involves modifying thermosensitive polymers with light-responsive functional groups. Light can directly influence photo-responsive functional groups of polymers and affect their thermal properties (e.g., LCST). For example, He et al. synthesized copolymers composed of a poly-[2-(dimethylamino)ethyl methacrylate] block and a PNIPAM block containing azobenzene substrates. UV or visible light-induced photo-isomerization of azobenzene groups and significantly influenced the cloud point of polymers [128]. In another study, poly(3-vinylbenzylideneoxazolones-co-dimethylacrylamide) was prepared, which contains benzylideneoxazolone as a light-responsive moiety. LCST of this polymer was influenced by light-induced photo-isomerization of benzylideneoxazolone. *Z* isomers showed lower LCST than *E* isomers [129]. Blasco et al. synthesized an azobenzene-containing star polymer AB3 which contains an azopolymer and three arms of poly(*N*,*N*-diethylacrylamide). This polymer formed stable micelles in water. UV irradiation induced a trans-to-cis isomerization of the azobenzene moiety and triggered the release of a model drug Nile Red from micelles [130].

The second approach to prepare thermal and light dual responsive systems is to incorporate light-responsive NPs in thermosensitive hydrogels. In response to light, NPs generate heat and trigger drug release from thermosensitive hydrogels. This topic was discussed in an earlier part of this chapter on hybrid systems. For example, chitosan-modified, chemically reduced graphene oxide NPs were incorporated in hydrogels and showed the ability to generate heat and trigger drug release in response to NIR light [131]. In another study, NIR irritation on silica core-gold shell NPs generated heat and resulted in the exposure of peptide targeting ligands which were otherwise masked by thermosensitive hydrogels coated on the surface of NPs [115].

6.8.3 Thermal and Ultrasound Dual Responsive Systems

For these types of delivery systems, externally applied ultrasound is utilized to trigger on-demand drug release. Paris et al. developed an ultrasound and thermal dual responsive drug delivery system based on polymer-grafted mesoporous silica NPs with pores.

They conjugated thermosensitive polymers onto the surface of mesoporous silica NPs. These polymers function as doorkeepers for the pores present on the surface of NPs. Drug molecules (fluorescein) can be loaded into pores at lower temperatures, when polymers remain at an open conformation. These conjugated polymers collapse at body temperature (37°C) which, in turn, closes the opening of pores and prevents the leakage of loaded drugs. Ultrasound treatment changes polymer hydrophobicity and conformation, then opens gates and triggers drug release [132]. Another example of ultrasound and thermal dual responsive system was reported by Chen et al. In their study, curcumin-loaded liposome bubbles were incorporated into chitosan-β-glycerophosphate thermosensitive hydrogels. Ultrasound treatment can effectively trigger the release of curcumin from the system [103].

6.8.4 Thermal and Redox Dual Responsive Systems

Nanocarriers with cleavable groups (such as disulfide bonds) have been extensively used to trigger drug release in response to intracellular reductive environments. Disulfide bonds are stable outside of cells with a low level of reducing glutathione (GSH), but are cleavable in the cytosol with a high level of GSH. Tian et al. synthesized disulfide cross-linked poly(vinylcaprolactam) nanogels. When the temperature changed from body temperature (37°C) to hyperthermia (41°C), nanogels underwent hydrophilic/hydrophobic transition, and switched from a hydrophilic and extended state to a collapsed state. This transition enhanced cellular uptake and drug release. Once inside cells, the cleavage of disulfide bonds can facilitate the dissociation of nanogels and further enhance drug release. The nanogels loaded with doxorubicin showed minimal drug leakage (15% in 48 h) in physiological conditions. A treatment of GSH and hyperthermia triggered quick drug release (92% in 12 h). Hyperthermia also enhanced anticancer efficacy of this nanogel's formulation [133].

6.8.5 Thermal, pH, and Redox Triple Responsive Systems

Several studies reported the design of pH, redox, and thermal triple responsive drug delivery systems. For example, Qiao et al. synthesized triple responsive nanogels composed of monomethyl oligo(ethylene glycol) acrylate, ortho ester-containing acrylic monomer, and 2-(5,5-dimethyl-1,3-dioxan-2-yloxy) ethyl acrylate. Bis(2-acryloyloxyethyl) disulfide was used as a cross-linking agent. This thermosensitive nanogel is degradable in either acidic and/or reductive environments. The loading and release of three drug molecules (i.e., Nile Red, paclitaxel, and doxorubicin) from these nanogels were tested. Drug release was triggered by acid-mediated hydrolysis of polymers or DTT mediated cleavage of disulfide bond [134]. In another study, nanogels were prepared from copolymerization of N,N'-bis(acryloyl)cystamine, NIPAM, acrylic acid through in situ polymerization. Doxorubicin can be loaded in these nanogels under physiological conditions. Anionic groups of polymers can interact with doxorubicin through electrostatic interactions, and thus enhance doxorubicin loading and prolong release. The release of drugs can be triggered by heat, reductive conditions, and an acidic environment [135].

6.8.6 Thermal, pH, and Light Triple Responsive Systems

Delivery systems which are responsive to light, pH, and temperature have also been developed in several studies to enhance tumor targeting. For example, Qin et al. encapsulated amphiphilic chitosan derivative-coated single-wall carbon nanotubes (CNT) in a pH and thermal dual responsive nanogel, named CS/PNIPAAm@CNT. CNT was incorporated in the system as a NIR stimulus antenna which can generate heat in response to NIR irradiation. PNIPAAm functioned as the thermosensitive hydrogel, which can switch between a swelling and shrinking state depending upon the temperature. Doxorubicin was incorporated in the system as a therapeutic drug. The release of doxorubicin is also pH dependent because of the Schiff base bonds formed between chitosan primary amino groups and doxorubicin ketone groups. In vitro drug release showed that doxorubicin release was faster at 40°C than at 25°C and faster at pH 5.0 than at pH 7.4. In addition, NIR light can also trigger drug release from this system [136]. Yuan et al. also prepared light, pH, and temperature triple responsive polymeric micelles with amphiphilic block copolymers containing multiple responsive functional groups. The hydrophilic block, poly(2-(*N*,*N*-dimethylamino)ethyl methacrylate), is pH and temperature dual sensitive. The hydrophobic block, poly{6-[4-(4-

pyridyazo)phenoxy] hexyl methacrylate, contains light-responsive azobenzene functional groups. This polymer forms core–shell micelles in water. The shell of micelles can swell and shrink in repose to a change in temperature and/or pH. The azobenzene of the micelle core block underwent tarns-cis transitions in response to light [137].

6.9 Conclusion and Future Directions

In this chapter, we introduced various types of thermosensitive nanocarriers. Their design principles and potential applications were discussed. Combining the use of local hyperthermia with nanomedicines can improve therapeutic outcomes. Nanocarriers with appropriate thermo-sensitivity are required. Ideally, nanocarriers should be stable at body temperature with drug release triggered by tolerable transit local heat (e.g., 40°C). On the other hand, a sharp transition around LCST is also preferred for the purpose of controlling drug release precisely. Materials or systems with a narrow range of transition temperatures and sharp phase transition behavior are preferred. Polymers with a wide distribution of molecular weights tend to have a broad and obscure phase transition. Therefore, adopting novel synthetic methodologies for the production of polymers with narrower distributions of molecular weights is highly suggested. Clear and extremely thoughtful rationale design for requisite polymers is also important. The incorporation of heat generating NPs is another effective way to enhance thermo-sensitive responses. These NPs may also make the system responsive to other stimuli. In the hybrid systems, external stimuli such as light and/or an external magnetic field can be absorbed by NPs to generate heat and promote drug release. Due to nanocarriers' intrinsic link to biomedical applications, numerous studies were performed to develop and evaluate thermosensitive nanomedicines. Various thermosensitive materials and different delivery systems are reported. Persistent efforts in basic research have laid a foundation for translational research and product development. There are at least two successful cases implementing thermosensitive nanomedicines: a thermosensitive liposome formulation of doxorubicin, Thermo DOX; and a Pluronic micelle formulated doxorubicin. Both of these are currently under clinical trials [24, 41]. However, many of the proof-of-concept studies actively reported are not feasible for clinical applications. For example, the commonly studied poly(*N*-isopropylacrylamide)-based materials are less likely to be developed into medicine because they are toxic and non-biodegradable. Therefore, safe and biocompatible materials are more promising for future applications. Lastly, manufacturing products in a simplistic and economical manner is another challenge for nanomedicine development. Many nanomedicines developed in research laboratories are extremely complex and rather difficult to be developed into clinically available products. In addition, the preparation method alone is too complicated, unreliable, or unaffordable. The future efforts should be directed towards developing nanocarriers which can be produced in large scale with relatively simple methods and good reproducibility.

Acknowledgments

This work was supported by the following grants: NIH-SC3 grant, NSF-PREM grant, Hampton University Faculty Research Grant, and Auburn University Start-up grant.

References

1. A. Dabbagh, B. J. Abdullah, H. Abdullah, M. Hamdi, N. H. Kasim, Triggering mechanisms of thermosensitive nanoparticles under hyperthermia condition, *J. Pharm. Sci.*, 104 (2015) 2414–2428.
2. X. Yin, A. S. Hoffman, P. S. Stayton, Poly(N-isopropylacrylamide-co-propylacrylic acid) copolymers that respond sharply to temperature and pH, *Biomacromolecules*, 7 (2006) 1381–1385.
3. E. S. Gil, S. M. Hudson, Stimuli-responsive polymers and their bioconjugates, *Prog. Polymer Sci.*, 29 (2004) 1173–1222.
4. C. Wang, J. Mallela, U. S. Garapati, S. Ravi, V. Chinnasamy, Y. Girard, M. Howell, S. Mohapatra, A chitosan-modified graphene nanogel for noninvasive controlled drug release, *Nanomedicine*, 9 (2013) 903–911.
5. G. A. Kinberger, W. Cai, M. Goodman, Collagen mimetic dendrimers, *J. Am. Chem. Soc.*, 124 (2002) 15162–15163.

6. C. Kojima, S. Tsumura, A. Harada, K. Kono, A collagen-mimic dendrimer capable of controlled release, *J. Am. Chem. Soc.*, 131 (2009) 6052–6053.
7. C. Kojima, K. Irie, Synthesis of temperature-dependent elastin-like peptide-modified dendrimer for drug delivery, *Biopolymers*, 100 (2013) 714–721.
8. M. Talelli, W. E. Hennink, Thermosensitive polymeric micelles for targeted drug delivery, *Nanomedicine (Lond)*, 6 (2011) 1245–1255.
9. G. Bonacucina, M. Cespi, G. Mencarelli, G. Giorgioni, G. F. Palmieri, Thermosensitive self-assembling block copolymers as drug delivery systems, *Polymers*, 3(2011) 779–811.
10. J. E. Chung, M. Yokoyama, M. Yamato, T. Aoyagi, Y. Sakurai, T. Okano, Thermo-responsive drug delivery from polymeric micelles constructed using block copolymers of poly(N-isopropylacrylamide) and poly(butylmethacrylate), *J. Control. Release*, 62 (1999) 115–127.
11. C. Choi, S. Y. Chae, J. W. Nạh, Thermosensitive poly(N-isopropylacrylamide)-b-poly(ε{lunate}-caprolactone) nanoparticles for efficient drug delivery system, *Polymer*, 47 (2006) 4571–4580.
12. F. Kohori, K. Sakai, T. Aoyagi, M. Yokoyama, Y. Sakurai, T. Okano, Preparation and characterization of thermally responsive block copolymer micelles comprising poly(N-isopropylacrylamide-b-DL-lactide), *J. Control. Release*, 55 (1998) 87–98.
13. H. Wei, X. Z. Zhang, Y. Zhou, S. X. Cheng, R. X. Zhuo, Self-assembled thermoresponsive micelles of poly(N-isopropylacrylamide-b-methyl methacrylate), *Biomaterials*, 27 (2006) 2028–2034.
14. M. Nakayama, T. Okano, T. Miyazaki, F. Kohori, K. Sakai, M. Yokoyama, Molecular design of biodegradable polymeric micelles for temperature-responsive drug release, *J. Control. Release*, 115 (2006) 46–56.
15. S. Q. Liu, Y. W. Tong, Y. Y. Yang, Incorporation and in vitro release of doxorubicin in thermally sensitive micelles made from poly(N-isopropylacrylamide-co-N,N-dimethylacrylamide)-b-poly(D,L-lactide-co-glyco lide) with varying compositions, *Biomaterials*, 26 (2005) 5064–5074.
16. B. Liu, M. Yang, R. Li, Y. Ding, X. Qian, L. Yu, X. Jiang, The antitumor effect of novel docetaxel-loaded thermosensitive micelles, *Eur. J. Pharm. Biopharm.*, 69 (2008) 527–534.
17. M. D. C. Topp, P. J. Dijkstra, H. Talsma, J. Feijen, Thermosensitive micelle-forming block copolymers of poly(ethylene glycol) and poly(N-isopropylacrylamide), *Macromolecules*, 30 (1997) 8518–8520.

18. J. Virtanen, S. Holappa, H. Lemmetyinen, H. Tenhu, Aggregation in aqueous poly(N-isopropylacrylamide)-block-poly(ethylene oxide) solutions studied by fluorescence spectroscopy and light scattering, *Macromolecules*, 35 (2002) 4763–4769.
19. W. Zhang, L. Shi, K. Wu, Y. An, Thermoresponsive micellization of poly(ethylene glycol)-b-poly(N-isopropylacrylamide) in water, *Macromolecules*, 38 (2005) 5743–5747.
20. C.-Y. Hong, Y.-Z. You, C.-Y. Pan, Synthesis and characterization of well-defined diblock and triblock copolymers of poly(N-isopropylacrylamide) and poly(ethylene oxide), *J. Polymer Sci. Part A: Polymer Chem.*, 42 (2004) 4873–4881.
21. Y. Maeda, N. Taniguchi, I. Ikeda, Changes in the hydration state of a block copolymer of poly(n-isopropylacrylamide) and poly(ethylene oxide) on thermosensitive micellization in water, *Macromol. Rapid Commun.*, 22 (2001) 1390–1393.
22. A. V. Kabanov, E. V. Batrakova, V. Y. Alakhov, Pluronic block copolymers as novel polymer therapeutics for drug and gene delivery, *J. Control. Release*, 82 (2002) 189–212.
23. K. T. Oh, T. K. Bronich, A. V. Kabanov, Micellar formulations for drug delivery based on mixtures of hydrophobic and hydrophilic Pluronic block copolymers, *J. Control. Release*, 94 (2004) 411–422.
24. http://www.supratek.com/pipeline/products (2016).
25. D. Neradovic, C. F. van Nostrum, W. E. Hennink, Thermoresponsive polymeric micelles with controlled instability based on hydrolytically sensitive N-isopropylacrylamide copolymers, *Macromolecules*, 34 (2001) 7589–7591.
26. C. J. Rijcken, O. Soga, W. E. Hennink, C. F. van Nostrum, Triggered destabilisation of polymeric micelles and vesicles by changing polymers polarity: An attractive tool for drug delivery, *J. Control. Release*, 120 (2007) 131–148.
27. Y. Li, B. S. Lokitz, C. L. McCormick, RAFT synthesis of a thermally responsive ABC triblock copolymer incorporating N-acryloxysuccinimide for facile in situ formation of shell cross-linked micelles in aqueous media, *Macromolecules*, 39 (2006) 81–89.
28. Y. Li, B. S. Lokitz, S. P. Armes, C. L. McCormick, Synthesis of reversible shell cross-linked micelles for controlled release of bioactive agents, *Macromolecules*, 39 (2006) 2726–2728.
29. C. Chang, H. Wei, D. Q. Wu, B. Yang, N. Chen, S. X. Cheng, X. Z. Zhang, R. X. Zhuo, Thermo-responsive shell cross-linked PMMA-b-P(NIPAAm-co-NAS) micelles for drug delivery, *Int. J. Pharm.*, 420 (2011) 333–340.

30. Y. Shi, C. F. van Nostrum, W. E. Hennink, Interfacially hydrazone cross-linked thermosensitive polymeric micelles for acid-triggered release of paclitaxel, *ACS Biomater. Sci. Eng.*, 1 (2015) 393–404.

31. Y. Shi, R. M. Cardoso, C. F. van Nostrum, W. E. Hennink, Anthracene functionalized thermosensitive and UV-crosslinkable polymeric micelles, *Polymer Chem.*, 6 (2015) 2048–2053.

32. T. Ta, T. M. Porter, Thermosensitive liposomes for localized delivery and triggered release of chemotherapy, *J. Control. Release*, 169 (2013) 112–125.

33. B. Kneidl, M. Peller, G. Winter, L. H. Lindner, M. Hossann, Thermosensitive liposomal drug delivery systems: State of the art review, *Int. J. Nanomed.*, 9 (2014) 4387–4398.

34. M. B. Yatvin, J. N. Weinstein, W. H. Dennis, R. Blumenthal, Design of liposomes for enhanced local release of drugs by hyperthermia, *Science*, 202 (1978) 1290–1293.

35. Y. Nibu, T. Inoue, I. Motoda, Effect of headgroup type on the miscibility of homologous phospholipids with different acyl chain lengths in hydrated bilayer, *Biophys. Chem.*, 56 (1995) 273–280.

36. D. Needham, G. Anyarambhatla, G. Kong, M. W. Dewhirst, A new temperature-sensitive liposome for use with mild hyperthermia: Characterization and testing in a human tumor xenograft model, *Cancer Res.*, 60 (2000) 1197–1201.

37. C. D. Landon, J. Y. Park, D. Needham, M. W. Dewhirst, Nanoscale drug delivery and hyperthermia: The materials design and preclinical and clinical testing of low temperature-sensitive liposomes used in combination with mild hyperthermia in the treatment of local cancer, *Open Nanomed. J.*, 3 (2011) 38–64.

38. G. R. Anyarambhatla, D. Needham, Enhancement of the phase transition permeability of DPPC liposomes by incorporation of MPPC: A new temperature-sensitive liposome for use with mild hyperthermia, *J. Liposome Res.*, 9 (1999) 491–506.

39. G. Kong, G. Anyarambhatla, W. P. Petros, R. D. Braun, O. M. Colvin, D. Needham, M. W. Dewhirst, Efficacy of liposomes and hyperthermia in a human tumor xenograft model: Importance of triggered drug release, *Cancer Res.*, 60 (2000) 6950–6957.

40. J. K. Mills, D. Needham, Lysolipid incorporation in dipalmitoylphosphatidylcholine bilayer membranes enhances the ion permeability and drug release rates at the membrane phase transition, *Biochim. Biophys. Acta*, 1716 (2005) 77–96.

41. http://celsion.com/docs/technology_thermodox (2016).
42. M. C. Sandstrom, L. M. Ickenstein, L. D. Mayer, K. Edwards, Effects of lipid segregation and lysolipid dissociation on drug release from thermosensitive liposomes, *J. Control. Release*, 107 (2005) 131–142.
43. K. Kono, Thermosensitive polymer-modified liposomes, *Adv. Drug Deliv. Rev.*, 53 (2001) 307–319.
44. K. Kono, A. Henmi, H. Yamashita, H. Hayashi, T. Takagishi, Improvement of temperature-sensitivity of poly(N-isopropylacrylamide)-modified liposomes, *J. Control. Release*, 59 (1999) 63–75.
45. E. P. Lukashev, R. Govindjee, M. Kono, T. G. Ebrey, Y. Sugiyama, Y. Mukohata, pH dependence of the absorption spectra and photochemical transformations of the archaerhodopsins, *Photochem. Photobiol.*, 60 (1994) 69–75.
46. H. Hayashi, K. Kono, T. Takagishi, Temperature sensitization of liposomes using copolymers of N-isopropylacrylamide, *Bioconjug. Chem.*, 10 (1999) 412–418.
47. H. D. Han, B. C. Shin, H. S. Choi, Doxorubicin-encapsulated thermo-sensitive liposomes modified with poly(N-isopropylacrylamide-co-acrylamide): Drug release behavior and stability in the presence of serum, *Eur. J. Pharm. Biopharm.*, 62 (2006) 110–116.
48. K. Kono, R. Nakai, K. Morimoto, T. Takagishi, Thermosensitive polymer-modified liposomes that release contents around physiological temperature, *Biochim. Biophys. Acta*, 1416 (1999) 239–250.
49. Y. G. Takei, T. Aoki, K. Sanui, N. Ogata, Y. Sakurai, T. Okano, Dynamic contact angle measurement of temperature-responsive surface properties for poly(N-isopropylacrylamide) grafted surfaces, *Macromolecules*, 27 (1994) 6163–6166.
50. T. Yakushiji, K. Sakai, A. Kikuchi, T. Aoyagi, Y. Sakurai, T. Okano, Graft architectural effects on thermoresponsive wettability changes of poly(N-isopropylacrylamide)-modified surfaces, *Langmuir*, 14 (1998) 4657–4662.
51. K. Yoshino, A. Kadowaki, T. Takagishi, K. Kono, Temperature sensitization of liposomes by use of N-isopropylacrylamide copolymers with varying transition endotherms, *Bioconjug. Chem.*, 15 (2004) 1102–1109.
52. K. Kono, T. Murakami, T. Yoshida, Y. Haba, S. Kanaoka, T. Takagishi, S. Aoshima, Temperature sensitization of liposomes by use of thermosensitive block copolymers synthesized by living cationic polymerization: Effect of copolymer chain length, *Bioconjug. Chem.*, 16 (2005) 1367–1374.

53. L. Paasonen, B. Romberg, G. Storm, M. Yliperttula, A. Urtti, W. E. Hennink, Temperature-sensitive poly(N-(2-hydroxypropyl)methacrylamide mono/dilactate)-coated liposomes for triggered contents release, *Bioconjug. Chem.*, 18 (2007) 2131–2136.

54. K. Kono, T. Ozawa, T. Yoshida, F. Ozaki, Y. Ishizaka, K. Maruyama, C. Kojima, A. Harada, S. Aoshima, Highly temperature-sensitive liposomes based on a thermosensitive block copolymer for tumor-specific chemotherapy, *Biomaterials*, 31 (2010) 7096–7105.

55. B. M. Dicheva, A. L. Seynhaeve, T. Soulie, A. M. Eggermont, T. L. Ten Hagen, G. A. Koning, Pharmacokinetics, Tissue distribution and therapeutic effect of cationic thermosensitive liposomal doxorubicin upon mild hyperthermia, *Pharm. Res.*, 33 (2016) 627–638.

56. P. Kulshrestha, M. Gogoi, D. Bahadur, R. Banerjee, In vitro application of paclitaxel loaded magnetoliposomes for combined chemotherapy and hyperthermia, *Colloids Surf. B Biointerfaces*, 96 (2012) 1–7.

57. T. Tagami, W. D. Foltz, M. J. Ernsting, C. M. Lee, I. F. Tannock, J. P. May, S. D. Li, MRI monitoring of intratumoral drug delivery and prediction of the therapeutic effect with a multifunctional thermosensitive liposome, *Biomaterials*, 32 (2011) 6570–6578.

58. S. M. Park, M. S. Kim, S. J. Park, E. S. Park, K. S. Choi, Y. S. Kim, H. R. Kim, Novel temperature-triggered liposome with high stability: Formulation, in vitro evaluation, and in vivo study combined with high-intensity focused ultrasound (HIFU), *J. Control. Release*, 170 (2013) 373–379.

59. M. Kimura, M. Kato, T. Muto, K. Hanabusa, H. Shirai, Temperature-sensitive dendritic hosts: Synthesis, characterization, and control of catalytic activity, *Macromolecules*, 33 (2000) 1117–1119.

60. Y. Z. You, C. Y. Hong, C. Y. Pan, P. H. Wang, Synthesis of a dendritic core–shell nanostructure with a temperature-sensitive shell, *Adv. Mater.*, 16 (2004) 1953–1957.

61. Z. Wang, B. Zhou, D. Liu, X.-P. Fan, Thermosensitive poly(N-isopropylacrylamide)-g-polyamidoamine dendrimer derivatives: Preparation and the drug release behaviors, *Macromol. Symp.*, 297 (2010) 158–166.

62. J. Xu, S. Luo, W. Shi, S. Liu, Two-stage collapse of unimolecular micelles with double thermoresponsive coronas, *Langmuir*, 22 (2006) 989–997.

63. Z. Yang, W. Zhang, J. Zou, W. Shi, Synthesis and thermally responsive characteristics of dendritic poly(ether-amide) grafting with PNIPAAm and PEG, *Polymer*, 48 (2007) 931–938.

64. L. Zhu, G. Zhu, M. Li, E. Wang, R. Zhu, X. Qi, Thermosensitive aggregates self-assembled by an asymmetric block copolymer of dendritic polyether and poly(N-isopropylacrylamide), *Eur. Polymer J.*, 38 (2002) 2503–2506.

65. Z. Ge, S. Luo, S. Liu, Syntheses and self-assembly of poly(benzyl ether)-b-poly(N-isopropylacrylamide) dendritic–linear diblock copolymers, *J. Polymer Sci. Part A: Polymer Chem.*, 44 (2006) 1357–1371.

66. H. I. Lee, J. A. Lee, Z. Poon, P. T. Hammond, Temperature-triggered reversible micellar self-assembly of linear-dendritic block copolymers, *Chem. Commun. (Camb.)*, 32 (2008) 3726–3728.

67. Y. S. Kim, E. S. Gil, T. L. Lowe, Synthesis and characterization of thermoresponsive-co-biodegradable linear–dendritic copolymers, *Macromolecules*, 39 (2006) 7805–7811.

68. Y. Haba, A. Harada, T. Takagishi, K. Kono, Rendering poly(amidoamine) or poly(propylenimine) dendrimers temperature sensitive, *J. Am. Chem. Soc.*, 126 (2004) 12760–12761.

69. K. Kono, E. Murakami, Y. Hiranaka, E. Yuba, C. Kojima, A. Harada, K. Sakurai, Thermosensitive molecular assemblies from poly(amidoamine) dendron-based lipids, *Angew. Chem.*, 123 (2011) 6456–6460.

70. Y. Tono, C. Kojima, Y. Haba, T. Takahashi, A. Harada, S. Yagi, K. Kono, Thermosensitive properties of poly(amidoamine) dendrimers with peripheral phenylalanine residues, *Langmuir*, 22 (2006) 4920–4922.

71. Y. Haba, C. Kojima, A. Harada, K. Kono, Control of temperature-sensitive properties of poly(amidoamine) dendrimers using peripheral modification with various alkylamide groups, *Macromolecules*, 39 (2006) 7451–7453.

72. Y. Haba, C. Kojima, A. Harada, K. Kono, Comparison of thermosensitive properties of poly(amidoamine) dendrimers with peripheral n-isopropylamide groups and linear polymers with the same groups, *Angew. Chem. Int. Ed.*, 46 (2007) 234–237.

73. S. V. Aathimanikandan, E. N. Savariar, S. Thayumanavan, Temperature-sensitive dendritic micelles, *J. Am. Chem. Soc.*, 127 (2005) 14922–14929.

74. C. Kojima, K. Yoshimura, A. Harada, Y. Sakanishi, K. Kono, Temperature-sensitive hyperbranched poly(glycidol)s with oligo(ethylene glycol) monoethers, *J. Polymer Sci. Part A: Polymer Chem.*, 48 (2010) 4047–4054.

75. C. Kojima, K. Yoshimura, A. Harada, Y. Sakanishi, K. Kono, Synthesis and characterization of hyperbranched poly(glycidol) modified with pH- and temperature-sensitive groups, *Bioconjug. Chem.*, 20 (2009) 1054–1057.

76. X. Li, Y. Haba, K. Ochi, E. Yuba, A. Harada, K. Kono, PAMAM dendrimers with an oxyethylene unit-enriched surface as biocompatible temperature-sensitive dendrimers, *Bioconjug. Chem.*, 24 (2013) 282–290.

77. Y. Guo, Y. Zhao, J. Zhao, M. Han, A. Zhang, X. Wang, Codendrimer from polyamidoamine (PAMAM) and oligoethylene dendron as a thermosensitive drug carrier, *Bioconjug. Chem.*, 25 (2014) 24–31.

78. Y. Zhao, L. Shen, Y. Wan, X. Zhu, Z. Wang, Thermosensitivity of low generation poly(amidoamine) dendrimers with enriching peripheral functional groups, *Colloids Surf. A: Physicochem. Eng. Aspects*, 403 (2012) 164–168.

79. R. Liu, M. Sun, X. Liu, A. Fan, Z. Wang, Y. Zhao, Interplay of stimuli-responsiveness, drug loading and release for a surface-engineered dendrimer delivery system, *Int. J. Pharm.*, 462 (2014) 103–107.

80. T. Koga, M. Iimura, N. Higashi, Novel peptide-shelled dendrimer with dramatically changeable thermo-responsive character, *Macromol. Biosci.*, 12 (2012) 1043–1047.

81. C. Kojima, U. H. Sk, D. Fukushima, K. Irie, N. Akazawa, M. Umeda, T. Niidome, Effect of main chain conformation on thermosensitivity in elastin-like peptide-grafted polylysine, *RSC Adv.*, 5 (2015) 104900–104906.

82. F. Meng, Z. Zhong, J. Feijen, Stimuli-responsive polymersomes for programmed drug delivery, *Biomacromolecules*, 10 (2009) 197–209.

83. O. Onaca, R. Enea, D. W. Hughes, W. Meier, Stimuli-responsive polymersomes as nanocarriers for drug and gene delivery, *Macromol. Biosci.*, 9 (2009) 129–139.

84. T. Anajafi, S. Mallik, Polymersome-based drug-delivery strategies for cancer therapeutics, *Ther. Deliv.*, 6 (2015) 521–534.

85. R. Cheng, F. Meng, S. Ma, H. Xu, H. Liu, X. Jing, Z. Zhong, Reduction and temperature dual-responsive crosslinked polymersomes for targeted intracellular protein delivery, *J. Mater. Chem.*, 21 (2011) 19013–19020.

86. S. Ma, M. Xiao, R. Wang, Formation and structural characteristics of thermosensitive multiblock copolymer vesicles, *Langmuir*, 29 (2013) 16010–16017.

87. F. Liu, V. Kozlovskaya, S. Medipelli, B. Xue, F. Ahmad, M. Saeed, D. Cropek, E. Kharlampieva, Temperature-sensitive polymersomes for controlled delivery of anticancer drugs, *Chem. Mater.*, 27 (2015) 7945–7956.

88. W. Agut, A. Brulet, C. Schatz, D. Taton, S. Lecommandoux, pH and temperature responsive polymeric micelles and polymersomes by self-assembly of poly[2-(dimethylamino)ethyl methacrylate]-b-poly(glutamic acid) double hydrophilic block copolymers, *Langmuir*, 26 (2010) 10546–10554.

89. Z. Y. Qiao, R. Ji, X. N. Huang, F. S. Du, R. Zhang, D. H. Liang, Z. C. Li, Polymersomes from dual responsive block copolymers: Drug encapsulation by heating and acid-triggered release, *Biomacromolecules*, 14 (2013) 1555–1563.

90. S. Qin, Y. Geng, D. E. Discher, S. Yang, Temperature-controlled assembly and release from polymer vesicles of poly(ethylene oxide)-block-poly(n-isopropylacrylamide), *Adv. Mater.*, 18 (2006) 2905–2909.

91. X. Chen, X. Ding, Z. Zheng, Y. Peng, Thermosensitive cross-linked polymer vesicles for controlled release system, *New J. Chem.*, 30 (2006) 577–582.

92. E. Amstad, S.-H. Kim, D. A. Weitz, Photo- and thermoresponsive polymersomes for triggered release, *Angew. Chem. Int. Ed.*, 51 (2012) 12499–12503.

93. C. Gong, T. Qi, X. Wei, Y. Qu, Q. Wu, F. Luo, Z. Qian, Thermosensitive polymeric hydrogels as drug delivery systems, *Curr. Med. Chem.*, 20 (2013) 79–94.

94. Y. Lee, S. Y. Park, C. Kim, T. G. Park, Thermally triggered intracellular explosion of volume transition nanogels for necrotic cell death, *J. Control. Release*, 135 (2009) 89–95.

95. W. McPhee, K. C. Tam, R. Pelton, Poly(N-isopropylacrylamide) lattices prepared with sodium dodecyl sulfate, *J. Colloid Interface Sci.*, 156 (1993) 24–30.

96. D. Kuckling, C. D. Vo, H. J. P. Adler, A. Völkel, H. Cölfen, Preparation and characterization of photo-cross-linked thermosensitive PNIPAAm nanogels, *Macromolecules*, 39 (2006) 1585–1591.

97. Q. Wang, H. Xu, X. Yang, Y. Yang, Drug release behavior from in situ gelatinized thermosensitive nanogel aqueous dispersions, *Int. J. Pharm.*, 361 (2008) 189–193.

98. Z. Jian, Q. Zhiyu, G. Yueqing, In vivo anti-tumor efficacy of docetaxel-loaded thermally responsive nanohydrogel, *Nanotechnology*, 20 (2009) 325102.

99. J. Zhang, H. Chen, L. Xu, Y. Gu, The targeted behavior of thermally responsive nanohydrogel evaluated by NIR system in mouse model, *J. Control. Release*, 131 (2008) 34–40.

100. M. Fathi, A. A. Entezami, A. Ebrahimi, S. K. D., Synthesis of thermosensitive nanohydrogels by crosslinker free method based on N-isopropylacrylamide: Applicable in the naltrexone sustained release, *Macromol. Res.*, 21 (2013) 17–26.

101. J. Jing, D. Alaimo, E. De Vlieghere, C. Jerome, O. De Wever, B. G. De Geest, R. Auzely-Velty, Tunable self-assembled nanogels composed of well-defined thermoresponsive hyaluronic acid-polymer conjugates, *J. Mater. Chem. B*, 1 (2013) 3883–3887.

102. E. Ruel-Gariepy, G. Leclair, P. Hildgen, A. Gupta, J. C. Leroux, Thermosensitive chitosan-based hydrogel containing liposomes for the delivery of hydrophilic molecules, *J. Control. Release*, 82 (2002) 373–383.

103. D. Chen, H. Yu, H. Mu, J. Wei, Z. Song, H. Shi, R. Liang, K. Sun, W. Liu, Novel chitosan derivative for temperature and ultrasound dual-sensitive liposomal microbubble gel, *Carbohydr. Polymers*, 94 (2013) 17–23.

104. A. A. Date, A. Shibata, M. Goede, B. Sanford, K. La Bruzzo, M. Belshan, C. J. Destache, Development and evaluation of a thermosensitive vaginal gel containing raltegravir+efavirenz loaded nanoparticles for HIV prophylaxis, *Antiviral. Res.*, 96 (2012) 430–436.

105. M. Shen, Y. Y. Xu, Y. Sun, B. S. Han, Y. R. Duan, Preparation of a thermosensitive gel composed of a mPEG-PLGA-PLL-cRGD nanodrug delivery system for pancreatic tumor therapy, *ACS Appl. Mater. Interfaces*, 7 (2015) 20530–20537.

106. Q. Q. Yin, L. Wu, M. L. Gou, Z. Y. Qian, W. S. Zhang, J. Liu, Long-lasting infiltration anaesthesia by lidocaine-loaded biodegradable nanoparticles in hydrogel in rats, *Acta Anaesthesiol. Scand.*, 53 (2009) 1207–1213.

107. H.-Y. Huang, S.-H. Hu, C.-S. Chian, S.-Y. Chen, H.-Y. Lai, Y.-Y. Chen, Self-assembling PVA-F127 thermosensitive nanocarriers with highly sensitive magnetically-triggered drug release for epilepsy therapy in vivo, *J. Mater. Chem.*, 22 (2012) 8566–8573.

108. L. Jianbo, Q. Yang, R. Jie, Y. Weizhong, S. Donglu, Magnetocaloric effect in magnetothermally-responsive nanocarriers for hyperthermia-triggered drug release, *Nanotechnology*, 23 (2012) 505706.

109. C. Echeverria, C. Mijangos, UCST-like hybrid PAAm-AA/Fe_3O_4 microgels. Effect of Fe_3O_4 nanoparticles on morphology, thermosensitivity and elasticity, *Langmuir*, 27 (2011) 8027–8035.

110. S. Kaamyabi, D. Habibi, M. M. Amini, Preparation and characterization of the pH and thermosensitive magnetic molecular imprinted nanoparticle polymer for the cancer drug delivery, *Bioorg. Med. Chem. Lett.*, 26 (2016) 2349–2354.

111. Y.-H. Lien, J.-H. Wu, J.-W. Liao, T.-M. Wu, In vitro evaluation of the thermosensitive and magnetic nanoparticles for the controlled drug delivery of vitamin D3, *Macromol. Res.*, 21 (2013) 511–518.

112. J.-Y. Kim, W. I. Choi, M. Kim, G. Tae, Tumor-targeting nanogel that can function independently for both photodynamic and photothermal therapy and its synergy from the procedure of PDT followed by PTT, *J. Control. Release*, 171 (2013) 113–121.

113. T. Shang, C.-D. Wang, L. Ren, X.-H. Tian, D.-H. Li, X.-B. Ke, M. Chen, A.-Q. Yang, Synthesis and characterization of NIR-responsive Aurod@ pNIPAAm-PEGMA nanogels as vehicles for delivery of photodynamic therapy agents, *Nanoscale Res. Lett.*, 4 (2013) 1–8.

114. X. Lian, J. Jin, J. Tian, H. Zhao, Thermoresponsive nanohydrogels cross-linked by gold nanoparticles, *ACS Appl. Mater. Interfaces*, 2 (2010) 2261–2268.

115. A. Barhoumi, W. Wang, D. Zurakowski, R. S. Langer, D. S. Kohane, Photothermally targeted thermosensitive polymer-masked nanoparticles, *Nano Lett.*, 14 (2014) 3697–3701.

116. S. Kazakov, K. Levon, Liposome-nanogel structures for future pharmaceutical applications, *Curr. Pharm. Des.*, 12 (2006) 4713–4728.

117. B. Zhang, J. Chen, Y. Lu, J. Qi, W. Wu, Liposomes interiorly thickened with thermosensitive nanogels as novel drug delivery systems, *Int. J. Pharm.*, 455 (2013) 276–284.

118. J. S. Lee, W. Zhou, F. Meng, D. Zhang, C. Otto, J. Feijen, Thermosensitive hydrogel-containing polymersomes for controlled drug delivery, *J. Control. Release*, 146 (2010) 400–408.

119. C. Cheng, H. Wei, B. X. Shi, H. Cheng, C. Li, Z. W. Gu, S. X. Cheng, X. Z. Zhang, R. X. Zhuo, Biotinylated thermoresponsive micelle

self-assembled from double-hydrophilic block copolymer for drug delivery and tumor target, *Biomaterials*, 29 (2008) 497–505.

120. X. J. Loh, J. del Barrio, P. P. Toh, T. C. Lee, D. Jiao, U. Rauwald, E. A. Appel, O. A. Scherman, Triply triggered doxorubicin release from supramolecular nanocontainers, *Biomacromolecules*, 13 (2012) 84–91.
121. C. Y. Chen, T. H. Kim, W. C. Wu, C. M. Huang, H. Wei, C. W. Mount, Y. Tian, S. H. Jang, S. H. Pun, A. K. Jen, pH-dependent, thermosensitive polymeric nanocarriers for drug delivery to solid tumors, *Biomaterials*, 34 (2013) 4501–4509.
122. S. Zschoche, J. C. Rueda, M. Binner, H. Komber, A. Janke, K.-F. Arndt, S. Lehmann, B. Voit, Reversibly switchable pH- and thermoresponsive core–shell nanogels based on poly(NiPAAm)-graft-poly(2-carboxyethyl-2-oxazoline)s, *Macromol. Chem. Phys.*, 213 (2012) 215–226.
123. D. Kuckling, C. D. Vo, S. E. Wohlrab, Preparation of nanogels with temperature-responsive core and pH-responsive arms by photo-cross-linking, *Langmuir*, 18 (2002) 4263–4269.
124. D. An, D. Zhao, X. Li, X. Lu, G. Qiu, K. J. Shea, Synthesis of surfactant-free hydroxypropylcellulose nanogel and its dual-responsive properties, *Carbohydr. Polym.*, 134 (2015) 385–389.
125. W.-H. Chiang, V. T. Ho, W.-C. Huang, Y.-F. Huang, C.-S. Chern, H.-C. Chiu, Dual stimuli-responsive polymeric hollow nanogels designed as carriers for intracellular triggered drug release, *Langmuir*, 28 (2012) 15056–15064.
126. M. Ahmed, R. Narain, Intracellular delivery of DNA and enzyme in active form using degradable carbohydrate-based nanogels, *Mol. Pharm.*, 9 (2012) 3160–3170.
127. D. Chen, H. Yu, K. Sun, W. Liu, H. Wang, Dual thermoresponsive and pH-responsive self-assembled micellar nanogel for anticancer drug delivery, *Drug Deliv.*, 21 (2014) 258–264.
128. J. He, L. Tremblay, S. Lacelle, Y. Zhao, How can photoisomerization of azobenzene induce a large cloud point temperature shift of PNIPAM?, *Polymer Chem.*, 5 (2014) 5403–5411.
129. A. Balamurugan, H.-I. Lee, New benzylidene oxazolone derived polymeric photoswitches for light-induced tunable thermoresponsive behaviors, *Polymer Chem.*, 5 (2014) 6426–6430.
130. E. Blasco, B. V. K. J. Schmidt, C. Barner-Kowollik, M. Pinol, L. Oriol, Dual thermo- and photo-responsive micelles based on miktoarm star polymers, *Polymer Chem.*, 4 (2013) 4506–4514.

131. C. Wang, J. Mallela, U. S. Garapati, S. Ravi, V. Chinnasamy, Y. Girard, M. Howell, S. Mohapatra, A chitosan-modified graphene nanogel for noninvasive controlled drug release, *Nanomed. Nanotechnol. Biol. Med.*, 9 (2013) 903–911.

132. J. L. Paris, M. V. Cabañas, M. Manzano, M. Vallet-Regí, Polymer-grafted mesoporous silica nanoparticles as ultrasound-responsive drug carriers, *ACS Nano*, 9 (2015) 11023–11033.

133. Y. Tian, Y. Wang, S. Shen, X. Jiang, Y. Wang, W. Yang, Temperature and redox dual-responsive biodegradable nanogels for optimizing antitumor drug delivery, *Particle Particle Syst. Characterization*, 32 (2015) 1092–1101.

134. Z. Y. Qiao, R. Zhang, F. S. Du, D. H. Liang, Z. C. Li, Multi-responsive nanogels containing motifs of ortho ester, oligo(ethylene glycol) and disulfide linkage as carriers of hydrophobic anti-cancer drugs, *J. Control. Release*, 152 (2011) 57–66.

135. Y. Zhan, M. Goncalves, P. Yi, D. Capelo, Y. Zhang, J. Rodrigues, C. Liu, H. Tomas, Y. Li, P. He, Thermo/redox/pH-triple sensitive poly(N-isopropylacrylamide-co-acrylic acid) nanogels for anticancer drug delivery, *J. Mater. Chem. B*, 3 (2015) 4221–4230.

136. Y. Qin, J. Chen, Y. Bi, X. Xu, H. Zhou, J. Gao, Y. Hu, Y. Zhao, Z. Chai, Near-infrared light remote-controlled intracellular anti-cancer drug delivery using thermo/pH sensitive nanovehicle, *Acta Biomater.*, 17 (2015) 201–209.

137. W. Yuan, W. Guo, H. Zou, J. Ren, Tunable thermo-, pH- and light-responsive copolymer micelles, *Polymer Chem.*, 4 (2013) 3934–3937.

Chapter 7

Magnetically Responsive Nanomedicine

Meng Zhang, Ergang Liu, and Yongzhuo Huang

State Key Laboratory of Drug Research,
Shanghai Institute of Materia Medica, Chinese Academy of Sciences,
501 Hai-ke Rd, Shanghai 201203, China

liuergang@hotmail.com, yzhuang@simm.ac.cn

Cancer is a major threat to human health due to its complexities. Magnetic nanoparticles have been widely studied for delivery of therapeutic agents and generation of magnetic hyperthermia for tumor treatment due to their unique properties. This chapter gives an overview of magnetically responsive nanoparticles/nanomedicines on aspects of their magnetic properties, synthetic methods, surface modification, and biomedical applications.

7.1 Introduction

Cancer severely threatens human health. The complexities of cancer have significantly contributed to the difficulty of anti-cancer

Stimuli-Responsive Nanomedicine
Edited by Lin Zhu

ISBN 978-981-4800-70-9 (Hardcover), 978-0-429-29529-4 (eBook)
www.jennystanford.com

treatment. With the remarkable progress in molecular biology and nanotechnology during the past decades, advances have been made in the translation of multifunctional nanostructures into biomedical applications, which help to overcome the shortcomings related to traditional diagnosis and therapy [1]. The concept of nanotechnology was first proposed by the physicist Richard Feynman in 1959 [2], and its application in biomedical sciences has led to the formation of a multidisciplinary field termed nanomedicine [3]. Compared to conventional cancer treatment, nanomedicine can achieve timed, controlled and targeted drug delivery, and overcome multidrug resistance caused by repeated dosing [4].

Magnetic nanoparticles, especially superparamagnetic iron oxide nanoparticles (SPIONs), have attracted more and more researchers to investigate their applications in cancer therapy due to their unique magnetic properties and good biocompatibility [5]. The SPIONs used in the biomedical investigation are usually categorized into two groups: the ones with a hydrodynamic size larger than 50 nm and the ultrafine ones less than 30 nm [6]. The applications of SPIONs have been expanding widely from early use of cell sorting, magnetic resonance imaging (MRI) to current use of magnetic targeted drug delivery and hyperthermia. These applications of SPIONs benefit from their special physical natures, such as size, shape, and crystal structure dependent magnetic properties (saturation magnetization, remanent magnetization, and coercivity), and achieve substantial improvement in cancer treatment and diagnosis [7]. However, the potential adverse effects should be considered for the clinical trials [8]. Surface modification with various inorganic or organic materials can prevent the aggregation of SPIONs in aqueous solutions, reduce their in vivo side effects, and improve stability by maintaining the size and shape of nanoparticles. On the other hand, surface modification may compromise their ability to respond to the external magnetic field, which means the saturation magnetization would be reduced. The SPION products with desired properties can be obtained by optimizing the conditions such as synthetic strategies, kinds of polymeric materials and the thickness of the surface coating [9].

For imaging purpose, magnetic nanoparticles can combine with other imaging modalities such as fluorescence imaging [10–12], near-infrared imaging (NIR) [13, 14], computed tomography (CT) [15], ultrasound imaging [16], positron emission tomography (PET) [17, 18], and single-photon emission computed tomography (SPECT) [19], to achieve multifunctional imaging [20, 21]. Concurrently, magnetic nanoparticles have also been used to deliver chemotherapeutics [22], nucleic acids [23–32], monoclonal antibodies [33–36], and agents with photodynamic [37–39], photothermal [40, 41], or sonodynamic properties [42] for cancer combination therapy. In addition, magnetic nanoparticles have been investigated in cancer hyperthermia because they can convert the energy obtained from an external electromagnetic field into heat. Hyperthermia is rarely used alone, and the combination of magnetic hyperthermia with other therapeutic strategies, such as chemotherapy [43, 44], gene therapy [45], radiotherapy [46], and phototherapy [47], have been widely reported.

This chapter discusses the current magnetically responsive nanoparticles according to their magnetic properties, synthetic methods, surface modification, and biomedical applications based on these nano-magnetic systems.

7.2 Properties of Magnetic Nanoparticles

Nanoparticles refer to the particles in the size range of 1–1,000 nm. The small size and large surface-to-volume ratio grant nanomaterials with unique size-dependent properties which are different from those of the bulk matters [48]. Magnetic nanoparticles are commonly used as MRI contrast agents for tumor diagnosis, drug carriers for tumor targeting, and therapeutic agents for cancer hyperthermia [49]. The understanding of the crystal structure, synthetic strategies, size-magnetism relationship has been growing [50], which provides helpful information for preparation of the nanoparticles with the desired properties by controlling the composition, size, and shape. So far, several magnetic nanoparticles have been approved by FDA as contrast agents [49].

The magnetic properties of the nanoparticles are governed by their composition and structure [50]. Magnetic nanoparticles usually have a solid magnetic core that is often composed of metal oxides such as iron, cobalt, and/or nickel oxides. The electrons on 3d atomic orbitals are restrained to the periphery of the nucleus, thus generating a local atomic magnetic moment [51]. Exchange of the electrons between adjacent atoms renders the local magnetic moment aligned parallel. In a sufficiently strong external magnetic field, all the atomic magnetic moment in local region align in an identical direction, which enables the metallic nanoparticles to become magnetic [51, 52]. This mechanism lays the foundation for applications of magnetic nanoparticles. Saturation magnetization, remanent magnetization, and coercivity are the key parameters to measure the magnetic properties [53]. Saturation magnetization is the maximal ability of magnetic response under an external magnetic field, while remanent magnetization is the magnetic capacity after the withdraw of external magnetic field, and coercivity is the required external magnetic field to offset the remanent magnetization [54]. Unlike the bulk matters that contain many non-parallel magnetic domains, a magnetic nanocrystal below the critical size contains only one magnetic domain, and the magnetic moments of all atoms in the nanoparticle can be vectorially summed as a large magnetic moment [48]. Upon withdraw of the external magnetic field, the related magnetic orientation force would be offset by the thermal fluctuations of the nanoparticles, resulting in zero remanent magnetization and coercivity [56]. Superparamagnetism can prevent the nanoparticles against coagulation after removal of the external magnetic field, and improve the stability and dispersity of the nanoparticles in physiological environments [57]. Of note, superparamagnetism of magnetic nanoparticles is size dependent, of which the nano-magnetite less than 30 nm are superparamagnetic [55] and displays an on-and-off state in the presence or absence of an external magnetic field [5].

Magnetic nanoparticles can convert the energy obtained from an electromagnetic field into heat through two ways: Neel relaxation and Brownian relaxation [58]. Neel relaxation originates from the orientation changes of the magnetic moment of nanoparticles, along with the external magnetic field [59]. The time of Neel relaxation τ_N is determined by the following equation:

$$\tau_N = \tau_0 \exp\left(\frac{KV}{k_B T}\right), \tag{7.1}$$

Here, $\tau_0 = 10^{-9}$ s, K is the anisotropy constant, V is the volume of magnetic nanoparticle, k_B is the Boltzmann constant, and T is the Kelvin temperature [48]. Neel relaxation is related to the size of magnetic nanoparticles [60]. The smaller the particle size is, the less energy is needed to change the direction of magnetic moment, and thereby Neel relaxation is an important parameter [48].

Brownian relaxation depends on the friction between the magnetic nanoparticles and their surrounding liquid where they rotate to adapt to the changing external magnetic field [61]. The Brownian relaxation time τ_B can be expressed by the following equation:

$$\tau_B = \frac{3\eta V_H}{k_B T}, \tag{7.2}$$

where η is the viscosity of the surrounding liquid, V_H is the hydrodynamic volume of the particle, k_B is the Boltzmann constant, and T is Kelvin temperature [48]. The Brownian relaxation mechanism is associated not only with the particle size but also with the viscosity of the surrounding liquid [60]. The high viscosity retards the rotation of nanoparticles. Neel relaxation generally occurs in the small magnetic nanoparticles, while Brownian relaxation is more prominent for large magnetic nanoparticles [62]. For hyperthermia treatment, Neel relaxation is more important, because magnetic nanoparticles may be prevented from the free rotation once entering the cells where the cytosol viscosity increases, leading to the suppressed Brownian relaxation [48]. The general relaxation time can be calculated by the combination of the above two relaxation times:

$$\tau = \frac{\tau_B \tau_N}{\tau_B + \tau_N} \tag{7.3}$$

Under an alternating magnetic field, energy from the external magnetic field will convert to heat when the outer field oscillates faster than the intrinsic relaxation of magnetic nanoparticles [51]. Energy transfer efficiency, called specific absorption rate (SAR), is defined as the electromagnetic energy absorbed by per unit mass of magnetic materials, which is a parameter to evaluate the thermal ability of magnetic materials under an alternating magnetic field. It can be expressed by the following formulation:

$$\mathrm{SAR} = \pi\mu_0\chi_0 H_0^2 f \frac{2\pi f\tau}{1+(2\pi f\tau)^2}, \tag{7.4}$$

where μ_0 is vacuum permeability, χ_0 is equilibrium susceptibility, H_0 is radiofrequency (RF) field amplitude and f is the external RF field frequency [53].

SAR is highly dependent on the size, size distribution, shape, surface modification and saturation magnetization of the magnetic nanoparticles [63]. It is also associated with the amplitude H and frequency f of the external magnetic field [64]. Multi-domain magnetic nanoparticles present a low SAR, while superparamagnetic nanoparticles offer a high SAR in a low amplitude [65]. In clinical trials, the safe exposure dose of RF frequency (the $H\cdot f$ factor) should be less than 5×10^9 $\mathrm{Am^{-1}\ s^{-1}}$ [48]. Thus, instead of increasing the frequency, it is more safe and feasible to acquire a high SAR by controlling the physical and chemical properties of the magnetic nanoparticles [66].

Moreover, magnetic nanoparticles are commonly used as MRI contrast agents and offer tissue imaging with higher temporal and spatial resolution [67–69]. MRI is based on the response of hydrogen protons in a magnetic field to an RF pulse [69]. Under an external magnetic field, hydrogen protons can align with the field and reach a steady state. When an additional RF pulse is applied, the protons will adsorb the RF energy and lose alignment with magnetic field, and realignment of the protons generates signals that can be recorded by MRI instrument. Generally, MRI relies on the proton density variances among different organs [69]. On this account, magnetic nanoparticles accumulated in the target tissue can shorten the T_2 relaxation time of protons, thus enhancing the resolution of MR imaging [70]. The size and surface properties of magnetic nanoparticles are key parameters that determine magnetic properties and subsequently affect the contrast enhancing effect [71]. For example, small size is favorable for T_2-MRI. MR imaging results can be improved by tuning the parameters of the magnetic nanoparticles such as the hydrophilic surface coating [49].

The unique properties of magnetic nanoparticles render them as the promising drug carriers, MRI contrast agents, and magnetic hyperthermia agents.

7.3 Synthetic Methods

Nanomaterials can be produced via either top-down (e.g., physical nanofabrication) or bottom-up methods (e.g., chemical synthesis) [49]. For magnetically responsive nanomedicines, the particle size is important, and the small size is favorable. Thereby, the chemical methods that can produce tiny and monodispersed magnetic nanostructures are preferred for preparing magnetic nanoparticles [54]. The commonly used chemical methods include co-precipitation, thermal decomposition, and microemulsion, which are depicted in the following sections.

7.3.1 Co-Precipitation

Co-precipitation is the most commonly used and the most efficient method to fabricate magnetic nanoparticles [72]. A typical preparation of SPIONs involves alkaline precipitation of $FeCl_3$ and $FeCl_2$. Generally, the magnetite (Fe_3O_4) can be produced when the stoichiometric ratio of ferrous and ferric ions is 1:2 in the alkaline solution (pH > 7) under the protection of inert gas at room temperature or elevated temperature. The partial reduction of aqueous ferric salt using Na_2SO_3 or partial oxidation of ferrous salt using ion nitrates can result in the formation of Fe_3O_4 nanoparticles in alkaline medium. It has been recently reported that in situ chemical oxidative polymerization of pyrrole along with the reduction of ferric chloride was used to achieve a one-step synthesis of polypyrrole-based magnetic nanoparticles [73]. The alkaline solutions of carbonates, hydroxides, ammonia, and amines are the commonly used precipitation agents [72]. However, Fe_3O_4 is not stable and will transform into γ-Fe_2O_3 in the presence of oxygen.

In a co-precipitation method, the concentration of metal ions and base solutions, composition of base solutions, digestion time, and type and degree of the mixing of the metal ions and base solutions are the important parameters for controlling the synthesis of SPIONs [72]. An optimized condition can result in good reproducibility of magnetic nanoparticle fabrication [74]. The co-precipitation process involves two stages: the rapid formation of crystal nucleus and the subsequent growth stage of pre-formed nucleus [75]. The growth of nanocrystals is governed by both thermodynamic and kinetic factors, but the kinetic factors are the predominant contributor to the crystal formation in co-precipitation. The improper conditions usually lead to the heterogeneously sized nanoparticles. This is a major shortcoming of the co-precipitation methods [55].

Magnetic crystals produced by co-precipitation are unstable and liable to aggregate at neutral pH. They can be better dispersed in acidic or basic conditions. However, the pH cannot be less than 3 because of acidolysis. It has been well established that using organic materials for stabilization can produce the monodispersed magnetic nanoparticles [74]. For example, PVA, PEG, chitosan, dextrin, and hyaluronic acid have been widely used to stabilize the magnetic nanoparticles [76]. However, the strategical selection of materials used for surface modification is important because the surface properties significantly affect the in vivo pharmacokinetic behavior of the modified magnetic nanoparticles [55].

7.3.2 Thermal Decomposition

Thermal decomposition is another commonly used method to fabricate iron oxide nanoparticles with high-quality nanocrystals and oxides [74]. In this process, organometallic iron compounds such as iron pentacarbonyl or acetylacetonate are often used, and the preparation involves a heating process in the organic solvents that contain stabilizing agents under the protection of inert gas. The nanoparticles prepared by the thermal decomposition are usually uniform in size and shape [74]. Fatty acids, especially oleic acid, and hexadecylamine are often used as the stabilizing agents [77]. Depending on the original valency of the metals in the precursors, different procedures of thermal decomposition are applied. For example, the zero-valency metals must experience two steps to form the magnetic nanoparticles, while the cationic metals only need one step [78–80]. In principle, the ratios of the starting reagents, including organometallic compounds, surfactants, and solvents, are the decisive parameters for controlling the size and morphology of magnetic nanoparticles. Furthermore, the reaction temperature, reaction time, as well as aging period may also be crucial for tuning the size and morphology [74].

7.3.3 Microemulsion

Microemulsions, composed of aqueous and oil phases and surfactants, are the thermodynamically stable isotropic systems [81–83]. Generally, microemulsions have two categories: normal microemulsion (o/w) and reverse microemulsion (w/o) [82]. Each microdroplet can serve as a nanoreactor, with the reactants dissolved in different droplets of microemulsion, and the reactions are triggered when two microemulsions mixed together. The size controllable nanoparticles can be produced along with the fusion of the microdroplets [84–88]. For instance, the Fe_3O_4 nanoparticles were prepared via the mixing of the reverse microemulsion containing ferric and ferrous salts (nFe^{3+}:nFe^{2+} = 2:1) in the aqueous cores and the reverse microemulsion containing NaOH [89]. Similarly, Fe^{3+} salts and $NaBH_4$ were dissolved into two identical w/o microemulsions, respectively, and they were mixed for reaction to synthesize the Fe_3O_4 nanoparticles [90]. It was also reported that the highly monodispersed iron oxide nanoparticles were prepared by using the w/o microdroplets, in which iron salts were dissolved inside the aqueous cores of the reverse microemulsion and precipitated when mixed with the microemulsion containing alkali agents, thus forming nanoparticles with the desired size [91].

7.4 Modification and Functionalization of Magnetic Nanoparticles

Magnetic nanoparticles are thermodynamically unstable because they are in a high-energy state, having large surface Gibbs free energy and dipole-dipole interactions [92]. For biological applications, the physical stability of the magnetic nanoparticles should be improved [93]. Moreover, magnetic nanoparticles are inclined to bind with a variety of biomolecules like plasma proteins, making it easy to be recognized by the reticuloendothelial system (RES) and rapidly cleared from the blood circulation [94].

Surface modification of magnetic nanoparticles can improve the biocompatibility and aqueous dispersion, and prolong the blood circulation, as well as enhance drug loading efficiency and control drug release [95]. Modification methods commonly involve the use of inorganic, organic, and hybrid materials (Figure 7.1).

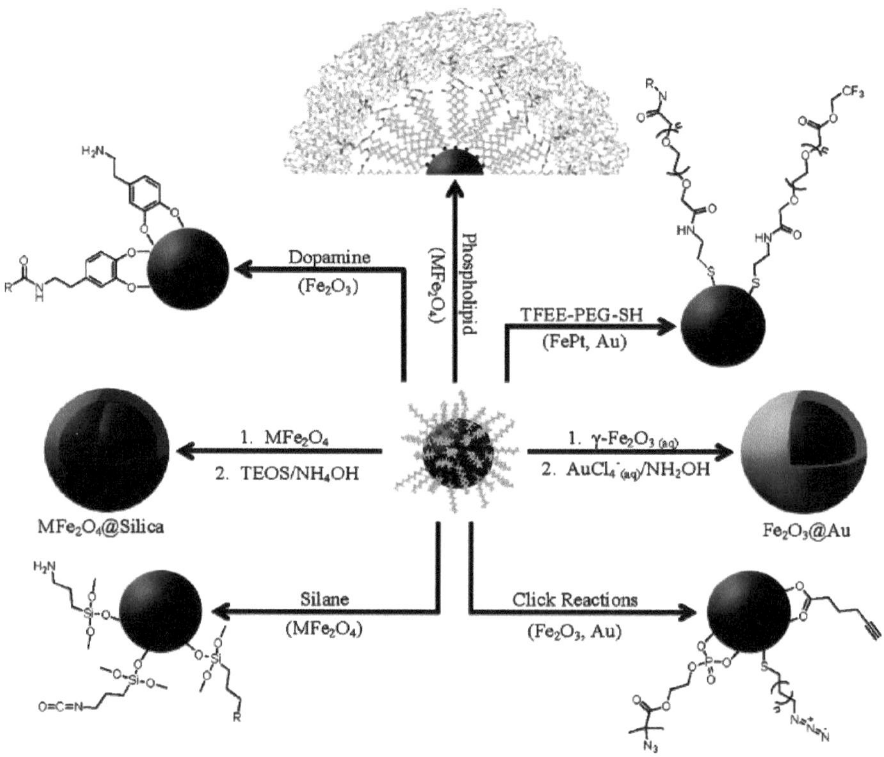

Figure 7.1 Magnetic nanoparticles with various shells to maintain stability. Reprinted with permission from [96].

7.4.1 Inorganic Modification

Using inorganic materials to modify magnetic nanoparticles has been widely investigated in the early years. The magnetic nanoparticles coated with small inorganic molecules often have smaller hydrodynamic diameters than the polymer-coated ones. The inorganic material-modified magnetic nanoparticles have been applied in the fields of surface plasmon resonance (SPR) biosensor [97], magnetic recording [98], separation [99], and DNA testing [100].

7.4.1.1 Inorganic metal composite

Gold, silver, and other noble metals are often used to modify magnetic nanoparticles. Mendez-Garza et al. produced α-Fe_2O_3 spindle-type particles in the presence of phosphate ions by hydrolysis of ferric salts [97]. The reaction was carried out in the dihydrogen atmosphere at a high temperature to reduce iron oxide powders to α-Fe. Subsequently, the thus-formed magnetic nanoparticles were coated by a layer of gold. The formation of the core@ shell structure was driven by a redox-transmetalation reaction between iron (0) at the surface of particles and Au^{3+} salt under the protection of N_2. The Fe@Au core–shell nanoparticles were further modified with the water-soluble, thiolated fluorescent dye labeled-PEG (Figure 7.2).

An immunosensor was previously developed based on the Ag/Au coated Fe_3O_4 nanoparticles [101]. In this study, the Fe_3O_4 magnetic cores were synthesized through a chemical co-precipitation method. Then by using a reverse micelle method, the Fe_3O_4/Ag seeds were produced by adding silver nitrate with excess acetone to precipitate the product. The Fe_3O_4/Ag seeds were further used for the synthesis of Fe_3O_4/Ag/Au nanocomposite by reaction with Au^{3+} in the solution containing hexadecyltrimethyl ammonium bromide (CTAB) and vitamin C. The Fe_3O_4/Ag/Au nanocomposite-based SPR biosensors had great protein immobilization capability, with advantages of high sensitivity and good biocompatibility (Figure 7.3).

Another strategy is to use tiny inorganic nanoparticles to modify the Fe_3O_4 nanoparticles. In a study, dopamine (DA)-coated SPIONs (Fe_3O_4@DA) were synthesized via a modified co-precipitation method. The gold nanoparticles (2–3 nm) could be easily conjugated to the Fe_3O_4@DA nanoparticles by the coordination interaction between the gold nanoparticles and the amino groups of dopamine, thus yielding the Au–Fe_3O_4 hybrid nanoparticles. These hybrid nanoparticles showed good dispersibility in aqueous solution, and were easily functionalized with a targeted peptide and fluorescein isothiocyanate (FITC) for liver cancer cell (BEL-7402) imaging, providing a multi-functional nano-platform [102].

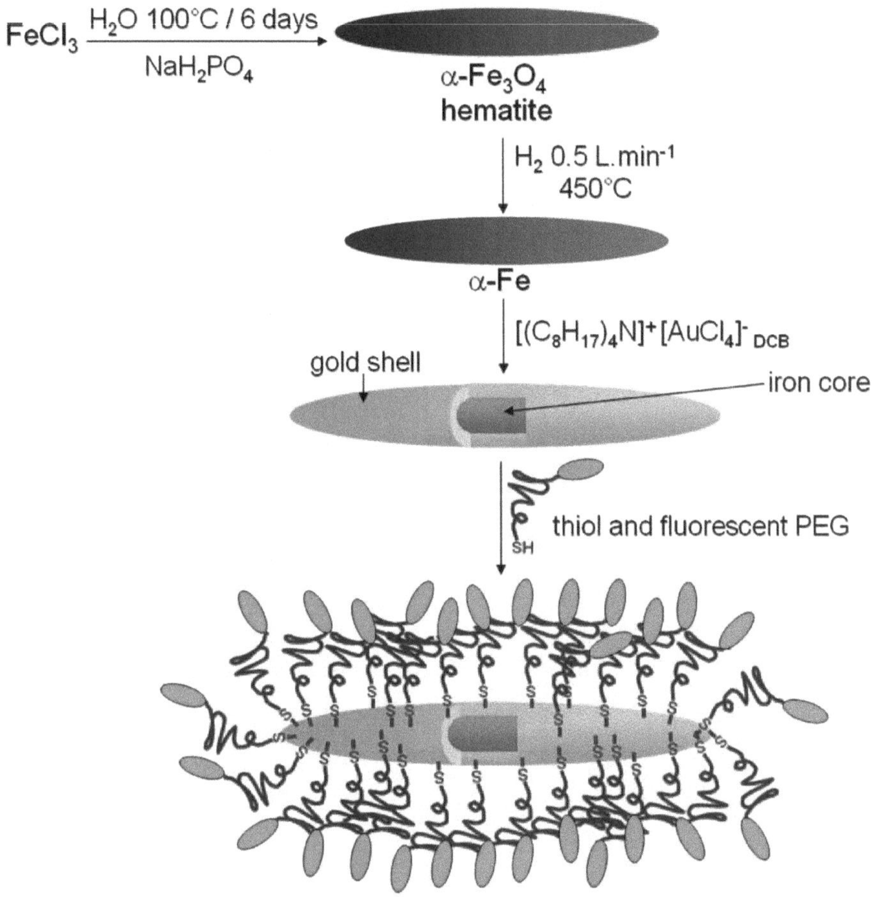

Figure 7.2 The general procedure to obtain the PEGylated iron@gold (core@shell) NPs. Reprinted with permission from [97].

7.4.1.2 SiO_2 modification

SiO_2 is another common coating material for magnetic nanoparticles during surface modification. The SiO_2 coatings are usually used to improve the stability and biocompatibility, and reduce the side toxicity of magnetic nanoparticles. Their hydroxyl-rich silicon surface makes it easy to introduce other functional groups like NH_2, which benefit the conjugation with antibodies or other biomolecules to achieve specific targeting [103].

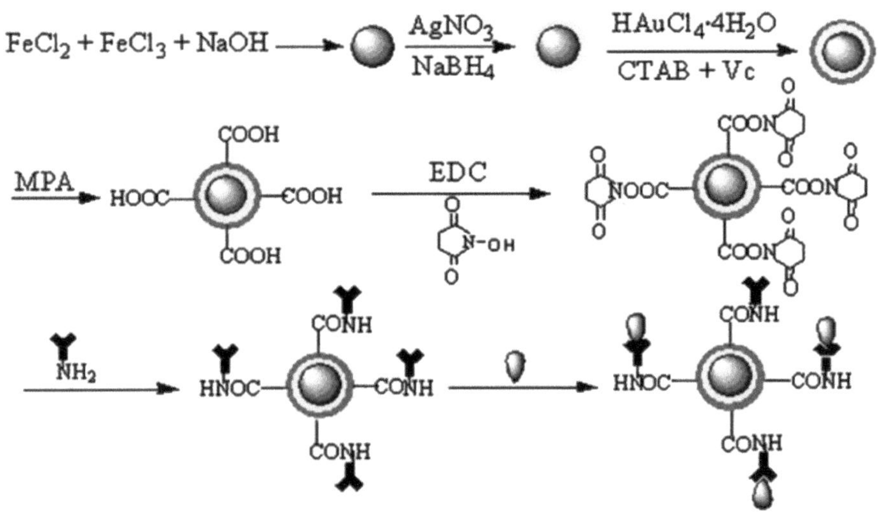

Figure 7.3 The scheme of the immobilization of antibody on the surface of the Au film. (-◎-) Fe_3O_4/Ag/Au nanocomposites, (◎) Fe_3O_4 nanoparticles (Y) antibody, and (♦) antigen. Reprinted with permission from [101].

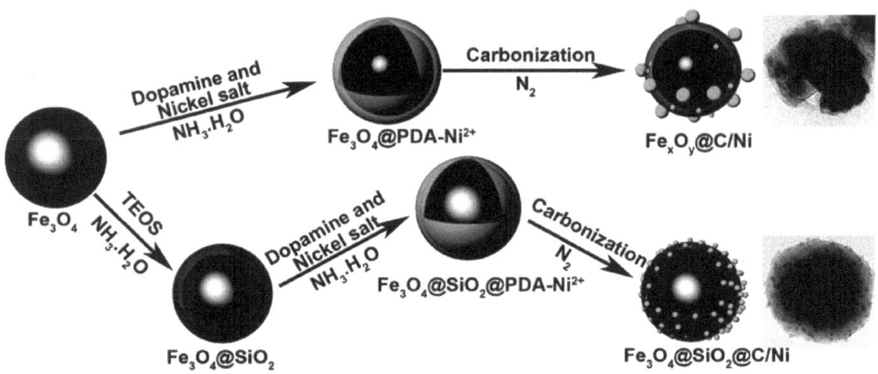

Figure 7.4 The synthetic procedures of $Fe_3O_4@SiO_2@C/Ni$ and $Fe_xO_y@C/Ni$ composites fabrication. Reprinted with permission from [105]. Copyright (2016) Royal Society of Chemistry.

The Fe_3O_4/SiO_2 nanoparticles were often synthesized through a sol-gel method. During the procedure, the addition of tetraethoxysilane (TEOS), the reactants of SiO_2, in aqueous solution could be optimized for preparing the magnetic nanoparticles with a narrow distribution and the proper thickness of the SiO_2 shell [104]. The Fe_3O_4@SiO_2 nanoparticles were synthesized via an ultrasound-assisted surfactant-template process with coprecipitation or microemulsion synthesized Fe_3O_4 as the core and TEOS as the silica shell [105–107]. The prepared Fe_3O_4@SiO_2 nanospheres could still maintain the uniform core–shell structure in the following process, but not the Fe_3O_4 nanoparticles without the SiO_2 layer (Figure 7.4) [105]. Zhang et al. coated a Ni^{2+}-polydopamine (PDA) complex onto the Fe_3O_4@SiO_2 surface, where the Ni^{2+} could be reduced to nickel nanoparticles by carbonization. In the process, the intermediate layer of SiO_2 between Fe_3O_4 and Ni^{2+}-PDA ensured the reduction of Ni^{2+} by PDA by preventing the electron transfer from PDA to Fe_3O_4. The thickness of SiO_2 could tune the size and density of nickel nanoparticles. The hybrid composites could more effectively catalyze the reduction of 4-nitrophenol and effectively separate the His-tagged proteins via the interaction between Ni and histidine [105].

7.4.2 Organic Modification

7.4.2.1 Modification by small organic molecules

Magnetic nanoparticles can be easily functionalized by surface decorations with hydrophilic, hydrophobic, or amphiphilic materials. Non-polymeric organic molecular stabilizers including ethanol [108], organic acids (alkane sulfonic acids, alkanoic acids, oleic acid) [109, 110], silane [111] and thiol are commonly used for the surface modification of magnetic nanoparticles. Furthermore, surfactants such as sodium oleate, dodecylamine, and sodium carboxymethylcellulose, phosphoryl vitamin B, and β-cyclodextrin are often used to improve the aqueous dispersity of magnetic nanoparticles.

By these surface modifications, hydrophobic nanoparticles can be converted into hydrophilic ones, and vice versa. The conversion can be achieved via two methods: surface coating and ligand exchange reaction. For the surface coating, for example, Yang et al. used the host–guest interaction between α-cyclodextrin and oleic acid to introduce a hydrophilic coating onto the hydrophobic magnetic nanoparticles [112].

The ligand exchange reactions are another common strategy to get ultrafine and water-dispersible superparamagnetic nanoparticles. This method generally includes two steps. First, the hydrophobic magnetic nanoparticles are synthesized in organic solvents using hydrophobic materials (e.g., oleic acid) as a dispersant. Second, the hydrophobic dispersant can be replaced by the ligand exchange to create a hydrophilic surface. For example, the ligand exchange between oleic acid and dopamine and Tiron (4,5-dihydroxy-1,3-benzenedisulfonic acid disodium salt) could convert the hydrophobic nanoparticles into the hydrophilic type [113]. Sun et al. used 11-amino undecanoic acid tetramethylammonium, a bipolar surface active agent, to replace oleic acid coating on the surface of magnetic nanoparticles, resulting in the hydrophilic magnetic nanoparticles [114].

It was reported that 2,3-dimercaptosuccinic acid (DMSA) served as a ligand exchange agent to convert the monodisperse hydrophobic magnetic nanoparticles into the hydrophilic nanoparticles [115]. The DMSA-modified magnetic nanoparticles were stable in the water and PBS and could be further engineered to a high-performance antibody probe, and used as an MRI agent for the diagnosis of breast cancer cells.

7.4.2.2 Modification by polymers

Although small molecular ligands have the advantage of not causing dramatic changes of the nanoparticle morphology, the surface modification using small molecules may not provide sufficient stabilization for the coated magnetic nanoparticles. To address the issues, the magnetic nanoparticles may be surface modified by macromolecules, mostly polymers including natural and synthetic polymers. Methods for polymeric modification include the in situ coating, post-synthesis adsorption, and post-synthesis grafting. The in situ coating is a process that the polymers are attached to the surface of nanoparticles during the formation of magnetic nanoparticles. The post-synthesis adsorption or grafting is to physically adsorb or chemically graft the polymers onto the surface of nanoparticles after the preparation of magnetic nanoparticles. Natural polymers mainly include dextran, starch, chitosan, chitin, and hyaluronic acid. Synthetic polymers mainly include polyethylene glycol (PEG), polyethyleneimine (PEI), polyvinyl alcohol (PVA), poly (N-isopropylacrylamide), polyvinyl pyrrolidone (PVP), polyacrylic acid (PAA), and their copolymers.

Chitosan is a class of nontoxic, biocompatible, cationic, and hydrophilic polymers, which have been actively investigated for drug delivery. Due to the abundance of nature, biocompatibility, and ease of functionalization, chitosan and its derivatives have been widely used as surface modifiers to prepare polymeric nanoparticles. A cellular imaging nanoprobe was designed by the surface modification of the γ-Fe_3O_4 nanoparticles with the FITC-labeled chitosan [116]. This magneto-fluorescent nanoagent showed high affinity with the cancer cells (SMMC-7721).

Dextran is a group of polymers widely used in the magnetic nanocomposite coatings due to its biocompatibility and polar interactions (chelation and hydrogen bonding). The polar interactions render high affinity between dextran and iron oxide surface. So far, the most widely applied magnetic nanocomposites in clinical are products stabilized by dextran and its derivatives [117]. Osborne et al. developed a simple, rapid, and reproducible microwave-assisted method for the synthesis of biocompatible iron oxide nanoparticles [118]. The prepared nanoparticles exhibited superparamagnetic properties and, with the dextran coating, their water dispersibility and biocompatibility were significantly improved, benefiting their in vivo application.

PEG, the synthetic linear polymer, is featured by good biocompatibility, and often used as the end grafting materials. The hydrophilic PEG chain can improve the stability and blood circulation of the magnetic nanoparticles by shielding the nanoparticles from the RES capture. It has been reported that a biocompatible and biodegradable Fe_3O_4-PEG-LAC-chitosan-PEI system could deliver siRNA to the human breast cancer (MCF-7) and leukemia (K562) cells. The system showed efficient and stable affinity for survivin siRNA and significant cytotoxicity to MCF-7 and K562 cells [119].

Hyaluronic acids (HA), the acidic mucopolysaccharide, show a variety of important physiological functions such as regulation of vascular wall permeability, lubrication of joints, and wound-healing [120]. HA is also a ligand of tumor-associated CD44 receptor. So far, HA has been widely used to modify magnetic nanoparticles. The HA-modified Fe_3O_4 NPs can specifically bind with and then be endocytosed by the CD44 overexpressing cancer cells, and serve as efficient probes for tumor-targeted MRI [114].

Moreover, PEI-based polymers have been used for gene delivery due to their ability to compact DNA via electrostatic interactions and the "proton sponge effect" for the endosomal escape and trafficking their cargo DNA into the nucleus. With the tunable amine-based conjugation chemistry, the PEI-stabilized Fe_3O_4 NPs have been prepared and used for gene delivery [121].

Magnetic nanoparticles are one of the most well-established nanomaterials and often used as alternative contrast enhancing agents. Xu and coworkers summarized the development of two straightforward applications of biofunctional magnetic nanoparticles—manipulating proteins and cells. The applications of magnetic nanoparticles depend on the molecular interactions between the functional molecules on the magnetic nanoparticles and the intended biological targets. Therefore, the strategies for surface functionalization of magnetic nanoparticles were examined. Because protein interactions are critical for biological functions, the use of magnetic nanoparticles for manipulating proteins should be an attractive application. The important "targeting" function is achieved by the introduction of the targeting ligands via physical binding or chemical conjugation with the surface functional groups onto the magnetic nanoparticles. Amino, carboxyl, aldehyde, and mercapto groups are commonly used anchors for chemical modification. Targeting ligand for magnetic nanoparticles can be categorized into five groups: (1) small molecules (e.g., folic acid, methotrexate); (2) targeting peptides (e.g., tumor-homing peptides); (3) aptamers (e.g., Thrm-A and Thrm-B DNA aptamers binding to human alpha-thrombin); (4) proteins (e.g., transferrin, and luteinizing hormone releasing hormone (LHRH)); and (5) antibodies (like herceptin, rituximab) (Figure 7.5).

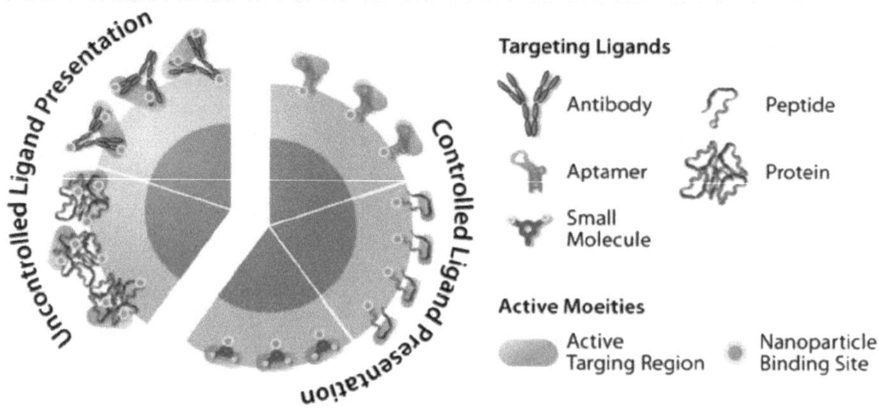

Figure 7.5 Illustration of the supermolecular assembly and presentation of targeting antibodies, proteins, peptides, aptamers, and small molecules on the surfaces of magnetic nanoparticles. Reprinted with permission from [93].

7.5 Biomedical Applications

7.5.1 Drug Delivery

Clinically used chemotherapeutics and many other promising drug candidates confront with the same major problems of poor targeting and severe side effects [122, 123]. Magnetic nanoparticles have attracted more and more attention [124] and their multiple functions have made them one of the promising smart systems for targeted drug delivery [55, 125]. The magnetic nanocrystals, referred to unmodified magnetic nanoparticles, are inclined to aggregate because of their excessive surface energy. The surface modification can effectively improve the stability of the magnetic nanoparticles. A great deal of organic and inorganic materials have been explored as coating agents for surface functionalization of magnetic nanoparticles [126].

The tiny magnetic nanoparticles may also be encapsulated inside a larger nanoparticle, rendering the thus-prepared nanoparticles the magnetic targeting ability [127]. These modified magnetic nanoparticles can be the excellent nanocarriers for delivery of various drug molecules.

7.5.1.1 Magnetically responsive passive targeting

Tumors are usually characterized by the leaky vasculature due to their malignant growth. The newly formed tumor vessels are abnormal in architecture with many gaps between endothelial cells. These gaps lead to the leakage and accumulation of nanoparticles or macromolecules into the tumor tissues. This is the so-called enhanced permeability and retention (EPR) effect [128]. Based on the EPR effect, nano-drug delivery systems have been designed to improve cancer treatment outcomes by enhancing drug accumulation in tumor.

Magnetic nanoparticles can be encapsulated in the lipid membrane or aqueous core of liposomes to form the so-called magnetoliposomes [129]. Many advances have been achieved in the field of magnetoliposomes [130, 131]. Recently, the combination therapy efficacy of gemcitabine-containing magnetoliposomes and oxaliplatin-containing magnetoliposomes was investigated in treating breast tumor. The combination strategy significantly inhibited the growth of the MCF-7 breast cancer in vitro and in vivo, by activating the apoptotic signaling pathway [132]. Moreover, the release of the encapsulated drugs can be effectively controlled by the magnetic nanoparticles under an external magnetic field [133]. The amine of doxorubicin can be protonated under acidic conditions, and thereby through a pH gradient method, doxorubicin was encapsulated in the aqueous core of magnetoliposomes containing the hydrophobic magnetic nanoparticles inside the bilayer membranes. In a specific time frame, the Dox-loaded magnetoliposomes displayed very minor cytotoxicity, similar to drug-free magnetoliposomes, demonstrating little drug leakage from the core of magnetoliposomes. However, the Dox-loaded magnetoliposomes showed significant cytotoxicity when exposed to an external alternating magnetic field, due to the triggered drug release [134].

A novel supermolecular delivery system with the sequential targeting ability for magneto-chemotherapy has been designed [135]. The delivery system was a type of double-emulsion capsule (DEC), composed of magnetic nanoparticles and polyvinyl alcohol polymers with or without PEG-PLGA (named P-DEC or DEC). The paclitaxel-HA conjugates (supermolecules) were encapsulated in the aqueous cores of nanocapsules. The drug conjugates were prepared via the chelation of paclitaxel with cyclodextrin decorated on HA. The DEC and P-DEC system showed the sensitivity to high-frequency magnetic field. After systemic administration, the delivery system accumulated at the tumor site via the EPR effect, followed by the disintegration of the nanocapsules and drug release under a high-frequency alternating magnetic field. Via CD44 receptor-mediated endocytosis, the drug conjugates could enter and kill the cancer cells efficiently (Figure 7.6).

Recently, a Janus nanocomposite combining a magnetic Fe_3O_4 head with a mesoporous silica body was prepared [136]. The system showed a "nano-bullet" type with doxorubicin encapsulated in the silica pores. The Janus nanocomposites remarkably suppressed the growth of subcutaneous and orthotropic liver tumor in a mouse model and significantly reduced the systemic toxicity (Figure 7.7).

Using magnetic nanoparticles to deliver anticancer vaccines has also been studied. For example, the magnetic nanoparticle-mediated delivery of mucin-1 (MUC1) glycopeptide, a tumor-associated antigen, could effectively stimulate B cells to produce the desired level of antibodies [137].

7.5.1.2 Magnetically responsive active targeting

To enhance the accumulation of therapeutic agents in the tumor site, the active targeting strategies have been developed by conjugating the ligands or antibodies on the nanoparticles for recognition of the receptors or antigens which are overexpressed by the tumor cells.

The hydrophobic oleic acid-modified magnetic nanoparticles and PTX were encapsulated in the micelles composed of PLGA-PEG and PCL-PEG. The anti-cancer efficacy of the drug delivery systems was evaluated on the mice bearing CT26 colon tumor. The combination of RGD-mediated targeting and magnetic nanoparticle-mediated targeting significantly improved the therapeutic efficacy [138].

Porous materials such as mesoporous silica can modify magnetic nanoparticles. The therapeutic agents thus can be encapsulated in the mesoporous silica coating layer on magnetic nanoparticles. The folate receptor-targeted, magnetic-mesoporous silica system has been studied for co-delivery of doxorubicin and VEGF shRNA [139]. In this system, doxorubicin was encapsulated in the pores of magnetic-mesoporous silica

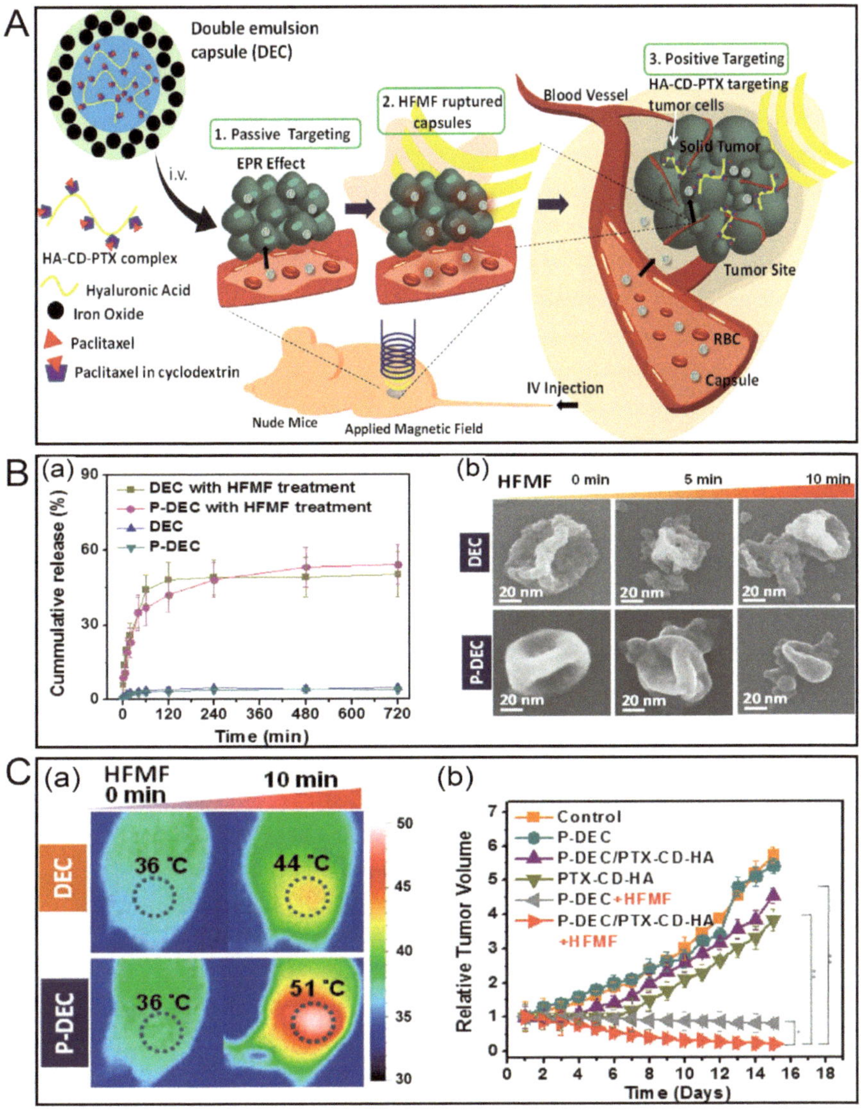

Figure 7.6 (A) The scheme of the supermolecular delivery system with sequential targeting ability. (B) The cumulative release of supermolecules from the DEC or P-DEC with/without the high-frequency magnetic field (HFMF) treatment and the morphology changes of DEC and P-DEC with the HFMF treatment. (C) The magneto-chemotherapy efficacy of the supermolecular delivery system. Reprinted with permission from [135].

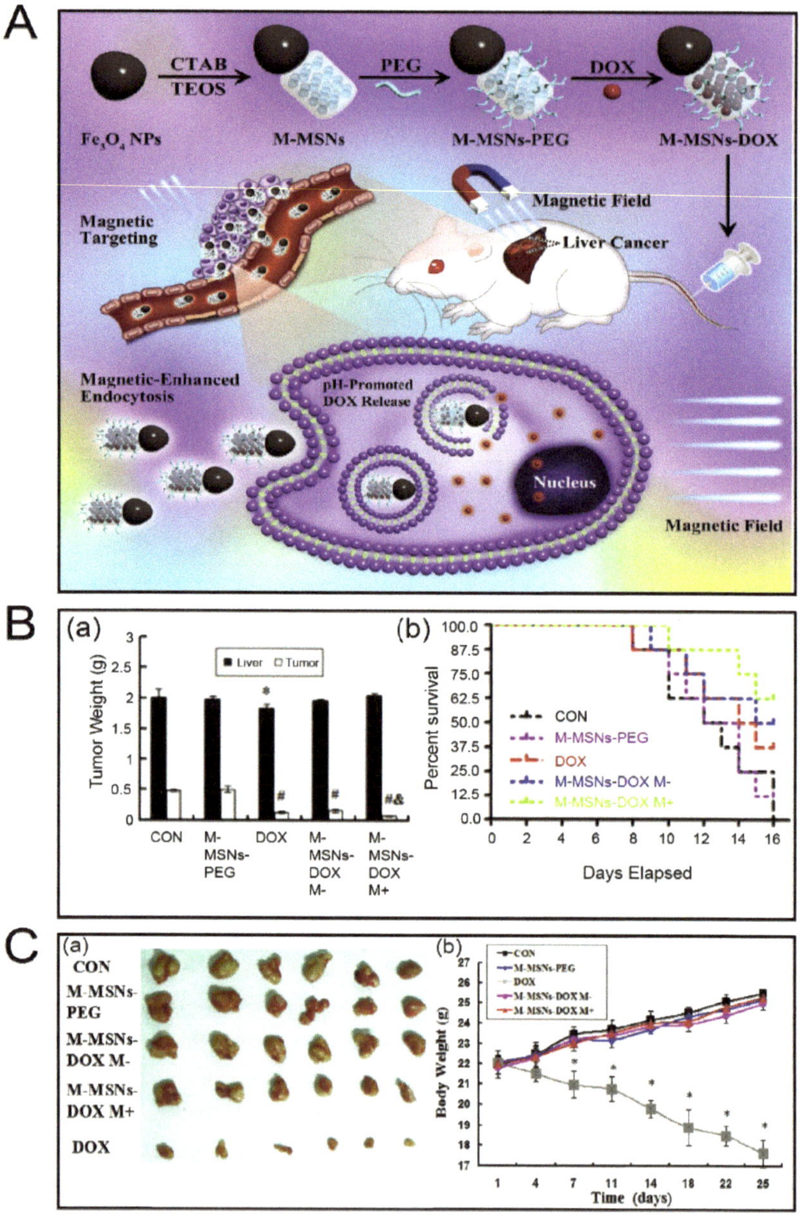

Figure 7.7 (A) The scheme of the synthesis and delivery of Janus nanocomposites. (B) The therapeutic efficacy of Janus nanocomposites in the orthotropic liver tumor in mice. (C) The therapeutic efficacy of Janus nanocomposites in the subcutaneous liver tumor in mice. Reprinted with permission from [136].

particles produced by a sol-gel procedure. The folate-modified PEI was then capped on the surface of doxorubicin-loaded magnetic-mesoporous silica particles as a blocker against drug leakage. The surface coated PEI was further used to bind with and compact the VEGF shRNA, thus forming the drug and gene co-delivery system. The system was effectively endocytosed by HeLa cells and down-regulated VEGF expression at both gene and protein levels, and thereby significantly inhibited the growth of cancer cells (Figure 7.8).

In addition, utilizing cationic materials to modify magnetic nanoparticles can deliver anionic nucleic acid therapeutics through electrostatic interactions that facilitate drug loading and cellular uptake. PEI is the most commonly used cationic polymer for surface coating of magnetic nanoparticles. PEI-coated magnetic nanoparticles can deliver gene drugs [140, 141]. L. F. Wang et al. designed a CD44 receptor-mediated magnetic system to deliver miRNA-128 for gene therapy of glioblastoma. The chondroitin sulfate, a ligand of CD44 receptor, was conjugated to the PEI polymers. The chondroitin sulfate-conjugated PEI was used to coat on the surface of magnetic nanoparticles to form the magneto-gene delivery system. The system showed higher gene expression and less cytotoxicity in 293T, CRL5802, and U87-MG cells compared to PEI- and the commercial magnetofection agent PolyMag-based gene delivery systems. High tumor accumulation was observed in the regions where an external magnetic field was applied on the mice bearing the U87-xenografted tumor [141].

The lactoferrin (Lf)-modified magnetic double emulsion nanocapsules (Lf-MDCs) were prepared with polyvinyl alcohol (PVA), polyacrylic acid (PAA) and iron oxide nanoparticles for glioma therapy [142]. The hydrophilic doxorubicin (Dox) and the hydrophobic curcumin (Cur) were encapsulated in the aqueous core and hydrophobic shell, respectively. The dual-drug-loaded nanocapsules were significantly endocytosed by the RG2 cells under the magnetic field and lactoferrin-mediated low-density lipoprotein-related protein-1 (LRP-1) targeting, resulting in the synergistic cytotoxicity. Moreover, after an intravenous injection, the dual-drug-loaded nanocapsules accumulated at the glioma under an external magnetic field and efficiently suppressed the tumor growth in the subcutaneous glioma-bearing mice (Figure 7.9).

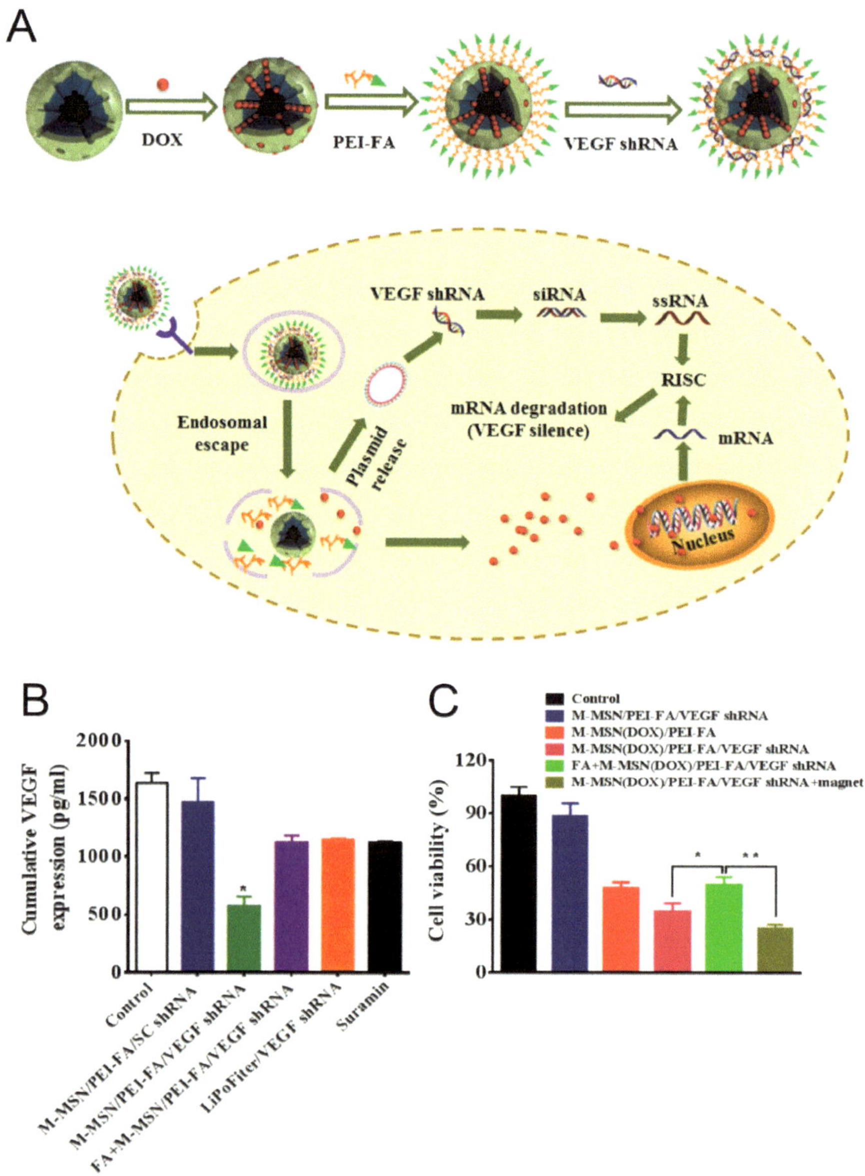

Figure 7.8 (A) The scheme of the magnetic-mesoporous silica drug delivery system for co-delivery of doxorubicin and VEGF shRNA. (B) The knockdown of the VEGF protein in HeLa cells. (C) In vitro cytotoxicity of the system in HeLa cells. Reprinted with permission from [139]. Copyright (2016) American Chemical Society.

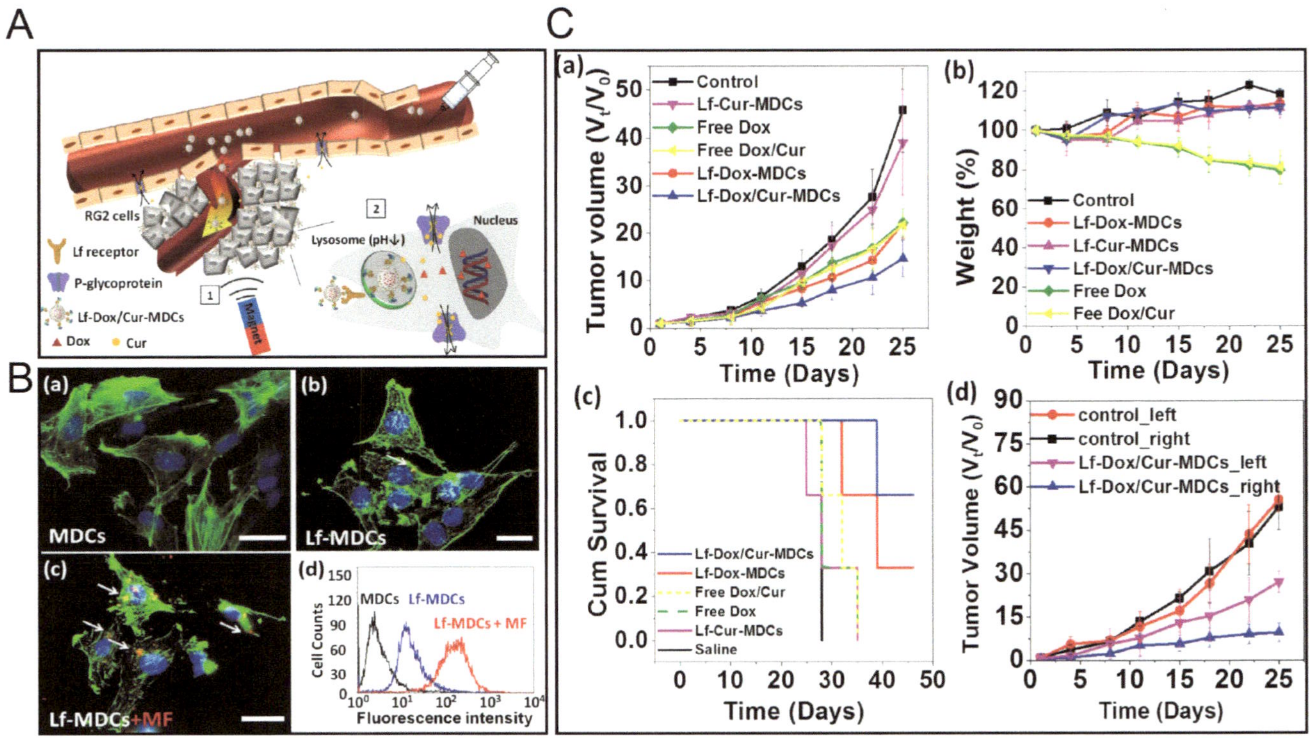

Figure 7.9 (A) Lactoferrin-modified magnetic double emulsion capsules (Lf-MDCs) co-loading Dox and Cur were designed for synergistic therapy. (B) Cellular uptake micrographs of (a) MDCs, (b) Lf-MDCs, and (c) Lf–MDCs in RG2 cells under an external magnetic field for 2 h. (C) The therapeutic efficacy of Lf-MDCs. Reprinted with permission from [142].

L. Yang et al. prepared a magnetic nanoparticle system for delivery of doxorubicin using a ligand (IGF-1) targeting the growth factor-1 receptor (IGF-1R). The system could penetrate the stroma and reach the tumor cells via IGF-1R mediated endocytosis on a pancreatic tumor xenografted animal model [143].

Moreover, magnetic nanoparticles can be used to control the death of cancer cells by modulating apoptosis. Extrinsic apoptosis signaling pathways initiated by death receptors are widely investigated to treat cancers. Clustering of death receptors is the main activator of extrinsic apoptosis signaling pathways. For example, TNF-related apoptosis inducing ligand (TRAIL) can significantly induce the formation of death receptor clusters and cell apoptosis. Anti-death receptor 4 (DR4) antibody conjugated zinc-doped iron oxide magnetic nanoparticles could effectively turn on the apoptosis signaling in the DR4-overexpressed DLD-1 colon cancer cells, and under an external magnetic field they could as well induce morphologic changes in a zebrafish model expressing zebrafish ovarian TNF receptor, which functioned similarly to human DR4 [144, 145]. In this system, the antibody-modified magnetic nanoparticles could bind to DR4 on the cancer cells, and DR4 thus formed the receptor clusters under the applied magnetic field to activate the apoptosis signaling of cancer cells.

7.5.1.3 Multi-stimuli-responsive drug delivery

Apart from the abovementioned magnetic drug delivery systems, the prodrug-type and microenvironment-responsive magnetic drug delivery systems have also been developed by chemical conjugation of drug molecules with nanoparticles via a responsive linkage against pH, enzyme, redox, or heat. In theory, prodrug-like systems do not display pharmacological activity, unless the systems are in response to the disease microenvironments or external stimuli.

7.5.1.3.1 Magnetic and pH dual-responsive targeting

The tumor microenvironment is acidic compared to the normal tissues due to the active metabolism of cancer cells and accumulation of acidic metabolites [146, 147]. The endosomes and lysosomes of tumor cells are even more acidic [148]. Thus, acid-cleavable bonds can trigger local drug release in the tumor microenvironment.

Among the acid cleavable bonds, hydrazone bonds are the most commonly used linkage for drug delivery due to their reducible yield and simple synthesis [149–151]. Other widely used acid cleavable bonds include imines, acetals/ketals, oximes, 1,3,5-triazaadamantanes, and boronate esters [152–154]. The magnetic pH-responsive drug delivery system based on magnetic nanocubes and hydrazone bonds was designed by K. Y. Cai and coworkers [155]. The magnetic nanocubes were coated with poly (methyl methacrylate) on the surface and then the hydrazine tailor was introduced to the poly (methyl methacrylate). The hydrazine tailor was changed into the hydrazone bond through reaction with the carbonyl group of doxorubicin. The system showed the pH-responsive drug release and cytotoxicity profiles. The in vivo results showed that the system could accumulate at the tumor site and dramatically inhibit the tumor growth with the application of an external magnetic field (Figure 7.10) [155].

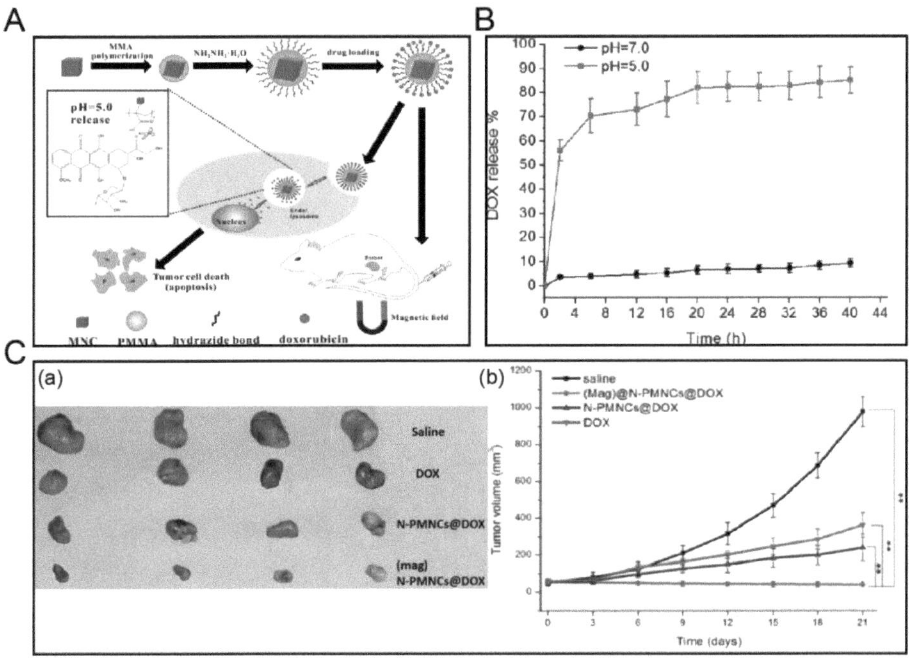

Figure 7.10 (A) The schematic illustration of the magnetic pH-responsive drug delivery system for tumor targeting. (B) In vitro doxorubicin release profiles at pH 7 and 5. (C) The anti-tumor efficacy in mice. Reprinted with permission from [155]. Copyright (2014) American Chemical Society.

7.5.1.3.2 Magnetic and redox dual-responsive targeting

Cancer cells usually contain high levels of glutathione in the intracellular compartments to rescue them from chemotherapy, resulting in drug resistance [156, 157]. High levels of glutathione create a reducing microenvironment in cancer cells that can be exploited to trigger drug release. Disulfide-based magnetic delivery systems have been used in the delivery of chemotherapeutics and genes [76, 158, 159]. A system composed of magnetic nanoparticles coated by disulfide-cross-linking PEI has been designed to respond to the intracellular reducing environment for gene delivery [160]. Another example was the redox-responsive magnetic nanoparticles that were used to deliver O6-benzylguanine (BG) for inhibition of DNA repair protein O6-methylguanine-DNA methyltransferase (MGMT). MGMT is associated with the resistance of tumor cells to temozolomide (TMZ) [159]. By convection-enhanced delivery, the system efficiently distributed into the brain in an orthotopic glioblastoma multiforme (GBM) xenografted murine model. The combination of this system and TMZ effectively prolonged the median overall survival. In addition, the tissues in the inflammatory regions can also produce abundant hydrogen peroxide and reactive oxygen species, by which the oxidization-responsive delivery of anti-inflammatory agents could be achieved [161].

7.5.1.3.3 Magnetic and enzyme dual-responsive targeting

On the basis of the overexpression of tumor-associated enzymes in the tumor microenvironment, the enzyme-activatable prodrug-type systems were developed to convert nontoxic drugs to toxic ones in the malignant tumor tissues for improving the therapeutic efficacy [162]. Matrix metalloproteinases (MMPs) are a family of tumor-associated enzymes, which are overexpressed in many types of solid malignant tumors and can selectively cleave their substrates [163]. Daldrup-Link et al. designed a multifunctional theranostic nanoparticle system composed of the magnetic nanoparticles, ferumoxytol, and MMP14-sensitive peptide-azademethylcolchicine (ICT) conjugates [164]. The system showed remarkable cytotoxicity in the MMP14 positive breast cancer cells, but not in the MMP14 negative fibroblasts. After intravenous administration, the nanoparticles significantly accumulated in the tumor and induced tumor cell necrosis (Figure 7.11).

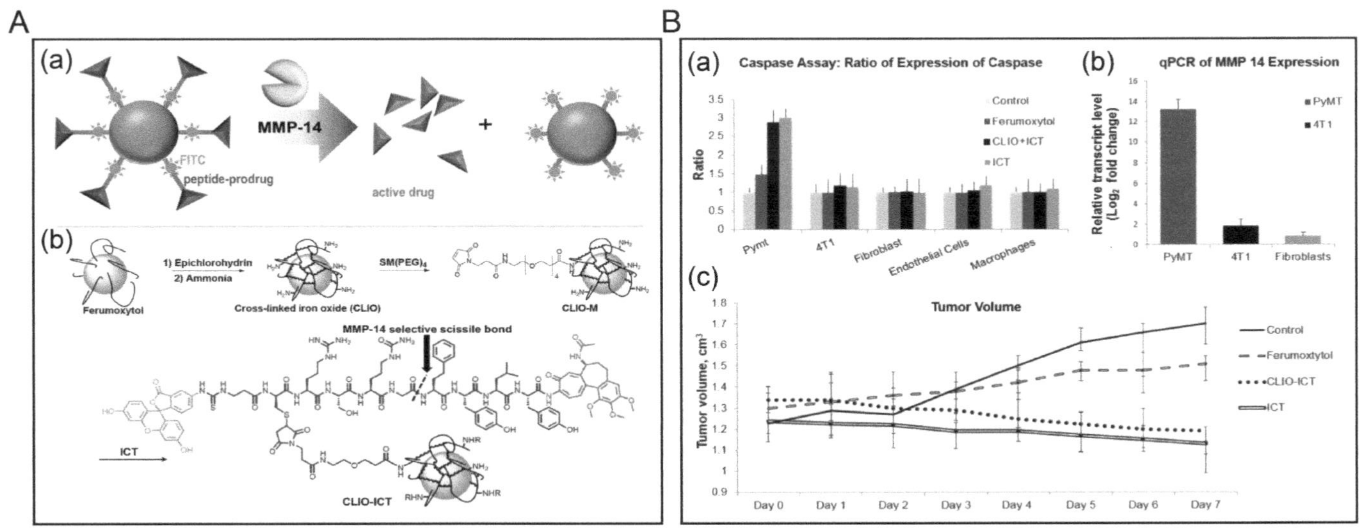

Figure 7.11 (A) The schematic illustration of the MMP-14 responsive magnetic nanoparticles. (B) The antitumor activity. Reprinted with permission from [164].

Furthermore, a lysosomally cleavable tetrapeptide linker was conjugated to the gemcitabine-loaded magnetic nanoparticles that were modified with the urokinase plasminogen activator receptor targeting ligand [165]. The nanoparticles showed the pH- and lysosomal enzyme-dependent gemcitabine release. Intravenous injection of the system significantly inhibited the tumor growth in the orthotopic human pancreatic tumor xenografted mice. In addition, the phosphatases and esterases have also been investigated as the stimuli to enhance the magnetic nanoparticle-based drug delivery [166–168].

7.5.1.3.4 Magnetic and thermo-responsive targeting

Thermosensitive magnetic drug delivery systems are advantageous for the ability of converting electromagnetic energy to heat under an alternating magnetic field [169, 170]. For example, the thermo-responsive magnetic micelles were designed for controlled release of the platinum complex [171]. The micelles were composed of the p (NIPAAm-co-AAm)-b-PCL and hydrophobic Pt (IV) complex. Furthermore, the hydrophobic Fe_3O_4 nanoparticles were encapsulated in the hydrophobic cores of the micelles. The in vitro hard X-ray fluorescence microscopy results showed that the magnetic micelles accumulated in the specific subcellular regions. The in vivo results showed that the magnetic micelles could target the tumor site under an external magnetic field.

In addition, the glucose-responsive magnetically targeted drug delivery system has also been developed for the treatment of type II diabetes [172].

7.5.2 Magnetic Hyperthermia

Hyperthermia refers to the use of heat produced from various heating strategies (e.g., microwave, ultrasound, radio wave, electromagnetic field, and infrared radiator) to kill cancer cells and impede tumor growth, which is based on the mechanism that cancer cells are more sensitive to high temperature than normal tissue cells [173, 174]. Magnetic nanoparticles are the appealing heat generators that produce heat through the internal dipole rotation and physical particle rotation when they are exposed to an external alternating magnetic field [175, 176]. Jordan et al. in 1993 revealed that the superparamagnetic crystal suspension could absorb the energy of an oscillating magnetic field and convert into heat, and based on that, the concept of magnetic hyperthermia was proposed [177]. Compared with the conventional hyperthermia methods such as phototherapy and radiofrequency ablation, magnetic hyperthermia elicits fewer side effects because it may achieve local heating in the tumor.

SAR is a crucial parameter of magnetic nanoparticles for clinical trials; higher SAR values imply lower dose. The factors governing SAR include the composition, size, and shape of the magnetic nanoparticles, as well as the external magnetic field. Guardia et al. prepared the magnetic nanocubes between 13 and 40 nm via a one-pot synthesis [178]. When the size was 19 nm, the magnetic nanocubes had the highest SAR and showed the hyperthermia effect in KB cells. The oleic acid-based magnetic nanocubes with an average size of 30 nm, coated with chitosan-DOPA, had higher specific loss power (SLP) than the commercial product, Feridex, and exhibited superior magnetic heating ability. The chitosan-DOPA conjugated magnetic nanocubes offered higher in vitro cell apoptosis and in vivo anti-cancer activity on the A549 tumor xenografted mice than Feridex (Figure 7.12) [179].

Magnetic hyperthermia is commonly combined with other currently used treatment methods. K. S. Kim et al. designed the multifunctional photosensitizer-conjugated hyaluronic acid-coated magnetic nanoparticles (AHP@MNPs) [180]. The surface coating was processed through electrostatic interactions and hydrophobic interactions. The CD44 receptor-targeting ligand hyaluronic acid enhanced the uptake efficiency of AHP@MNPs in the tumor cells via the receptor-mediated endocytosis. The hyperthermia induced by the combination use of Fe_3O_4 and photosensitizer significantly enhanced the anti-tumor effect than those induced by either Fe_3O_4 or photosensitizer on an animal tumor model.

Recently, magnetic nanoparticles have also been studied for the photothermal therapy. Wilhelm et al. developed a magnetic nanocube system with photothermal capacity. The system could simultaneously respond to an alternating magnetic field and 808 nm near-infrared laser irradiation. The dual action effectively induced the apoptosis of cancer cells. When the single-mode treatment (either magnetic or laser) was applied, the solid tumor was partially inhibited. But the complete inhibition of the tumor growth was achieved by the dual mode treatment (Figure 7.13) [47].

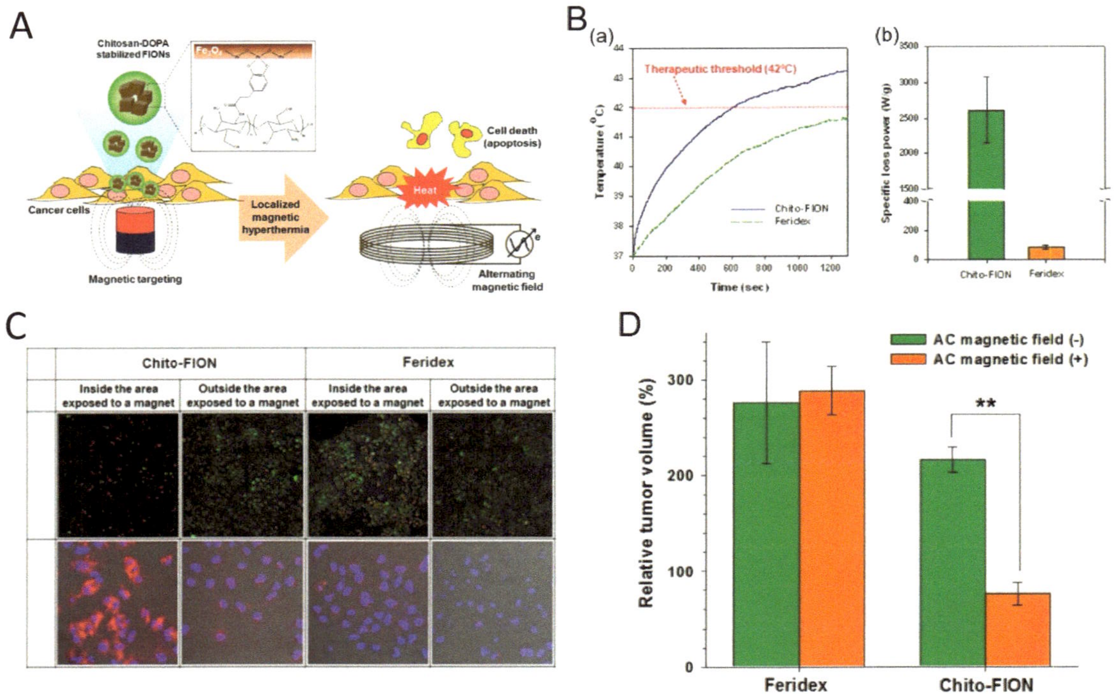

Figure 7.12 (A) The schematic illustration of chitosan-DOPA conjugated magnetic nanocubes for the localized magnetic hyperthermia. (B) The temperature change under an alternating magnetic field and SLP values. (C) The live and dead cells staining of A549 cells after treatment by chitosan-DOPA conjugated magnetic nanocubes and Feridex with/without an external magnetic field. (D) The hyperthermia-induced anticancer activity of chitosan-DOPA conjugated magnetic nanocubes. Reprinted with permission from [179]. Copyright (2012) American Chemical Society.

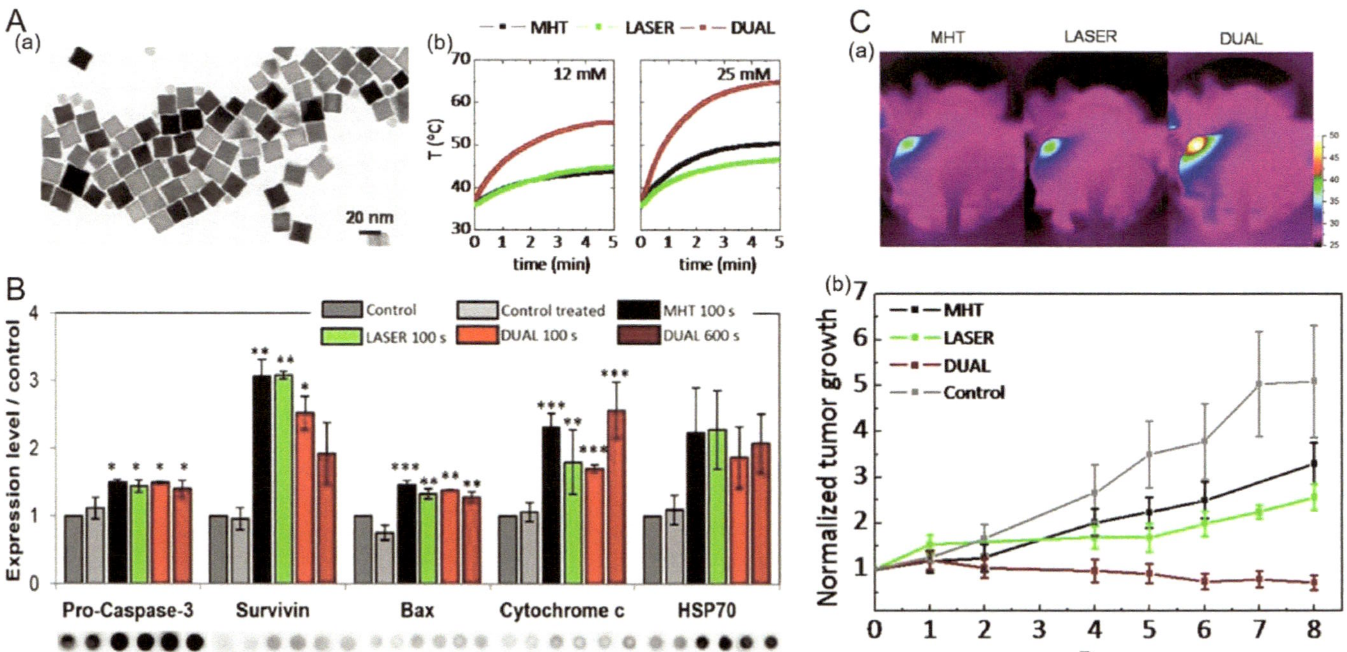

Figure 7.13 (A) The transmission electron microscopy image of the magnetic nanocubes and the temperature curves of magnetic nanocubes (Fe: 12 or 25 mM) under magnetic field, laser, or both. (B) The apoptosis of cancer cells under magnetic field, laser, or both. (C) The in vivo tumor inhibitory effects of nanocubes. Reprinted with permission from [47]. Copyright (2016) American Chemical Society.

It was reported that hyperthermia could cause up-regulation of the expression of heat shock proteins, thus developing the resistance against the treatment. The combination of heat shock protein inhibitors and magnetic hyperthermia could restore the sensitivity of cancer cells to hyperthermia [181].

7.6 Conclusions

Magnetic nanoparticle-based systems display great potential to treat various diseases. Although many preclinical studies of magnetically responsive nanomedicines have been reported, there are a few problems that need to be addressed prior to their clinical translation. The future interests of magnetic nanoparticle-based systems will be on the development of new synthetic methods to reproducibly and economically prepare magnetic nanoparticles and the effective surface modification strategies to improve the nanomedicines' magnetic and physicochemical properties. Moreover, how to improve the specificity and efficiency of magnetically responsive drug delivery is a key issue for the improved treatment outcomes and reduced adverse effects. Hyperthermia is a promising application for magnetic anoparticles, particularly when combined with other types of therapies. Furthermore, the smart nano systems that can simultaneously respond to more than one stimuli besides magnetic field, i.e. dual or multi responsive/targeting, will improve the specificity and therapeutic outcomes of magnetically responsive nanomedicines.

References

1. H. Kratz, D. Eberbeck, S. Wagner, M. Taupitz, J. Schnorr, Synthetic routes to magnetic nanoparticles for MPI, *Biomedizinische Technik. Biomedical Engineering*, 58(6) (2013) 509–515.
2. O. L. Gobbo, K. Sjaastad, M. W. Radomski, Y. Volkov, A. Prina-Mello, Magnetic nanoparticles in cancer theranostics, *Theranostics*, 5(11) (2015) 1249–1263.
3. J. R. Heath, Nanotechnologies for biomedical science and translational medicine, *Proceedings of the National Academy of Sciences*, 112(47) (2015) 14436–14443.

4. A. M. El-Toni, M. A. Habila, J. P. Labis, Z. A. ALOthman, M. Alhoshan, A. A. Elzatahry, F. Zhang, Design, synthesis and applications of core–shell, hollow core, and nanorattle multifunctional nanostructures, *Nanoscale,* 8(5) (2016) 2510–2531.

5. G. Kandasamy, D. Maity, Recent advances in superparamagnetic iron oxide nanoparticles (SPIONs) for in vitro and in vivo cancer nanotheranostics, *International Journal of Pharmaceutics,* 496(2) (2015) 191–218.

6. S. Laurent, J.-L. Bridot, L. V. Elst, R. N. Muller, Magnetic iron oxide nanoparticles for biomedical applications, *Future Medicinal Chemistry,* 2(3) (2010) 427–449.

7. S. Behrens, I. Appel, Magnetic nanocomposites, *Current Opinion in Biotechnology,* 39 (2016) 89–96.

8. V. Valdiglesias, N. Fernandez-Bertolez, G. Kilic, C. Costa, S. Costa, S. Fraga, M. J. Bessa, E. Pasaro, J. P. Teixeira, B. Laffon, Are iron oxide nanoparticles safe? Current knowledge and future perspectives, *Journal of Trace Elements in Medicine and Biology: Organ of the Society for Minerals and Trace Elements,* 38 (2016) 53–63.

9. H. Zhao, H.-J. Cui, M.-L. Fu, A general and facile method for improving carbon coat on magnetic nanoparticles with a thickness control, *Journal of Colloid and Interface Science,* 461 (2016) 20–24.

10. D. Wang, J. Zhou, R. Chen, R. Shi, G. Zhao, G. Xia, R. Li, Z. Liu, J. Tian, H. Wang, Z. Guo, H. Wang, Q. Chen, Controllable synthesis of dual-MOFs nanostructures for pH-responsive artemisinin delivery, magnetic resonance and optical dual-model imaging-guided chemo/photothermal combinational cancer therapy, *Biomaterials,* 100 (2016) 27–40.

11. J. Cai, B. Gu, F. Cao, S. Liu, A transferrin-target magnetic/fluorescent dual-mode probe significantly enhances the diagnosis of non-small cell lung cancer, *Oncotarget,* 7(26) (2016) 40047–40059.

12. H. Wang, S. Zhou, Magnetic and fluorescent carbon-based nanohybrids for multi-modal imaging and magnetic field/NIR light responsive drug carriers, *Biomaterials Science,* 4(7) (2016) 1062–1073.

13. Y.-K. Huang, C.-H. Su, J.-J. Chen, C.-T. Chang, Y.-H. Tsai, S.-F. Syu, T.-T. Tseng, C.-S. Yeh, Fabrication of silica-coated hollow carbon nanospheres encapsulating Fe_3O_4 cluster for magnetical and MR imaging guided NIR light triggering hyperthermia and US imaging, *ACS Applied Materials & Interfaces,* 8(23) (2016) 14470–14480.

14. H. Vu-Quang, M. S. Vinding, T. Nielsen, M. G. Ullisch, N. C. Nielsen, J. Kjems, Theranostic tumor targeted nanoparticles combining

drug delivery with dual near infrared and 19 F magnetic resonance imaging modalities, *Nanomedicine: Nanotechnology, Biology and Medicine*, 12(7) (2016) 1873–1884.

15. R. Geevarghese, R. O'Gorman Tuura, D. Lumsden, M. Samuel, K. Ashkan, Registration accuracy of CT/MRI fusion for localisation of deep brain stimulation electrode position: An imaging study and systematic review, *Stereotactic and Functional Neurosurgery*, 94(3) (2016) 159–163.
16. P. Yang, F. Wang, X. Luo, Y. Zhang, J. Guo, W. Shi, C. Wang, Rational design of magnetic nanorattles as contrast agents for ultrasound/magnetic resonance dual-modality imaging, *ACS Applied Materials & Interfaces*, 6(15) (2014) 12581–12587.
17. A. Kiani, A. Esquevin, N. Lepareur, P. Bourguet, F. Le Jeune, J.-Y. Gauvrit, Main applications of hybrid PET-MRI contrast agents: A review, *Contrast media & Molecular Imaging*, 11(2) (2015) 92–98.
18. K. Gariani, J. Gariani, G. Amzalag, B. M. Delattre, O. Ratib, V. Garibotto, Hybrid PET/MRI as a tool to detect brown adipose tissue: Proof of principle, *Obesity Research & Clinical Practice*, 9(6) (2015) 613–617.
19. D. Cheng, X. Li, C. Zhang, H. Tan, C. Wang, L. Pang, H. Shi, Detection of vulnerable atherosclerosis plaques with a dual-modal single-photon-emission computed tomography/magnetic resonance imaging probe targeting apoptotic macrophages, *ACS Applied Materials & Interfaces*, 7(4) (2015) 2847–2855.
20. G. Trencsényi, S. Barna, I. Garai, Hybrid imaging: Clinical evidence, opportunities, *Orvosi Hetilap*, 156(52) (2015) 2110–2115.
21. K. Roy, R. K. Kanwar, J. R. Kanwar, LNA aptamer based multi-modal, Fe_3O_4-saturated lactoferrin (Fe 3 O 4-bLf) nanocarriers for triple positive (EpCAM, CD133, CD44) colon tumor targeting and NIR, MRI and CT imaging, *Biomaterials*, 71 (2015) 84–99.
22. D. H. Kim, Y. Guo, Z. Zhang, D. Procissi, J. Nicolai, R. A. Omary, A. C. Larson, Temperature-sensitive magnetic drug carriers for concurrent gemcitabine chemohyperthermia, *Advanced Healthcare Materials*, 3(5) (2014) 714–724.
23. T. Li, X. Shen, Y. Geng, Z. Chen, L. Li, S. Li, H. Yang, C. Wu, H. Zeng, Y. Liu, Folate-functionalized magnetic-mesoporous silica nanoparticles for drug/gene codelivery to potentiate the antitumor efficacy, *ACS Applied Materials & Interfaces*, 8(22) (2016) 13748–13758.
24. T. Zhang, G. Qu, Magnetic nanosphere-guided site-specific delivery of vascular endothelial growth factor gene attenuates restenosis

in rabbit balloon-injured artery, *Journal of Vascular Surgery*, 63(1) (2016) 226–33 e1.

25. R. Wang, Y. Hu, N. Zhao, F. J. Xu, Well-defined peapod-like magnetic nanoparticles and their controlled modification for effective imaging guided gene therapy, *ACS Applied Materials & Interfaces*, 8(18) (2016) 11298–11308.
26. S. Majidi, F. Zeinali Sehrig, M. Samiei, M. Milani, E. Abbasi, K. Dadashzadeh, A. Akbarzadeh, Magnetic nanoparticles: Applications in gene delivery and gene therapy, *Artificial Cells, Nanomedicine, and Biotechnology*, 44(4) (2016) 1186–1193.
27. W. Huang, Z. Liu, G. Zhou, A. Tian, N. Sun, Magnetic gold nanoparticle-mediated small interference RNA silencing Bag-1 gene for colon cancer therapy, *Oncology Reports*, 35(2) (2016) 978–984.
28. W. Huang, Z. Liu, G. Zhou, J. Ling, A. Tian, N. Sun, Silencing Bag-1 gene via magnetic gold nanoparticle-delivered siRNA plasmid for colorectal cancer therapy in vivo and in vitro, *Tumour Biology: The Journal of the International Society for Oncodevelopmental Biology and Medicine*, 37(8) (2016) 10365–10374.
29. X. Hou, H. Zhang, H. Li, D. Zhang, Magnetic albumin immuno-nanospheres as an efficient gene delivery system for a potential use in lung cancer: Preparation, in vitro targeting and biological effect analysis, *Journal of Drug Targeting*, 24(3) (2016) 247–256.
30. S. Xiao, R. Castro, J. Rodrigues, X. Shi, H. Tomas, PAMAM Dendrimer/pDNA functionalized-magnetic iron oxide nanoparticles for gene delivery, *Journal of Biomedical Nanotechnology*, 11(8) (2015) 1370–1384.
31. W. Siene Ng, K. Binley, B. Song, J. E. Morgan, Use of magnetic nanoparticles and oscillating magnetic field for non-viral gene transfer into mouse cornea, *Lancet*, 385 Suppl 1 (2015) S75.
32. X. Zhao, H. Cui, W. Chen, Y. Wang, B. Cui, C. Sun, Z. Meng, G. Liu, Morphology, structure and function characterization of PEI modified magnetic nanoparticles gene delivery system, *PloS One*, 9(6) (2014) e98919.
33. M. A. Abakumov, N. V. Nukolova, M. Sokolsky-Papkov, S. A. Shein, T. O. Sandalova, H. M. Vishwasrao, N. F. Grinenko, I. L. Gubsky, A. M. Abakumov, A. V. Kabanov, V. P. Chekhonin, VEGF-targeted magnetic nanoparticles for MRI visualization of brain tumor, *Nanomedicine: Nanotechnology, Biology, and Medicine*, 11(4) (2015) 825–833.
34. D. Valdepérez, P. del Pino, L. Sánchez, W. J. Parak, B. Pelaz, Highly active antibody-modified magnetic polyelectrolyte capsules, *Journal of Colloid and Interface Science*, 474 (2016) 1–8.

35. S.-C. Wu, Y.-J. Chen, H.-C. Wang, M.-Y. Chou, T.-Y. Chang, S.-S. Yuan, C.-Y. Chen, M.-F. Hou, J. T.-A. Hsu, Y.-M. Wang, Bispecific antibody conjugated manganese-based magnetic engineered iron oxide for imaging of HER2/neu-and EGFR-expressing tumors, *Theranostics*, 6(1) (2016) 118–130.

36. M. A. Shevtsov, B. P. Nikolaev, V. A. Ryzhov, L. Y. Yakovleva, Y. Y. Marchenko, M. A. Parr, V. I. Rolich, A. L. Mikhrina, A. V. Dobrodumov, E. Pitkin, Ionizing radiation improves glioma-specific targeting of superparamagnetic iron oxide nanoparticles conjugated with cmHsp70. 1 monoclonal antibodies (SPION–cmHsp70. 1), *Nanoscale*, 7(48) (2015) 20652–20664.

37. K. S. Kim, J. Y. Kim, J. Y. Lee, S. Matsuda, S. Hideshima, Y. Mori, T. Osaka, K. Na, Stimuli-responsive magnetic nanoparticles for tumor-targeted bimodal imaging and photodynamic/hyperthermia combination therapy, *Nanoscale*, 8(22) (2016) 11625–11634.

38. H. Basoglu, M. D. Bilgin, M. M. Demir, Protoporphyrin IX-loaded magnetoliposomes as a potential drug delivery system for photodynamic therapy: Fabrication, characterization and in vitro study, *Photodiagnosis and Photodynamic Therapy*, 13 (2016) 81–90.

39. H. Unterweger, D. Subatzus, R. Tietze, C. Janko, M. Poettler, A. Stiegelschmitt, M. Schuster, C. Maake, A. R. Boccaccini, C. Alexiou, Hypericin-bearing magnetic iron oxide nanoparticles for selective drug delivery in photodynamic therapy, *International Journal of Nanomedicine*, 10 (2015) 6985–6996.

40. A. Espinosa, M. Bugnet, G. Radtke, S. Neveu, G. A. Botton, C. Wilhelm, A. Abou-Hassan, Can magneto-plasmonic nanohybrids efficiently combine photothermia with magnetic hyperthermia?, *Nanoscale*, 7(45) (2015) 18872–18877.

41. L. Ji, T. Yan, Y. Li, J. Gao, Q. Wang, L. Hu, D. Wu, Q. Wei, B. Du, Preparation of Au-polydopamine functionalized carbon encapsulated Fe_3O_4 magnetic nanocomposites and their application for ultrasensitive detection of carcino-embryonic antigen, *Scientific Reports*, 6 (2016) 21017.

42. S. Shen, L. Wu, J. Liu, M. Xie, H. Shen, X. Qi, Y. Yan, Y. Ge, Y. Jin, Core–shell structured Fe_3O_4@TiO_2-doxorubicin nanoparticles for targeted chemo-sonodynamic therapy of cancer, *International Journal of Pharmaceutics*, 486(1) (2015) 380–388.

43. R. Yang, Q. Tang, F. Miao, Y. An, M. Li, Y. Han, X. Wang, J. Wang, P. Liu, R. Chen, Inhibition of heat-shock protein 90 sensitizes liver cancer

stem-like cells to magnetic hyperthermia and enhances anti-tumor effect on hepatocellular carcinoma-burdened nude mice, *International Journal of Nanomedicine*, 10 (2015) 7345–7358.

44. L. Zhao, M. Huo, J. Liu, Z. Yao, D. Li, Z. Zhao, J. Tang, In vitro investigation on the magnetic thermochemotherapy mediated by magnetic nanoparticles combined with methotrexate for breast cancer treatment, *Journal of Nanoscience and Nanotechnology,* 13(2) (2013) 741–745.
45. A. Ito, F. Matsuoka, H. Honda, T. Kobayashi, Heat shock protein 70 gene therapy combined with hyperthermia using magnetic nanoparticles, *Cancer Gene Therapy,* 10(12) (2003) 918–925.
46. H. Wang, X. Li, X. Xi, B. Hu, L. Zhao, Y. Liao, J. Tang, Effects of magnetic induction hyperthermia and radiotherapy alone or combined on a murine 4T1 metastatic breast cancer model, *International Journal of Hyperthermia*, 27(6) (2011) 563–572.
47. A. Espinosa, R. Di Corato, J. Kolosnjaj-Tabi, P. Flaud, T. Pellegrino, C. Wilhelm, Duality of iron oxide nanoparticles in cancer therapy: Amplification of heating efficiency by magnetic hyperthermia and photothermal bimodal treatment, *ACS Nano*, 10(2) (2016) 2436–2446.
48. A. Hervault, N. T. K. Thanh, Magnetic nanoparticle-based therapeutic agents for thermo-chemotherapy treatment of cancer, *Nanoscale*, 6(20) (2014) 11553–11573.
49. A. Singh, S. K. Sahoo, Magnetic nanoparticles: A novel platform for cancer theranostics, *Drug Discovery Today*, 19(4) (2014) 474–481.
50. K. C. Barick, S. Singh, N. V. Jadhav, D. Bahadur, B. N. Pandey, P. A. Hassan, pH-responsive peptide mimic shell cross—linked magnetic nanocarriers for combination therapy, *Advanced Functional Materials*, 22(23) (2012) 4975–4984.
51. Y. Akin, I. Obaidat, B. Issa, Y. Haik, $Ni_{1-x}Cr_x$ alloy for self controlled magnetic hyperthermia, *Crystal Research and Technology*, 44(4) (2009) 386–390.
52. G. Nedelcu, Magnetic nanoparticles impact on tumoral cells in the treatment by magnetic fluid hyperthermia, *Digest Journal of Nanomaterials and Biostructures*, 3(3) (2008) 103–107.
53. V. Nandwana, M. De, S. Chu, M. Jaiswal, M. Rotz, T. J. Meade, V. P. Dravid, Theranostic magnetic nanostructures (MNS) for cancer, *Nanotechnology-Based Precision Tools for the Detection and Treatment of Cancer*, Springer 166 (2015) 51–83.

54. C. Mirkin, T. J. Meade, S. H. Petrosko, A. H. Stegh, *Nanotechnology-Based Precision Tools for the Detection and Treatment of Cancer*, Springer (2015).

55. K. Ulbrich, K. I. Holá, V. Šubr, A. Bakandritsos, J. I. Tuček, R. Zbořil, Targeted drug delivery with polymers and magnetic nanoparticles: covalent and noncovalent approaches, release control, and clinical studies, *Chemical Reviews*, 116(9) (2016) 5338–5431.

56. H. K. Kim, L. T. Schelhas, S. Keller, J. L. Hockel, S. H. Tolbert, G. P. Carman, Magnetoelectric control of superparamagnetism, *Nano Letters*, 13(3) (2013) 884–888.

57. A. K. Guria, K. Dey, S. Sarkar, B. K. Patra, S. Giri, N. Pradhan, Tuning the growth pattern in 2D confinement regime of Sm_2O_3 and the emerging room temperature unusual superparamagnetism, *Scientific Reports*, 4 (2014) 6514.

58. C. S. Kumar, F. Mohammad, Magnetic nanomaterials for hyperthermia-based therapy and controlled drug delivery, *Advanced Drug Delivery Reviews*, 63(9) (2011) 789–808.

59. H. F. Rodrigues, F. M. Mello, L. C. Branquinho, N. Zufelato, E. P. Silveira-Lacerda, A. F. Bakuzis, Real-time infrared thermography detection of magnetic nanoparticle hyperthermia in a murine model under a non-uniform field configuration, *International Journal of Hyperthermia*, 29(8) (2013) 752–767.

60. A. K. Iyer, G. Khaled, J. Fang, H. Maeda, Exploiting the enhanced permeability and retention effect for tumor targeting, *Drug Discovery Today*, 11(17) (2006) 812–818.

61. J.-P. Fortin, F. Gazeau, C. Wilhelm, Intracellular heating of living cells through Néel relaxation of magnetic nanoparticles, *European Biophysics Journal*, 37(2) (2008) 223–228.

62. B. E. Kashevsky, S. B. Kashevsky, V. S. Korenkov, Y. P. Istomin, T. I. Terpinskaya, V. S. Ulashchik, Magnetic hyperthermia with hard-magnetic nanoparticles, *Journal of Magnetism and Magnetic Materials*, 380 (2015) 335–340.

63. R. Hergt, S. Dutz, Magnetic particle hyperthermia—biophysical limitations of a visionary tumour therapy, *Journal of Magnetism and Magnetic Materials*, 311(1) (2007) 187–192.

64. A. E. Deatsch, B. A. Evans, Heating efficiency in magnetic nanoparticle hyperthermia, *Journal of Magnetism and Magnetic Materials*, 354 (2014) 163–172.

65. J. P. May, S.-D. Li, Hyperthermia-induced drug targeting, *Expert Opinion on Drug Delivery*, 10(4) (2013) 511–527.

66. Y.-W. Jun, Y.-M. Huh, J.-S. Choi, J.-H. Lee, H.-T. Song, S. Kim, S. Kim, S. Yoon, K.-S. Kim, J.-S. Shin, Nanoscale size effect of magnetic nanocrystals and their utilization for cancer diagnosis via magnetic resonance imaging, *Journal of the American Chemical Society*, 127(16) (2005) 5732–5733.

67. Y. W. Jun, J. H. Lee, J. Cheon, Chemical design of nanoparticle probes for high-performance magnetic resonance imaging, *Angewandte Chemie International Edition*, 47(28) (2008) 5122–5135.

68. M. De, S. S. Chou, H. M. Joshi, V. P. Dravid, Hybrid magnetic nano-structures (MNS) for magnetic resonance imaging applications, *Advanced Drug Delivery Reviews*, 63(14) (2011) 1282–1299.

69. N. Lee, T. Hyeon, Designed synthesis of uniformly sized iron oxide nanoparticles for efficient magnetic resonance imaging contrast agents, *Chemical Society Reviews*, 41(7) (2012) 2575–2589.

70. P. F. Hahn, D. D. Stark, J. M. Lewis, S. Saini, G. Elizondo, R. Weissleder, C. Fretz, J. Ferrucci, First clinical trial of a new superparamagnetic iron oxide for use as an oral gastrointestinal contrast agent in MR imaging, *Radiology*, 175(3) (1990) 695–700.

71. Z. Karimi, L. Karimi, H. Shokrollahi, Nano-magnetic particles used in biomedicine: Core and coating materials, *Materials Science and Engineering: C*, 33(5) (2013) 2465–2475.

72. N. N. Ghosh, P. Pant, S. Bhuvaneswari, Chemical methodologies for preparation of micron and nanometer scale ferrites–a mini review of patents, *Recent Patents on Nanotechnology*, 2(1) (2008) 8–18.

73. M. F. Attia, N. Anton, I. U. Khan, C. A. Serra, N. Messaddeq, A. Jakhmola, R. Vecchione, T. Vandamme, One-step synthesis of iron oxide polypyrrole nanoparticles encapsulating ketoprofen as model of hydrophobic drug, *International Journal of Pharmaceutics*, 508(1) (2016) 61–70.

74. A. H. Lu, E. E. L. Salabas, F. Schüth, Magnetic nanoparticles: Synthesis, protection, functionalization, and application, *Angewandte Chemie International Edition*, 46(8) (2007) 1222–1244.

75. T. Sugimoto, Formation of monodispersed nano-and micro-particles controlled in size, shape, and internal structure, *Chemical Engineering & Technology*, 26(3) (2003) 313–321.

76. C. Boyer, M. R. Whittaker, V. Bulmus, J. Liu, T. P. Davis, The design and utility of polymer-stabilized iron-oxide nanoparticles for nanomedicine applications, *NPG Asia Materials*, 2(1) (2010) 23–30.

77. Y. Li, M. Afzaal, P. O'Brien, The synthesis of amine-capped magnetic (Fe, Mn, Co, Ni) oxide nanocrystals and their surface modification for aqueous dispersibility, *Journal of Materials Chemistry*, 16(22) (2006) 2175–2180.

78. T. Hyeon, S. S. Lee, J. Park, Y. Chung, H. B. Na, Synthesis of highly crystalline and monodisperse maghemite nanocrystallites without a size-selection process, *Journal of the American Chemical Society*, 123(51) (2001) 12798–12801.

79. S. Sun, H. Zeng, D. B. Robinson, S. Raoux, P. M. Rice, S. X. Wang, G. Li, Monodisperse MFe_2O_4 (M = Fe, Co, Mn) nanoparticles, *Journal of the American Chemical Society*, 126(1) (2004) 273–279.

80. S. Sun, H. Zeng, Size-controlled synthesis of magnetite nanoparticles, *Journal of the American Chemical Society*, 124(28) (2002) 8204–8205.

81. H. C. Vadlamudi, H. Narendran, T. Nagaswaram, G. Yaga, J. Thanniru, P. R. Yalavarthi, Microemulsions based transdermal drug delivery systems, *Current Drug Discovery Technologies*, 11(3) (2014) 169–180.

82. I. Sharifi, H. Shokrollahi, S. Amiri, Ferrite-based magnetic nanofluids used in hyperthermia applications, *Journal of Magnetism and Magnetic Materials*, 324(6) (2012) 903–915.

83. M. Eslamian, M. Shekarriz, Recent advances in nanoparticle preparation by spray and microemulsion methods, *Recent Patents on Nanotechnology*, 3(2) (2009) 99–115.

84. S. H. Kim, H. Jeong, J. Kim, I. S. Lee, Fabrication of supported AuPt alloy nanocrystals with enhanced electrocatalytic activity for formic acid oxidation through conversion chemistry of layer-deposited Pt^{2+} on Au nanocrystals, *Small*, 11(37) (2015) 4884–4893.

85. M. Sonmez, M. Georgescu, L. Alexandrescu, D. Gurau, A. Ficai, D. Ficai, E. Andronescu, Synthesis and applications of Fe_3O_4/SiO_2 core-shell materials, *Current Pharmaceutical Design*, 21(37) (2015) 5324–5335.

86. X. Yu, G. Cheng, M.-D. Zhou, S.-Y. Zheng, On-demand one-step synthesis of monodisperse functional polymeric microspheres with droplet microfluidics, *Langmuir*, 31(13) (2015) 3982–3992.

87. E. Song, W. Han, J. Li, Y. Jiang, D. Cheng, Y. Song, P. Zhang, W. Tan, Magnetic-encoded fluorescent multifunctional nanospheres for simultaneous multicomponent analysis, *Analytical Chemistry*, 86(19) (2014) 9434–9442.

88. X. Zhao, H. Zhao, H. Yuan, M. Lan, Multifunctional super-paramagnetic $Fe_3O_4@SiO_2$ core/shell nanoparticles: Design and

application for cell imaging, *Journal of Biomedical Nanotechnology*, 10(2) (2014) 262–270.

89. J. Tang, M. Myers, K. A. Bosnick, L. E. Brus, Magnetite Fe_3O_4 nanocrystals: Spectroscopic observation of aqueous oxidation kinetics, *The Journal of Physical Chemistry B*, 107(30) (2003) 7501–7506.
90. K. Kekalo, K. Koo, E. Zeitchick, I. Baker, Microemulsion synthesis of iron core/iron oxide shell magnetic nanoparticles and their physicochemical properties, MRS Proceedings, Cambridge Univ Press (2012), pp. mrsf11-1416-jj05-54.
91. A. K. Gupta, M. Gupta, Synthesis and surface engineering of iron oxide nanoparticles for biomedical applications, *Biomaterials*, 26(18) (2005) 3995–4021.
92. T. Hiemstra, Formation, stability, and solubility of metal oxide nanoparticles: Surface entropy, enthalpy, and free energy of ferrihydrite, *Geochimica et Cosmochimica Acta*, 158 (2015) 179–198.
93. O. Veiseh, J. W. Gunn, M. Zhang, Design and fabrication of magnetic nanoparticles for targeted drug delivery and imaging, *Advanced Drug Delivery Reviews*, 62(3) (2010) 284–304.
94. A. K. Gupta, R. R. Naregalkar, V. D. Vaidya, M. Gupta, Recent advances on surface engineering of magnetic iron oxide nanoparticles and their biomedical applications, *Nanomedicine*, 2(1) (2007) 23–39.
95. B. I. Kharisov, H. R. Dias, O. V. Kharissova, A. Vázquez, Y. Peña, I. Gómez, Solubilization, dispersion and stabilization of magnetic nanoparticles in water and non-aqueous solvents: Recent trends, *RSC Advances*, 4(85) (2014) 45354–45381.
96. A. H. Latham, M. E. Williams, Controlling transport and chemical functionality of magnetic nanoparticles, *Accounts of Chemical Research*, 41(3) (2008) 411–420.
97. J. Mendez-Garza, B. Wang, A. Madeira, C. Di Giorgio, G. Bossis, Synthesis and surface modification of spindle-type magnetic nanoparticles: Gold coating and PEG functionalization, *Journal of Biomaterials and Nanobiotechnology*, 4 (2013) 222–228.
98. Q. Dai, D. Berman, K. Virwani, J. Frommer, P.-O. Jubert, M. Lam, T. Topuria, W. Imaino, A. Nelson, Self-assembled ferrimagnet–polymer composites for magnetic recording media, *Nano Letters*, 10(8) (2010) 3216–3221.
99. S. Mohapatra, S. Sahu, S. Nayak, S. K. Ghosh, Design of Fe_3O_4@ SiO_2@ carbon quantum dot based nanostructure for fluorescence

sensing, magnetic separation, and live cell imaging of fluoride ion, *Langmuir*, 31(29) (2015) 8111–8120.

100. J. Göransson, T. Zardán Gómez De La Torre, M. Strömberg, C. Russell, P. Svedlindh, M. Strømme, M. Nilsson, Sensitive detection of bacterial DNA by magnetic nanoparticles, *Analytical Chemistry*, 82(22) (2010) 9138–9140.
101. J. Wang, D. Song, H. Zhang, J. Zhang, Y. Jin, H. Zhang, H. Zhou, Y. Sun, Studies of Fe_3O_4/Ag/Au composites for immunoassay based on surface plasmon resonance biosensor, *Colloids and Surfaces B: Biointerfaces*, 102 (2013) 165–170.
102. Z. Liang, X. Wu, Y. Xie, S. Liu, A facile approach to fabricate water-soluble Au-Fe_3O_4 nanoparticle for liver cancer cells imaging, *Chinese Journal of Chemistry*, 30(7) (2012) 1387–1392.
103. S. Ghosh, T. H. Fang, M. Uddin, K. Hidajat, Enantioselective separation of chiral aromatic amino acids with surface functionalized magnetic nanoparticles, *Colloids and Surfaces B: Biointerfaces*, 105 (2013) 267–277.
104. M. Abbas, B. P. Rao, M. N. Islam, S. Naga, M. Takahashi, C. Kim, Highly stable-silica encapsulating magnetite nanoparticles (Fe_3O_4/SiO_2) synthesized using single surfactantless-polyol process, *Ceramics International*, 40(1) (2014) 1379–1385.
105. Y. W. Zhang, M. Zhang, J. B. Yang, L. Ding, J. Zheng, J. L. Xu, S. L. Xiong, Formation of Fe_3O_4@SiO_2@C/Ni hybrids with enhanced catalytic activity and histidine-rich protein separation, *Nanoscale*, 8(35) (2016) 15978–15988.
106. M. Sonmez, M. Georgescu, L. Alexandrescu, D. Gurau, A. Ficai, D. Ficai, E. Andronescu, Synthesis and applications of Fe_3O_4/SiO_2 core-shell materials, *Current Pharmceutical Design*, 21(37) (2015) 5324–5335.
107. T. R. Santra S., Theodoropoulou N., Dobson J., Hebard A., Tan W., Synthesis and characterization of silica-coated iron oxide nanoparticles in microemulsion: The effect of nonionic surfactants, *Langmuir*, 17 (2001) 2900–2906.
108. Z. Zhang, C. Zhang, J. Sun, T. Kou, Q. Bai, Y. Wang, Y. Ding, Ultrafine nanoporous PdFe/Fe_3O_4 catalysts with doubly enhanced activities towards electro-oxidation of methanol and ethanol in alkaline media, *Journal of Materials Chemistry A*, 1(11) (2013) 3620–3628.
109. J. Mamani, A. Costa-Filho, D. Cornejo, E. Vieira, L. Gamarra, Synthesis and characterization of magnetite nanoparticles coated with lauric acid, *Materials Characterization*, 81 (2013) 28–36.

110. Q. Lan, C. Liu, F. Yang, S. Liu, J. Xu, D. Sun, Synthesis of bilayer oleic acid-coated Fe_3O_4 nanoparticles and their application in pH-responsive Pickering emulsions, *Journal of Colloid and Interface Science*, 310(1) (2007) 260–269.

111. X. Liu, J. Xing, Y. Guan, G. Shan, H. Liu, Synthesis of amino-silane modified superparamagnetic silica supports and their use for protein immobilization, *Colloids and Surfaces A: Physicochemical and Engineering Aspects*, 238(1) (2004) 127–131.

112. Y. Wang, J. F. Wong, X. Teng, X. Z. Lin, H. Yang, "Pulling" nanoparticles into water: Phase transfer of oleic acid stabilized monodisperse nanoparticles into aqueous solutions of α-cyclodextrin, *Nano Letters*, 3(11) (2003) 1555–1559.

113. K. V. Korpany, F. Habib, M. Murugesu, A. S. Blum, Stable water-soluble iron oxide nanoparticles using Tiron, *Materials Chemistry and Physics*, 138(1) (2013) 29–37.

114. M. Sun, A. Zhu, Q. Zhang, Q. Liu, A facile strategy to synthesize monodisperse superparamagnetic OA-modified Fe_3O_4 nanoparticles with PEG assistant, *Journal of Magnetism and Magnetic Materials*, 369 (2014) 49–54.

115. Z. Chen, Y. Zhang, S. Zhang, J. Xia, J. Liu, K. Xu, N. Gu, Preparation and characterization of water-soluble monodisperse magnetic iron oxide nanoparticles via surface double-exchange with DMSA, *Colloids and Surfaces A: Physicochemical and Engineering Aspects*, 316(1) (2008) 210–216.

116. Y. Q. Ge, Y. Zhang, S. Y. He, F. Nie, G. J. Teng, N. Gu, Fluorescence modified chitosan-coated magnetic nanoparticles for high-efficient cellular imaging, *Nanoscale Research Letters*, 4(4) (2009) 287–295.

117. C. Tassa, S. Y. Shaw, R. Weissleder, Dextran-coated iron oxide nanoparticles: a versatile platform for targeted molecular imaging, molecular diagnostics, and therapy, *Accounts of Chemical Research*, 44(10) (2011) 842–852.

118. E. A. Osborne, T. M. Atkins, D. A. Gilbert, S. M. Kauzlarich, K. Liu, A. Y. Louie, Rapid microwave-assisted synthesis of dextran-coated iron oxide nanoparticles for magnetic resonance imaging, *Nanotechnology*, 23(21) (2012) 215602.

119. S. Arami, M. Rashidi, M. Mahdavi, M. Fathi, A. Entezami, Synthesis and characterization of Fe_3O_4-PEG-LAC-chitosan-PEI nanoparticle as a survivin siRNA delivery system, *Human & Experimental Toxicology*, 36(3) (2016) 227–237.

120. G. Kogan, L. Šoltés, R. Stern, P. Gemeiner, Hyaluronic acid: A natural biopolymer with a broad range of biomedical and industrial applications, *Biotechnology Letters*, 29(1) (2007) 17–25.

121. W. Lu, M. Ling, M. Jia, P. Huang, C. Li, B. Yan, Facile synthesis and characterization of polyethylenimine-coated Fe_3O_4 superparamagnetic nanoparticles for cancer cell separation, *Molecular Medicine Reports*, 9(3) (2014) 1080–1084.

122. I. I. Lungu, M. Rădulescu, G. D. Mogoşanu, A. M. Grumezescu, pH sensitive core-shell magnetic nanoparticles for targeted drug delivery in cancer therapy, *Romanian Journal of Morphology and Embryology*, 57(1) (2016) 23–32.

123. S. Laurent, A. A. Saei, S. Behzadi, A. Panahifar, M. Mahmoudi, Superparamagnetic iron oxide nanoparticles for delivery of therapeutic agents: Opportunities and challenges, *Expert Opinion on Drug Delivery*, 11(9) (2014) 1449–1470.

124. M. R. Santoso, P. C. Yang, Magnetic nanoparticles for targeting and imaging of stem cells in myocardial infarction, *Stem Cells International*, 2016 (2016) 4198790.

125. H. Mok, M. Zhang, Superparamagnetic iron oxide nanoparticle-based delivery systems for biotherapeutics, *Expert Opinion on Drug Delivery*, 10(1) (2013) 73–87.

126. X. Mou, Z. Ali, S. Li, N. He, Applications of magnetic nanoparticles in targeted drug delivery system, *Journal of Nanoscience and Nanotechnology*, 15(1) (2015) 54–62.

127. T. K. Jain, J. Richey, M. Strand, D. L. Leslie-Pelecky, C. A. Flask, V. Labhasetwar, Magnetic nanoparticles with dual functional properties: Drug delivery and magnetic resonance imaging, *Biomaterials*, 29(29) (2008) 4012–4021.

128. Y. Matsumura, H. Maeda, A new concept for macromolecular therapeutics in cancer chemotherapy: Mechanism of tumoritropic accumulation of proteins and the antitumor agent smancs, *Cancer Research*, 46(12 Part 1) (1986) 6387–6392.

129. R. J. Hickey, A. S. Haynes, J. M. Kikkawa, S.-J. Park, Controlling the self-assembly structure of magnetic nanoparticles and amphiphilic block-copolymers: From micelles to vesicles, *Journal of the American Chemical Society*, 133(5) (2011) 1517–1525.

130. A. Wijaya, K. Hamad-Schifferli, High-density encapsulation of Fe_3O_4 nanoparticles in lipid vesicles, *Langmuir*, 23(19) (2007) 9546–9550.

131. C. Bonnaud, C. A. Monnier, D. Demurtas, C. Jud, D. Vanhecke, X. Montet, R. Hovius, M. Lattuada, B. Rothen-Rutishauser, A. Petri-Fink,

Insertion of nanoparticle clusters into vesicle bilayers, *ACS Nano*, 8(4) (2014) 3451–3460.

132. H. Ye, J. Tong, J. Liu, W. Lin, C. Zhang, K. Chen, J. Zhao, W. Zhu, Combination of gemcitabine-containing magnetoliposome and oxaliplatin-containing magnetoliposome in breast cancer treatment: A possible mechanism with potential for clinical application, *Oncotarget*, 7(28) (2016) 43762–43778.
133. E. Amstad, J. Kohlbrecher, E. Müller, T. Schweizer, M. Textor, E. Reimhult, Triggered release from liposomes through magnetic actuation of iron oxide nanoparticle containing membranes, *Nano Letters*, 11(4) (2011) 1664–1670.
134. Y. Chen, Y. Chen, D. Xiao, A. Bose, R. Deng, G. D. Bothun, Low-dose chemotherapy of hepatocellular carcinoma through triggered-release from bilayer-decorated magnetoliposomes, *Colloids and Surfaces B: Biointerfaces*, 116 (2014) 452–458.
135. I. C. Lin, J. H. Fang, C. T. Lin, S. Y. Sung, Y. L. Su, S. H. Hu, Enhanced targeted delivery of cyclodextrin-based supermolecules by core-shell nanocapsules for magnetothermal chemotherapy, *Macromolecular Bioscience*, 16(9) (2016) 1273–1286.
136. D. Shao, J. Li, X. Zheng, Y. Pan, Z. Wang, M. Zhang, Q.-X. Chen, W.-F. Dong, L. Chen, Janus "nano-bullets" for magnetic targeting liver cancer chemotherapy, *Biomaterials*, 100 (2016) 118–133.
137. S. Sungsuwan, Z. Yin, X. Huang, Lipopeptide-coated iron oxide nanoparticles as potential glycoconjugate-based synthetic anticancer vaccines, *ACS Applied Materials & Interfaces*, 7(31) (2015) 17535–17544.
138. N. Schleich, C. Po, D. Jacobs, B. Ucakar, B. Gallez, F. Danhier, V. Préat, Comparison of active, passive and magnetic targeting to tumors of multifunctional paclitaxel/SPIO-loaded nanoparticles for tumor imaging and therapy, *Journal of Controlled Release*, 194 (2014) 82–91.
139. T. T. Li, X. Shen, Y. Geng, Z. Chen, L. Li, S. Li, H. Yang, C. Wu, H. Zeng, Y. Y. Liu, Folate-functionalized magnetic-mesoporous silica nanoparticles for drug/gene co-delivery to potentiate the antitumor efficacy, *ACS Applied Materials & Interfaces*, 8(22) (2016) 13748–13758.
140. Y.-K. Kim, M. Zhang, J.-J. Lu, F. Xu, B.-A. Chen, L. Xing, H.-L. Jiang, PK11195-chitosan-graft-polyethylenimine-modified SPION as a mitochondria-targeting gene carrier, *Journal of Drug Targeting*, 24(4) (2015) 457–467.

141. Y.-L. Lo, H.-L. Chou, Z.-X. Liao, S.-J. Huang, J.-H. Ke, Y.-S. Liu, C.-C. Chiu, L.-F. Wang, Chondroitin sulfate-polyethylenimine copolymer-coated superparamagnetic iron oxide nanoparticles as an efficient magneto-gene carrier for microRNA-encoding plasmid DNA delivery, *Nanoscale*, 7(18) (2015) 8554–8565.

142. J. H. Fang, Y. H. Lai, T. L. Chiu, Y. Y. Chen, S. H. Hu, S. Y. Chen, Magnetic core–shell nanocapsules with dual-targeting capabilities and Co: Delivery of multiple drugs to treat brain gliomas, *Advanced Healthcare Materials*, 3(8) (2014) 1250–1260.

143. H. Zhou, W. Qian, F. M. Uckun, Z. Zhou, L. Wang, A. Wang, H. Mao, L. Yang, IGF-1 receptor targeted nanoparticles for image-guided therapy of stroma-rich and drug resistant human cancer, *Proceedings of SPIE–the International Society for Optical Engineering*, 17 (2016) 9836.

144. M. H. Cho, E. J. Lee, M. Son, J.-H. Lee, D. Yoo, J.-W. Kim, S. W. Park, J.-S. Shin, J. Cheon, A magnetic switch for the control of cell death signalling in in vitro and in vivo systems, *Nature Materials*, 11(12) (2012) 1038–1043.

145. J. Dobson, Cancer therapy: Death by magnetism, *Nature Materials*, 11(12) (2012) 1006–1008.

146. G. Helmlinger, F. Yuan, M. Dellian, R. K. Jain, Interstitial pH and pO_2 gradients in solid tumors in vivo: High-resolution measurements reveal a lack of correlation, *Nature Medicine*, 3(2) (1997) 177–182.

147. A. S. Trevani, G. Andonegui, M. Giordano, D. H. López, R. Gamberale, F. Minucci, J. R. Geffner, Extracellular acidification induces human neutrophil activation, *The Journal of Immunology*, 162(8) (1999) 4849–4857.

148. J. A. Mindell, Lysosomal acidification mechanisms, *Annual Review of Physiology*, 74 (2012) 69–86.

149. A. Ponta, Y. Bae, Tumor-preferential sustained drug release enhances antitumor activity of block copolymer micelles, *Journal of Drug Targeting*, 22(7) (2014) 619–628.

150. J. Zhang, R. Misra, Magnetic drug-targeting carrier encapsulated with thermosensitive smart polymer: Core–shell nanoparticle carrier and drug release response, *Acta Biomaterialia*, 3(6) (2007) 838–850.

151. X. Yang, J. J. Grailer, I. J. Rowland, A. Javadi, S. A. Hurley, V. Z. Matson, D. A. Steeber, S. Gong, Multifunctional stable and pH-responsive polymer vesicles formed by heterofunctional triblock copolymer for targeted anticancer drug delivery and ultrasensitive MR imaging, *ACS Nano*, 4(11) (2010) 6805–6817.

152. R. Tong, L. Tang, L. Ma, C. Tu, R. Baumgartner, J. Cheng, Smart chemistry in polymeric nanomedicine, *Chemical Society Reviews*, 43(20) (2014) 6982–7012.

153. J. Yu, X. Chu, Y. Hou, Stimuli-responsive cancer therapy based on nanoparticles, *Chemical Communications*, 50(79) (2014) 11614–11630.

154. H.-C. Wang, Y. Zhang, C. M. Possanza, S. C. Zimmerman, J. Cheng, J. S. Moore, K. Harris, J. S. Katz, Trigger chemistries for better industrial formulations, *ACS Applied Materials & Interfaces*, 7(12) (2015) 6369–6382.

155. X. Ding, Y. Liu, J. Li, Z. Luo, Y. Hu, B. Zhang, J. Liu, J. Zhou, K. Cai, Hydrazone-bearing PMMA-functionalized magnetic nanocubes as pH-responsive drug carriers for remotely targeted cancer therapy in vitro and in vivo, *ACS Applied Materials & Interfaces*, 6(10) (2014) 7395–7407.

156. J. M. Estrela, A. Ortega, E. Obrador, Glutathione in cancer biology and therapy, *Critical Reviews in Clinical Laboratory Sciences*, 43(2) (2006) 143–181.

157. F. Q. Schafer, G. R. Buettner, Redox environment of the cell as viewed through the redox state of the glutathione disulfide/glutathione couple, *Free Radical Biology and Medicine*, 30(11) (2001) 1191–1212.

158. J. H. Lee, K. Lee, S. H. Moon, Y. Lee, T. G. Park, J. Cheon, All-in-one target-cell-specific magnetic nanoparticles for simultaneous molecular imaging and siRNA delivery, *Angewandte Chemie*, 121(23) (2009) 4238–4243.

159. Z. R. Stephen, F. M. Kievit, O. Veiseh, P. A. Chiarelli, C. Fang, K. Wang, S. J. Hatzinger, R. G. Ellenbogen, J. R. Silber, M. Zhang, Redox-responsive magnetic nanoparticle for targeted convection-enhanced delivery of O 6-benzylguanine to brain tumors, *ACS Nano*, 8(10) (2014) 10383–10395.

160. A. H. Abouzeid, N. R. Patel, V. P. Torchilin, Polyethylene glycol-phosphatidylethanolamine (PEG-PE)/vitamin E micelles for co-delivery of paclitaxel and curcumin to overcome multi-drug resistance in ovarian cancer, *International Journal of Pharmaceutics*, 464(1–2) (2014) 178–184.

161. D. S. Wilson, G. Dalmasso, L. Wang, S. V. Sitaraman, D. Merlin, N. Murthy, Orally delivered thioketal nanoparticles loaded with TNF-α–siRNA target inflammation and inhibit gene expression in the intestines, *Nature Materials*, 9(11) (2010) 923–928.

162. J. Atkinson, C. Siller, J. Gill, Tumour endoproteases: The cutting edge of cancer drug delivery?, *British Journal of Pharmacology*, 153(7) (2008) 1344–1352.

163. L. M. Coussens, B. Fingleton, L. M. Matrisian, Matrix metalloproteinase inhibitors and cancer—trials and tribulations, *Science*, 295(5564) (2002) 2387–2392.

164. C. Ansari, G. A. Tikhomirov, S. H. Hong, R. A. Falconer, P. M. Loadman, J. H. Gill, R. Castaneda, F. K. Hazard, L. Tong, O. D. Lenkov, Development of novel tumor-targeted theranostic nanoparticles activated by membrane-type matrix metalloproteinases for combined cancer magnetic resonance imaging and therapy, *Small*, 10(3) (2014) 566–575.

165. G. Y. Lee, W. P. Qian, L. Wang, Y. A. Wang, C. A. Staley, M. Satpathy, S. Nie, H. Mao, L. Yang, Theranostic nanoparticles with controlled release of gemcitabine for targeted therapy and MRI of pancreatic cancer, *ACS Nano*, 7(3) (2013) 2078–2089.

166. J. R. Hwu, Y. S. Lin, T. Josephrajan, M.-H. Hsu, F.-Y. Cheng, C.-S. Yeh, W.-C. Su, D.-B. Shieh, Targeted paclitaxel by conjugation to iron oxide and gold nanoparticles, *Journal of the American Chemical Society*, 131(1) (2008) 66–68.

167. F. Cengelli, J. A. Grzyb, A. Montoro, H. Hofmann, S. Hanessian, L. Juillerat-Jeanneret, Surface-functionalized ultrasmall superparamagnetic nanoparticles as magnetic delivery vectors for camptothecin, *ChemMedChem*, 4(6) (2009) 988–997.

168. H. Wang, T. B. Shrestha, M. T. Basel, R. K. Dani, G.-M. Seo, S. Balivada, M. M. Pyle, H. Prock, O. B. Koper, P. S. Thapa, Magnetic-Fe/Fe_3O_4-nanoparticle-bound SN38 as carboxylesterase-cleavable prodrug for the delivery to tumors within monocytes/macrophages, *Beilstein Journal of Nanotechnology*, 3(1) (2012) 444–455.

169. K. Hayashi, K. Ono, H. Suzuki, M. Sawada, M. Moriya, W. Sakamoto, T. Yogo, High-frequency, magnetic-field-responsive drug release from magnetic nanoparticle/organic hybrid based on hyperthermic effect, *ACS Applied Materials & Interfaces*, 2(7) (2010) 1903–1911.

170. J. M. Shen, T. Yin, X. Z. Tian, F. Y. Gao, S. Xu, Surface charge-switchable polymeric magnetic nanoparticles for the controlled release of anticancer drug, *ACS Applied Materials & Interfaces*, 5(15) (2013) 7014–7024.

171. E. A. Vitol, E. A. Rozhkova, V. Rose, B. D. Stripe, N. R. Young, E. E. W. Cohen, L. Leoni, V. Novosad, Efficient cisplatin pro-drug delivery

visualized with sub-100 nm resolution: Interfacing engineered thermosensitive magnetomicelles with a living system, *Advanced Materials Interfaces*, 1(7) (2014) 1400182.

172. A. Sinha, A. Chakraborty, N. R. Jana, Dextran-gated, multifunctional mesoporous nanoparticle for glucose-responsive and targeted drug delivery, *ACS Applied Materials & Interfaces*, 6(24) (2014) 22183–22191.

173. Z. Behrouzkia, Z. Joveini, B. Keshavarzi, N. Eyvazzadeh, R. Z. Aghdam, Hyperthermia: How can it be used?, *Oman Medical Journal*, 31(2) (2016) 89.

174. H. Liu, J. Zhang, X. Chen, X.-S. Du, J.-L. Zhang, G. Liu, W.-G. Zhang, Application of iron oxide nanoparticles in glioma imaging and therapy: From bench to bedside, *Nanoscale*, 8(15) (2016) 7808–7826.

175. K. H. Bae, M. Park, M. J. Do, N. Lee, J. H. Ryu, G. W. Kim, C. Kim, T. G. Park, T. Hyeon, Chitosan oligosaccharide-stabilized ferrimagnetic iron oxide nanocubes for magnetically modulated cancer hyperthermia, *ACS Nano*, 6(6) (2012) 5266–5273.

176. M. Torres-Lugo, C. Rinaldi, Thermal potentiation of chemotherapy by magnetic nanoparticles, *Nanomedicine: Nanotechnology, Biology, and Medicine*, 8(10) (2013) 1689–1707.

177. A. Jordan, P. Wust, H. Fähling, W. John, A. Hinz, R. Felix, Inductive heating of ferrimagnetic particles and magnetic fluids: Physical evaluation of their potential for hyperthermia, *International Journal of Hyperthermia*, 25(7) (2009) 499–511.

178. P. Guardia, R. Di Corato, L. Lartigue, C. Wilhelm, A. Espinosa, M. Garcia-Hernandez, F. Gazeau, L. Manna, T. Pellegrino, Water-soluble iron oxide nanocubes with high values of specific absorption rate for cancer cell hyperthermia treatment, *ACS Nano*, 6(4) (2012) 3080–3091.

179. K. H. Bae, M. Park, M. J. Do, N. Lee, J. H. Ryu, G. W. Kim, C. Kim, T. G. Park, T. Hyeon, Chitosan oligosaccharide-stabilized ferrimagnetic iron oxide nanocubes for magnetically modulated cancer hyperthermia, *ACS Nano*, 6(6) (2012) 5266–5273.

180. K. S. Kim, J. Kim, J. Y. Lee, S. Matsuda, S. Hideshima, Y. Mori, T. Osaka, K. Na, Stimuli-responsive magnetic nanoparticles for tumor-targeted bimodal imaging and photodynamic/hyperthermia combination therapy, *Nanoscale*, 8(22) (2016) 11625–11634.

181. R. Yang, Q. Tang, F. Miao, Y. An, M. Li, Y. Han, X. Wang, J. Wang, P. Liu, R. Chen, Inhibition of heat-shock protein 90 sensitizes liver

cancer stem-like cells to magnetic hyperthermia and enhances anti-tumor effect on hepatocellular carcinoma-burdened nude mice, *International Journal of Nanomedicine*, 10 (2015) 7345–7358.

Chapter 8

Ultrasound-Responsive Nanomedicine

Tyrone M. Porter[a] and Jonathan A. Kopechek[b]

[a]*Mechanical Engineering and Biomedical Engineering, Boston University, Boston, Massachusetts 02215, USA*
[b]*Bioengineering, University of Louisville, Louisville, Kentucky 40292, USA*

tmp@bu.edu, jonathan.kopechek@louisville.edu

8.1 Introduction

Ultrasound is widely used throughout the world for sensing and imaging applications. In nature, ultrasound is used by animals such as bats, whales, and dolphins for communication and to locate prey. Over the past century, ultrasound technology has been developed for many industrial and military applications, including non-destructive evaluation of materials (to locate cracks in aircraft skin for example), underwater sensing (Sonar), and motion tracking. The development of ultrasound technology also led to advances in medical imaging applications. Ultrasound imaging was first developed in the 1950s by an engineer (Tom Brown)

Stimuli-Responsive Nanomedicine
Edited by Lin Zhu

ISBN 978-981-4800-70-9 (Hardcover), 978-0-429-29529-4 (eBook)
www.jennystanford.com

and an obstetrician (Ian Donald) in Scotland. Medical ultrasound is now widely used for imaging applications which include echocardiography (imaging heart anatomy and function), gynecological ultrasonography (including fetal imaging), and abdominal sonography.

There are several advantages to the use of ultrasound for medical imaging compared to other clinical imaging modalities such as magnetic resonance imaging (MRI), computed tomography (CT), or positron emission tomography (PET). Ultrasound imaging possesses excellent temporal resolution with frame rates that can exceed 100 frames per second, providing near real-time images. Ultrasound can also simultaneously provide both anatomical and physiologic (blood flow) information. Unlike CT and PET, ultrasound does not emit harmful ionizing radiation. Furthermore, ultrasound is portable and costs much less than most other clinical imaging technologies. However, clinical ultrasound has limited spatial resolution with minimum resolutions typically in the range of 0.1–1 mm, although recent advances in post-processing algorithms have significantly improved image quality. Ultrasound also has limited penetration of bone and gas, which presents a challenge for imaging the brain and lungs. Most organs and tissues, however, are readily imaged with ultrasound at penetration depths of at least 8 cm. Taken together, these features have led to the adoption of ultrasound as one of the most widespread clinical imaging modalities.

8.2 Principles of Diagnostic Ultrasound

Ultrasound is defined as pressure waves oscillating at frequencies above the audible range (>20 kHz). Clinical ultrasound pulses are generated by transducers consisting of multiple piezoelectric crystals, which convert electrical signals into pressure waves that propagate through the medium. As the ultrasound waves propagate, they reflect off any interfaces between media that have different acoustic impedances, sending reflected waves ("echoes") back to the transducer where they are converted back into electrical signals. The acoustic impedance of each medium is proportional to its density and speed of sound, which is a function of its compressibility. Thus, interfaces between tissues with different densities and sound speeds (such as bone and fat, for example) will produce reflections that provide contrast on ultrasound images. The locations and intensities of these reflections are converted into pixel intensities in order to generate 2-D or 3-D ultrasound images.

Clinical ultrasound systems generally transmit sinusoidal pulses consisting of positive (compression) and negative (rarefaction) pressure waves. The center frequency (i.e., "fundamental frequency") is selected based on the application. Clinical ultrasound imaging is most commonly performed at frequencies between 1–10 MHz. Tissue attenuation of ultrasound increases as a function of frequency, which negatively impacts the propagation depth of the transmitted and reflected waves. A lower center frequency provides lower spatial resolution but significantly improves tissue penetration, which can be important for imaging the heart or deep abdominal organs. On the other hand, a higher center frequency provides higher spatial resolution but poorer penetration and thus is better suited for superficial or intravascular imaging. The maximum allowable ultrasound intensities delivered by clinical imaging systems are limited by the FDA in order to minimize the risk of adverse bioeffects. However, in certain applications, the bioeffects induced at high ultrasound intensities can be utilized for therapeutic benefit.

8.3 Therapeutic Ultrasound

In addition to the widespread use of ultrasound for diagnostic imaging applications, therapeutic ultrasound applications have also been developed. Clinical therapeutic ultrasound applications primarily utilize high-intensity focused ultrasound (HIFU) to generate large pressure changes locally in tissue (Fig. 8.1). Extracorporeal shock wave lithotripsy has been used in patients since the 1980s to mechanically break up kidney and gall stones with minimal heating [1]. This non-invasive treatment is approved for clinical use in many countries, including the United States. More recently, therapeutic ultrasound has been developed for thermal ablation applications.

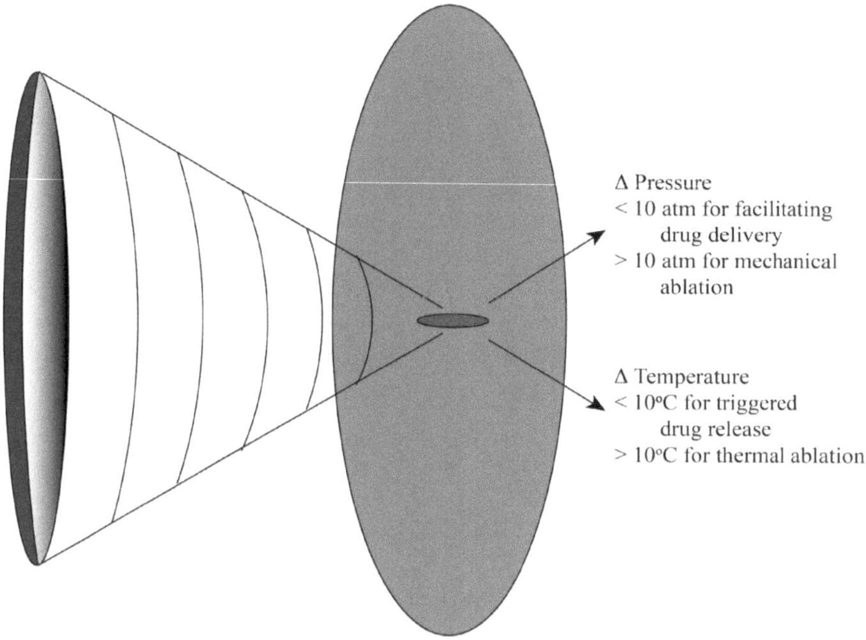

Figure 8.1 Focused ultrasound can modulate the local pressure in tissue by several atmospheres and elevate the local temperature by tens of degrees. These pressure and temperature changes can be used for various therapeutic applications, including triggered drug release and tumor ablation.

Long ultrasound pulses at very high pressures can generate significant heating at a target site, which occurs due to absorption of acoustic energy by tissue (Fig. 8.1). Magnetic resonance (MR) thermometry can be used to quantify temperature elevations in tissue and produce two-dimensional heat maps for image guidance [2]. MR thermometry has been leveraged for guiding the placement of the ultrasound focus in tissue before sonication and for feedback control of ultrasound-induced heating during the procedure [3–5]. This heating effect can be harnessed to non-invasively burn away tissue at a precise target site and has been used in patients to ablate many types of tumors [4, 6, 7]. Mild hyperthermia, which is ideal for heat-triggered release of drugs from thermosensitive liposomes (TSL), is achieved with lower acoustic pressures. For example, Dromi et al. used focused ultrasound and low acoustic pressures to heat implanted tumors by 4–5°C and trigger release of doxorubicin from lysolipid-containing TSL [8]. Designed predominately to release entrapped drugs when heated above 44°C, the threshold temperature and rate of drug release from TSL can be augmented with the inclusion of membrane-disruptive polymers or lysolipids in their formulation [9]. TSL can also be co-loaded with MR contrast agents (e.g., gadolinium) and drugs to enable noninvasive MR-based detection of heat-triggered release [10, 11]. Using the combination of MR-guided focused ultrasound and polymer-modified TSL, scientists have successfully delivered lethal dosages of doxorubicin locally to solid tumors and observed a positive response to therapy with negligible adverse effects [12]. Moreover, the complete regression of solid tumors in mice has been achieved with multiple treatments of ultrasound-mediated mild hyperthermia after systemic administration of TSL containing a copper-doxorubicin complex [13]. This result demonstrates the potential clinical benefit afforded by the use of noninvasive focused ultrasound for local heat-triggered release of chemotherapy within solid tumors.

Clinical use of HIFU therapy, which is often guided using MRI, is rapidly growing and has obtained regulatory approvals throughout the world, including Asia, Europe, and the United States, for multiple indications involving tumor ablation. Additional indications, including treatment of neurological conditions, are currently being tested in clinical trials [14–16].

Recent research efforts have focused on developing a new therapeutic ultrasound approach utilizing low acoustic pressures (as used in diagnostic imaging) for targeted drug and gene delivery applications. This non-invasive approach uses microbubbles as theranostic agents to deliver molecular therapeutics to a target site in response to ultrasound pulses. As the field of nanomedicine has emerged, it has intersected with therapeutic ultrasound in several areas. This chapter describes the recent developments and breakthroughs in ultrasound-responsive nanomedicine.

8.4 Ultrasound Contrast Agents

Microbubbles coated with a stabilizing shell have been used as ultrasound contrast agents in patients for decades, primarily to image blood flow and perfusion. Contrast-enhanced ultrasound is approved for tumor detection in Asia, Europe, and other countries, but in the United States, it is currently approved only to assess left ventricular function. Microbubbles, which are generally ~1–3 microns in diameter, consist of a gas core surrounded by a surfactant such as phospholipid, protein, or polymer. The gas core typically contains a perfluorocarbon or other inert compound with low solubility in water in order to prolong stability in circulation. Antibodies or peptides can be conjugated to the shell for targeted imaging of specific tissues [17–22]. For drug and gene delivery applications, however, even in the absence of antibody conjugation ultrasound can be focused to induce microbubble destruction at a target site.

Microbubbles respond more strongly to ultrasound compared to tissue due to the high compressibility of their gas core. As microbubbles are insonified, the ultrasound pressure waves cause compression and expansion of the microbubbles during each acoustic cycle (Fig. 8.2). When driven at relatively low-pressure amplitudes, oscillating microbubbles radiate pressure waves, which are used in diagnostic ultrasound to enhance contrast in the images. When driven at higher pressures, microbubbles undergo large amplitude oscillations, known as "cavitation." The oscillations become non-linear at moderate acoustic pressures and eventually the microbubble can collapse due to the inertia of the surrounding fluid ("inertial cavitation"). Cavitation can induce microstreaming and microjetting in the surrounding fluid [23–25], which can have profound effects on nearby structures. For example, ship propellers can be damaged by long-term cavitation activity occurring in near proximity. In biology, microstreaming and microjetting can stimulate bioeffects that can transiently enhance vascular permeability [26, 27] and can also transiently porate cell membranes, a phenomenon known as "sonoporation" [28, 29]. The small pores formed in cell membranes during sonoporation provide a temporary window for direct delivery of molecular therapeutics (such as drugs or genes) before the pores are repaired by the cell (Fig. 8.3) [30]. A range of pore sizes can occur, but pores with diameters of approximately 100–200 nm have been reported [31]. In addition to sonoporation, cavitation has also been shown to enhance endocytosis [32, 33].

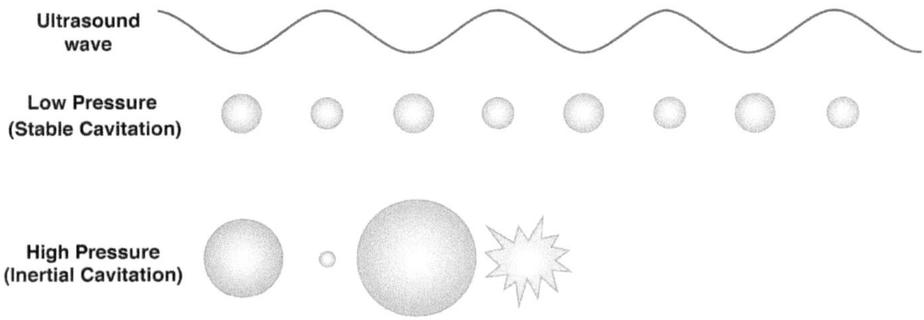

Figure 8.2 Ultrasound contrast agents respond to ultrasound wave, stably oscillating (stable cavitation) or rapidly expanding and collapsing (inertial cavitation). Stable cavitation is leveraged for diagnostic applications while inertial cavitation is typically used for therapeutic applications.

Researchers have leveraged the nonlinear oscillatory response of UCA and the associated stresses to increase the permeability of blood vessels for soluble molecules and drug-loaded colloids. An application that has garnered considerable attention is the use of focused ultrasound and ultrasound contrast agents to permeabilize the blood–brain barrier (BBB). The BBB protects the brain from pathogens and foreign compounds that gain access to the circulatory system. Focused ultrasound can be transmitted directly through the skull when operated at submegahertz frequencies, enabling noninvasive, reversible, safe and highly localized opening of the BBB when combined with UCA [34–36]. Several therapeutic agents have been delivered to brain tissue after ultrasound-mediated opening of the BBB, including proteins, small interfering RNA (siRNA), immune cells, stem cells, viral nanocarriers, and chemotherapeutic agents entrapped within nanoparticles [37–40]. Researchers have demonstrated the potential of this strategy for various applications, including treatment of brain tumors and neurodegenerative diseases [41–48].

In summary, cavitation-induced bioeffects can be utilized to significantly enhance delivery of molecular therapeutics to target cells.

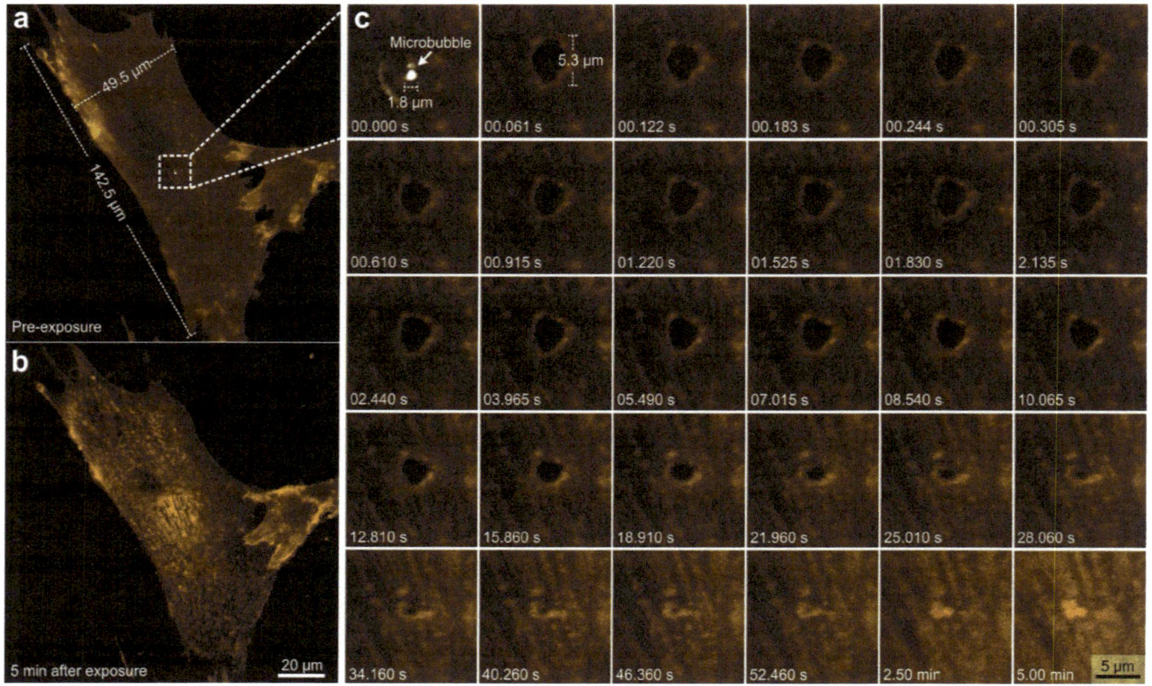

Figure 8.3 Real-time imaging of reversible cavitation-induced pore formation in a cell membrane captured with confocal microscopy. Confocal images of a fibroblast (a) before sonication with an ultrasound contrast agent (UCA) deposited on its membrane (b) and 5 min after sonication. A pore was formed in the cell membrane due to inertial cavitation nucleated by the UCA (c). The cell was able to recover and seal the pore over the course of 5 minutes. Reprinted from Ref. 30 with permission from Elsevier.

8.5 Microbubbles and Nanoparticles for Biomedical Ultrasound Applications

Significant research efforts have focused on engineering drug and gene carriers for intravenous delivery to target organs. These carriers must overcome several physiological barriers that hinder intravenous delivery of molecular therapeutics, which include (1) avoiding rapid clearance from circulation via the reticuloendothelial system, (2) avoiding enzymatic degradation in circulation, (3) escaping the vasculature at the target site, (4) crossing cellular membranes (typically via endocytosis), and (5) being released from intracellular vesicles in order to function inside the cell cytoplasm. Several novel formulations of nanoparticles have been developed to overcome these barriers.

8.5.1 Ultrasound and Lipid-Based Colloids for Localized Drug Delivery

Liposomes were first produced in the 1960s and have since been developed for many drug and gene delivery applications [49, 50]. Several liposome formulations have been approved in the United States and elsewhere for delivery of anti-cancer drugs such as doxorubicin [51]. Liposomes, which are typically less than 200 nm in diameter, consist of a phospholipid bilayer surrounding an aqueous core into which therapeutic molecules can be encapsulated. Targeting antibodies can be conjugated to the lipid shell but passive accumulation of liposomes has been observed in tumors, likely due to the enhanced permeability and retention (EPR) effect [52, 53]. Liposomes have been designed to release encapsulated payloads in response to various stimuli, including changes in pH, heat, and exposure to light [9, 54–58]. However, effective techniques to enhance delivery of liposomal payloads beyond the endothelium remain under investigation. To this end, recent efforts have focused on utilizing ultrasound to trigger drug release from liposomes. One approach utilizes low frequency ultrasound (LFUS) to enhance drug release from sono-sensitive liposomes. Enhanced delivery of drugs such as cisplatin and doxorubicin has been demonstrated using ultrasound at frequencies as low as 20 kHz [59, 60]. Although the mechanisms are not fully understood, it is thought that nucleation of cavitation may occur within the cell membrane at the ultrasound intensities used in these studies, which could permeabilize the lipid bilayer [61]. However, clinical ultrasound imaging systems generally operate at frequencies above 1 MHz and the onset of cavitation at these frequencies requires higher ultrasound intensities in the absence of cavitation nuclei (i.e., ultrasound contrast agents).

Another approach in development utilizes microbubbles to enhance liposomal delivery at the higher ultrasound frequencies (>1 MHz) used in diagnostic imaging applications. Drug-loaded liposomes can be attached to the microbubble shell using linkers such as biotin and streptavidin. Due to the size difference (radii of 50–100 nm for liposomes and >1000 nm for microbubbles) many liposomes can potentially be loaded onto each microbubble. Yu et al. [62], attached doxorubicin-loaded liposomes to polymer-shelled microbubbles, while Lentacker et al. [63], used a similar approach with lipid-shelled microbubbles. Both studies demonstrated ultrasound-enhanced drug release from the liposomes, which increased cytotoxicity in cancer cells. Liposome-microbubble complexes have also been tested in vivo. Yan et al. [64], delivered paclitaxel with liposomes attached to microbubbles and observed significantly enhanced drug targeting and tumor growth inhibition.

Nanobubbles, bubble liposomes (BLs) and echogenic liposomes (ELIPs) represent another formulation of sono-sensitive submicron colloids that are in development [65–67]. Whereas BL and ELIP formulations consist of a gas core coated with lipids, nanobubbles can be coated with lipids and/or polymers [68]. These largely submicron gas bodies can be detected with ultrasound imaging, and can be loaded with drugs or other therapeutic agents. The formulations appear to consist of a heterogeneous mixture of bubbles, liposomes, and bubble-containing liposomes. Nanobubbles, BLs and ELIPs have been used to deliver various therapeutic agents including plasmid DNA, morpholino oligomers, xenon gas for neuroprotection, nitric oxide, and thrombolytic drugs [65, 69–75].

Altogether, these studies demonstrate that ultrasound and microbubbles can be utilized to enhance targeted delivery of drugs or genes from lipid-based colloids. Efforts to move this therapy from preclinical research into clinical testing are ongoing.

8.5.2 Ultrasound and Polymeric Nanoparticles for Localized Drug Delivery

The recent development of polymeric nanoparticles has generated strong interest for drug and gene delivery applications due to their drug loading capacity and biological stability. Polymeric nanoparticles can be engineered to be biodegradable and biocompatible, to package water soluble or insoluble drugs, and to release their contents rapidly (i.e., minutes), gradually (i.e., days), or in response to a stimulus (i.e., acid, enzymatic cleavage, etc.). This versatility makes polymeric nanoparticles extremely appealing for drug delivery applications, whether systemic or localized. Ultrasound may be used as an external energy source to enhance drug release from polymeric nanoparticles as well as to enhance cell internalization of released drug or drug-loaded polymeric nanoparticles. In this section, we review studies reporting on the combinatorial strategy for localized drug delivery.

The interplay between ultrasound and micelles formed from Pluronic for chemotherapy applications has been studied extensively. Pluronic is a block copolymer composed of polyethylene oxide (PEO) and polypropylene oxide (PPO). When suspended in aqueous media, micelles assemble with PPO forming a hydrophobic core and PEO forming a hydrophilic corona. The hydrophobic core has been used to entrap hydrophobic and lipophilic drugs. It has been reported that doxorubicin release from Pluronic micelles could be enhanced with the application of ultrasound [76–78]. Additionally, ultrasound reportedly enhanced the cellular uptake of doxorubicin that had been encapsulated within Pluronic micelles [76, 78]. The authors proposed two mechanisms for the noted enhancement: (1) doxorubicin was released from the micelles by ultrasound and then internalized by cells, and (2) ultrasound increased the permeability of cell membranes to Pluronic micelles containing doxorubicin. There was strong evidence that inertial cavitation (i.e., rapid bubble expansion and inertia-driven collapse) played a critical role in ultrasound-mediated drug release and delivery to the cell cytoplasm. First, researchers detected acoustic emissions indicative of inertial cavitation activity that correlated with doxorubicin release from Pluronic micelles [79, 80]. Second, the researchers noted an inverse relationship between the ultrasound frequency and percent drug release and intracellular drug concentration after sonication [78]. This was most likely due to a reduction in inertial cavitation activity, which also has an inverse relationship with ultrasound frequency [81, 82]. Most of the in vitro studies were conducted with sub-megahertz ultrasound (i.e., 20–70 kHz), which is not ideal for in vivo applications. Recognizing this limitation, a series of experiments were conducted to examine the impact of megahertz frequency ultrasound on the delivery of Pluronic micelles to solid tumors [78, 83]. Studies show that 1 or 3 MHz ultrasound can induce release of doxorubicin from Pluronic micelles and increase doxorubicin delivery to malignant cells locally [78, 83], resulting in suppression of tumor growth [84]. These positive results prompted additional studies to evaluate the benefits of combining ultrasound with polymeric nanoparticles for local delivery of various therapeutic agents.

Cationic polymers, such as polyethylenimine (PEI), have been used for decades to package and transport oligonucleotides (i.e., DNA and RNA) to cells for transfection and modulation of protein expression [85, 86]. The resultant nanoparticles, termed polyplexes, protect the oligonucleotides from enzymatic degradation in blood and can facilitate transport across the cell membrane. However, cationic polymer polyplexes suffer from poor delivery efficiency and can have toxic effects on cells and tissue. Ultrasound combined with stabilized microbubbles has been utilized to increase the oligonucleotide delivery efficiency of cationic polymers, thereby reducing the polymer concentration required and associated toxicities. For example, Deshpande and Prausnitz reported that ultrasound combined with albumin-stabilized microbubbles increased the delivery of PEI/DNA polyplexes to cells by 55% [87]. In a separate study, Li et al. noted that ultrasound combined with a commercial ultrasound contrast agent enhanced PEI-mediated gene transfection of bone marrow stem cells by a factor of 38 [88]. Studies have shown that acoustic cavitation was most likely the primary mechanism for the observed enhancement in PEI/DNA delivery to cultured cells [89–91].

Researchers have leveraged cavitation-enhanced gene delivery to transfect cells and tissue in vivo with a plasmid expressing the reporter protein luciferase [92]. A strong luciferase signal was detected for up to 30 days from tissue treated with ultrasound after administration of PEI/DNA polyplexes [92]. Sustained protein expression can be difficult to achieve in vivo with non-viral gene delivery vehicles; thus, this result was extremely encouraging. Scientists had similar success delivering circulating polyplexes containing DNA coding for luciferase to brain tissue after opening the BBB temporarily with ultrasound and microbubbles [93]. The result of this work is particularly exciting given the difficulty of delivering almost any therapeutic agent across the BBB. Moreover, the study demonstrates the potential of ultrasound and cationic polyplexes for addressing devastating brain diseases that historically have been virtually untreatable. Potentially, gene delivery across vessel walls can be further enhanced by linking PEI/DNA polyplexes to microbubbles, thus ensuring the polyplexes and microbubbles are co-localized within the ultrasound beam [94].

In addition to DNA, scientists have also utilized ultrasound-mediated cavitation to enhance the delivery of short hairpin RNA (shRNA) complexed with PEI into solid tumors [95, 96]. By designing the shRNA to suppress the expression of oncogenes that promote proliferation and survival, successful delivery of the oligonucleotide can combat tumor growth and induce apoptosis (i.e., programmed cell death). Research in this area is in its infancy and most studies have been conducted with reporter proteins that do not impact cell biology. Additional studies are needed to examine the clinical utility of the strategy for delivery of DNA and RNA as well as the effect on cell biology and viability as a function of administered dose.

8.5.3 Photoacoustics and Multimodality Imaging with Nanoparticles

Ultrasound-responsive nanoparticles are also in development for multimodality imaging applications. For example, super-paramagnetic iron oxide (SPIO) nanoparticles have been loaded onto microbubbles for use as dual contrast agents in magnetic resonance and ultrasound imaging [97, 98]. In addition, nanoparticles containing optical contrast have been attached to microbubbles for dual optical and ultrasound contrast [99, 100].

Emerging multimodality imaging applications based on the photoacoustic effect have been recently developed. This phenomenon arises from the interaction of light and sound in tissue [101]. As pulsed or intensity-modulated light passes through tissue, the absorption of photons by endogenous chromophores (i.e., hemoglobin, fat, etc.) causes a slight increase in temperature and transient thermal expansion, which generates pressure waves that can be detected with ultrasound transducers (Fig. 8.4). Photoacoustic images can be generated using photoacoustic tomography (PAT), which spatially maps optical absorption in the biological media [102]. Photoacoustic microscopy (PAM) is a specific application of photoacoustic tomography that can generate high-resolution images (typically <20 μm per pixel) by coaxially focusing acoustic and optical beams [103]. This technology has generated significant interest due to the unique information that can be obtained, such as high sensitivity to hemoglobin oxygenation levels.

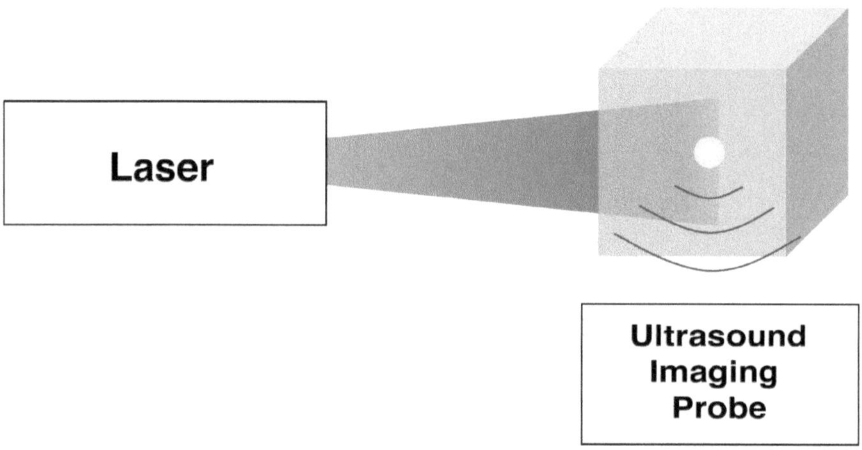

Figure 8.4 A schematic of the photoacoustic effect. A chromophore (i.e., compound or particle) is irradiated at its peak absorption wavelength resulting in rapid heating. A pressure wave is radiated during thermal expansion, which is received by an ultrasound imaging probe and used for image reconstruction.

The field of photoacoustic imaging (PAI) has expanded rapidly, but its limited imaging depth due to multiple scattering of the incident optical field poses a challenge to clinical translation.

Nanoparticles have been investigated as a possible solution to this problem. A major advantage of nanoparticles compared to endogenous chromophores (i.e., hemoglobin, lipids, fat, etc.) is the ability to tailor their optical properties. For example, nanoparticles engineered to absorb light at near-infrared wavelengths have been used to improve the photoacoustic imaging depth [104]. Additionally, the ability to fabricate nanoparticles with unique optical absorption spectra enables multispectral photoacoustic imaging [105]. Tuning the optical properties of nanoparticles is made possible with the use of noble metals such as gold for synthesis. When gold nanoparticles are illuminated, electrons at the surface oscillate collectively at a resonance frequency, which dramatically enhances optical absorption and scatter. This phenomenon, known as surface plasmon resonance (SPR), leads to a significant enhancement in the optical absorption coefficient of the nanoparticle [106]. The SPR frequency and the associated absorption spectrum depend on the physical dimensions of the gold nanoparticle [106]. A fundamental understanding of this relationship has allowed scientists to tune the optical absorption spectrum of gold nanoparticles through control of their shape and size [107]. An assortment of gold nanoparticles has been evaluated as PAI agents, ranging from nanoshells to nanorods. The performance of these nanoparticles in the context of PAI and potential applications are reviewed in the following paragraphs.

Nanoshells represent one class of photoacoustic contrast agents and were among the earliest to be developed. These are composed of a dielectric core surrounded by a nanometer-thick shell composed of an optically absorbing material, such as gold. By controlling the shell thickness, scientists have produced gold nanoshells with peak absorption between 600–900 nm [108, 109]. Gold nanoshells have been investigated for several applications, including imaging of cancer, atherosclerotic plaques, and rodent brain vasculature. In a study by Li et al., photoacoustic microscopy was employed to image extravasation and accumulation of gold nanoshells in murine colon carcinomas in vivo [73]. The authors found that gold nanoshells continued to accumulate in the mouse tumors for at least 6 h after injection. A study by Lee et al. investigated the release of a chemotherapeutic agent from hollow gold nanoshells with laser ablation thermal therapy and photoacoustic imaging [110]. Doxorubicin, a fluorescent chemotherapeutic drug, was loaded onto gold nanoshells, which also served to accelerate heating in laser ablation thermal therapy. Fluorescence imaging detected a significant increase in fluorescence intensity within laser-treated mouse tumors following injection of doxorubicin-loaded gold nanoshells. Furthermore, photoacoustic imaging revealed significantly enhanced heating in tumors with gold nanoparticles. This study demonstrated the potential utility of photoacoustic imaging to monitor nanoshell-enhanced laser therapy and drug delivery.

Gold nanoshells have also been used for PAI in blood vessels and organs. Rouleau et al. conjugated VCAM-1 antibody onto nanoshells for targeted detection of atherosclerotic plaque in mice [111]. The authors showed selective targeting of VCAM-1 antibodies, which enabled photoacoustic-based molecular imaging of atherosclerosis in mice. Several studies have shown enhanced optical absorption and photoacoustic signal from brain vasculature following injection of gold nanoshells in mice and rats (Fig. 8.5), with enhanced signal lasting on the order of hours [112–114]. These studies have helped to advance potential applications for contrast-enhanced photoacoustic imaging in the brain.

Gold nanorods, which are exceptional optical absorbers, have been evaluated for PAI. SPR is more pronounced for gold nanorods than spherical nanoparticles due to anisotropy in the physical dimensions, which enhances the absorption coefficient [107]. As a result of the high absorption coefficient, gold nanorods are very effective PAI agents. Gold nanorods have been detected in vitro and in vivo with PAI at concentrations as low as 1.25 pM [115]. The extremely low detection limit of gold nanorods makes PAI a viable modality for molecular imaging applications. Scientists have reported the photoacoustic detection of HER2/neu-targeted gold nanorods bound to a single prostate cancer cell [116] as well as ICAM-1 targeted gold nanorods bound to inflamed endothelial cells [117]. To enable multispectral photoacoustic imaging, gold nanorods with different absorption spectra can be synthesized by varying their physical dimensions, particularly the aspect ratio (length:width ratio) [106]. For example, gold nanoparticles synthesized to absorb at 785 and 1000 nm were utilized for multispectral PAI-based molecular imaging of the cancer biomarkers HER2 and EGFR simultaneously [118]. Lastly, gold nanorods have been utilized for PAI combined with surface enhanced Raman spectroscopy for cancer detection and resection guidance [119], or combined with magnetic resonance imaging and photothermal therapy for theranostic applications [120].

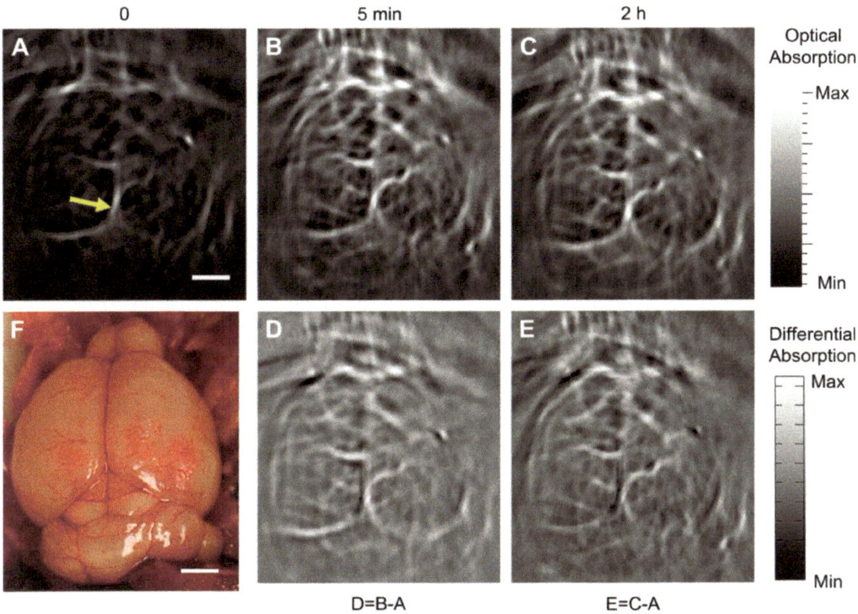

Figure 8.5 Intracranial vasculature was visualized with photoacoustic imaging after systemic administration of hollow gold nanospheres. The nanospheres had a mean diameter of 43–46 nm, a mean shell thickness of 2–4 nm, and the optical absorption peaked at 800 nm. The photoacoustic signal was weak before nanosphere administration (A), strong 5 minutes after injection (B), and endured for at least 2 h (C). Note that individual vessels were resolved by PAI. Differential photoacoustic images (D, E) were obtained by subtracting the pre-injection images from the post-injection images. Reprinted from Ref. 114 with permission from Elsevier.

The last gold nanoparticle that has been tested for PAI applications is gold nanocages. These nanoparticles are prepared by depositing gold on a silver sacrificial template [121]. Removal of the silver produces a gold nanostructure with a hollow interior and ultrathin porous walls. This process is very versatile and allows for producing nanocages with various geometries, such as nanorings, nanoboxes, and nanotubes [121]. The absorption spectrum can be tailored by controlling the size, shape, and wall thickness. Nanocages can be engineered to have an optical absorption cross-section that is superior to nanoshells and nanorods [107, 122], making these nanoparticles ideal for photoacoustic imaging and theranostic applications [123]. For example, gold nanocages have been utilized as photoacoustic agents to acquire tomographic images of blood flow in rat cerebral cortex [113]. Additionally, PAI has been used to track the accumulation of gold nanocages in sentinel lymph nodes after injection into the forepaw pad of rats [124]. This approach could be used to help guide biopsies of sentinel lymph nodes clinically in search of metastasized cancer cells. In a study conducted by Kim et al., gold nanocages were bioconjugated with a tumor specific ligand ([Nle^4,d-Phe^7]-α-melanocyte-stimulating hormone) to target melanoma cells [125]. Photoacoustic imaging was performed for up to 6 h after injection of the nanocages in melanoma tumor-bearing mice. The authors showed that targeted nanocages provided high specificity of melanomas using photoacoustic imaging, and the detection limit was over 2000 times better compared to single-walled carbon nanotubes. Lastly, it has been demonstrated that stem cells labeled with gold nanocages can be tracked in vivo with PAI [126]. This strategy could be of value for monitoring various cell-based therapies, including cell and tissue transplantation, regenerative medicine, and immunotherapy.

In summary, the development of novel multimodality and photoacoustic agents provides the potential to expand theranostic applications available for medicine in order to improve patient care.

8.6 Liquid Nanodroplets for Ultrasound Therapy

Liquid perfluorocarbon nanodroplets have been recently investigated as a new type of ultrasound-responsive agent. In contrast to gas-filled microbubbles, these nanodroplets contain a liquid perfluorocarbon core surrounded by a lipid or polymer shell. The liquid perfluorocarbon is inert, non-toxic, and has a low boiling point such that it can be vaporized into gas using ultrasound, and the resulting bubbles can be detected with ultrasound imaging (Fig. 8.6). This phenomenon, known as

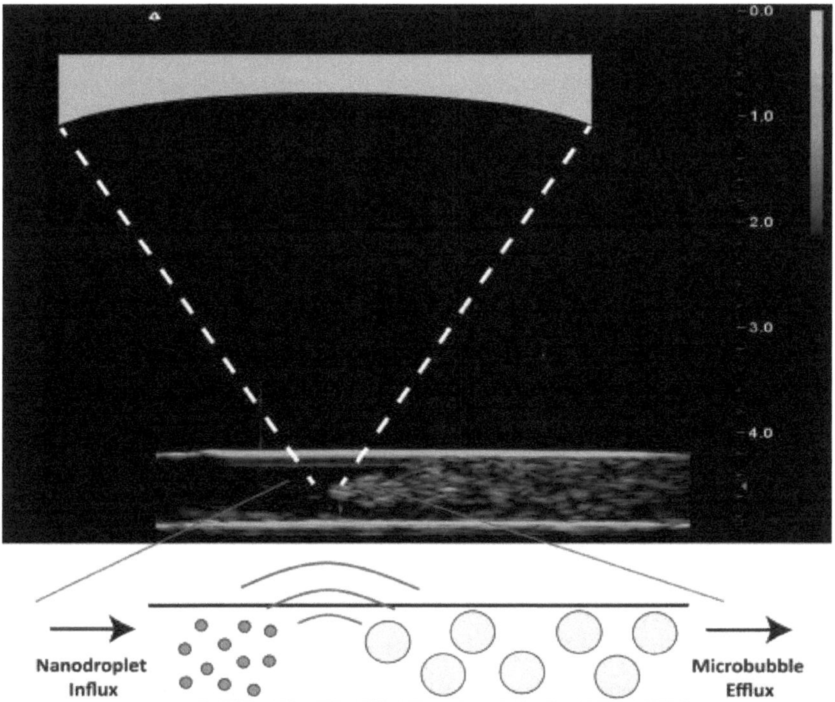

Figure 8.6 Liquid perfluorocarbon droplets are pumped through a dialysis tube. The droplets are incompressible and thus undetectable with ultrasound imaging. The droplets are vaporized within the tube by focused ultrasound, resulting in the formation of perfluorocarbon microbubbles that are compressible and detectable with ultrasound imaging.

"acoustic droplet vaporization," has been explored for multiple therapeutic ultrasound applications, including targeted delivery of drugs and genes [127]. The nanodroplets are generally 100–300 nm in diameter and have a much higher stability in circulation (several hours or more) compared to gas-filled microbubbles (typically less than 30 min) [128]. The stability and small size of the liquid nanodroplets allows them to accumulate in tumors through the EPR effect similar to other nanoparticles, in contrast to gas-filled microbubbles which are generally too large to extravasate from vessels. Furthermore, the fluorine content in perfluorocarbon nanodroplets can be detected with magnetic resonance imaging, thereby enabling their use as multimodality contrast agents [129]. Together, these properties have generated strong interest in developing perfluorocarbon nanodroplet emulsions as theranostic agents, and research efforts are actively growing in this area.

8.6.1 Nanodroplets for Ultrasound-Targeted Drug and Gene Delivery

Perfluorocarbon nanodroplets have been developed for ultrasound-targeted delivery of molecular therapeutics, including chemotherapeutics. Rapaport et al. developed perfluorocarbon nanodroplets with a block co-polymer shell to deliver doxorubicin and paclitaxel and successfully inhibit tumor growth in mice [130–133]. Fabiilli et al. produced double-emulsion perfluoro-carbon droplets for ultrasound-mediated delivery of chlorambucil and thrombin in vitro [134, 135]. Another formulation recently developed for drug and gene delivery, known as "eLiposomes," consists of perfluorocarbon nanodroplets encapsulated within the aqueous core of liposomes. These formulations have been used to deliver a fluorescent dye (calcein), a plasmid expressing green fluorescent protein (GFP), or doxorubicin to HeLa cells [136–138]. In another study, Burgess and Porter successfully delivered a GFP-targeted siRNA to GFP-expressing human breast carcinoma cells using ultrasound-activated perfluorocarbon nanodroplets emulsions [139].

8.6.2 Nanodroplets for Imaging and Tumor Ablation

Perfluorocarbon nanodroplets are also in development as contrast agents for ultrasound imaging [140–142]. Kripfgans et al. showed enhanced contrast in canine brain and kidneys following intravenous injection of perfluorocarbon droplets [143]. Sheeran et al. used a single ultrasound transducer to induce acoustic droplet vaporization and simultaneously image contrast generated by vaporized nanodroplets in rat kidneys [144]. In another study, Williams et al. demonstrated enhanced ultrasound contrast in mouse tumors using acoustic droplet vaporization 1 h after intravenous infusion of perfluorocarbon nanodroplets [145].

Tumor ablation therapy is another area where perfluorocarbon nanodroplets emulsions may have a positive impact. Clinical use of HIFU-mediated tumor ablation is currently limited by the long treatment times required (generally several hours per patient). After the perfluorocarbon nanodroplets accumulate in the tumor through the EPR effect, they can be activated using acoustic droplet vaporization to generate microbubbles inside the tumor, which can be utilized to accelerate HIFU-mediated heating and tumor ablation. Zhang et al. infused perfluorocarbon droplets intravenously to enhance HIFU-mediated ablation of canine liver [146]. In studies by Kopechek et al., perfluorocarbon nanodroplet emulsions ("phase-shift nanoemulsions") were used to accelerate HIFU-mediated heating and ablation in rabbit VX2 tumors 2 h after intravenous injection [147–149]. Moyer et al. used perfluorocarbon nanodroplets to enhance HIFU-mediated heating in rat liver [150], and Zhou et al. demonstrated enhanced HIFU-mediated ablation of tumors in mice using perfluorocarbon nanoemulsions [151]. Together, these studies demonstrate that perfluorocarbon nanodroplets can potentially overcome the challenges that currently limit clinical applications of HIFU-mediated tumor ablation.

8.7 Future Outlook

The field of ultrasound-responsive nanomedicine is rapidly growing. In preclinical studies, ultrasound and microbubbles have been utilized to enhance drug and gene delivery with liposomes, polymeric nanoparticles, and perfluorocarbon nanodroplets. Emulsions of perfluorocarbon nanodroplets have been developed to accelerate high-intensity focused ultrasound ablation of tumors. Furthermore, ultrasound and nanoparticles are being utilized in new multimodal imaging applications involving MRI and photoacoustics.

Each year, new clinical trials are initiated to test novel therapeutic ultrasound applications. It is highly likely that the fusion of nanomedicine with ultrasound imaging technologies will transform medicine and significantly improve the treatment of many diseases. Ultrasound imaging is well established, but the field of ultrasound-responsive nanomedicine has only recently emerged, and the future outlook for this field is very promising.

References

1. Lingeman JE, McAteer JA, Gnessin E, and Evan AP, Shock wave lithotripsy: Advances in technology and technique. *Nat Rev Urol*, 2009. 6(12): pp. 660–670.
2. Rieke V and Butts Pauly K, MR thermometry. *J Magn Reson Imaging*, 2008. 27(2): pp. 376–390.
3. Kennedy JE, High-intensity focused ultrasound in the treatment of solid tumours. *Nat Rev Cancer*, 2005. 5(4): pp. 321–327.
4. Jolesz FA, Hynynen K, McDannold N, and Tempany C, MR imaging-controlled focused ultrasound ablation: A noninvasive image-guided surgery. *Magn Reson Imaging Clin N Am*, 2005. 13(3): pp. 545–560.
5. Jolesz FA, MRI-guided focused ultrasound surgery. *Annu Rev Med*, 2009. 60: pp. 417–430.
6. Wu F, Wang ZB, Zhu H, Chen WZ, Zou JZ, Bai J, et al., Feasibility of US-guided high-intensity focused ultrasound treatment in patients with advanced pancreatic cancer: Initial experience. *Radiology*, 2005. 236(3): pp. 1034–1040.
7. Illing RO, Kennedy JE, Wu F, ter Haar GR, Protheroe AS, Friend PJ, et al., The safety and feasibility of extracorporeal high-intensity focused ultrasound (HIFU) for the treatment of liver and kidney tumours in a Western population. *Br J Cancer*, 2005. 93(8): pp. 890–895.
8. Dromi S, Frenkel V, Luk A, Traughber B, Angstadt M, Bur M, et al., Pulsed-high intensity focused ultrasound and low temperature-sensitive liposomes for enhanced targeted drug delivery and antitumor effect. *Clin Cancer Res*, 2007. 13(9): pp. 2722–2727.
9. Ta T and Porter TM, Thermosensitive liposomes for localized delivery and triggered release of chemotherapy. *J Control Release*, 2013. 169(1–2): pp. 112–125.
10. de Smet M, Heijman E, Langereis S, Hijnen NM, and Grull H, Magnetic resonance imaging of high intensity focused ultrasound mediated drug delivery from temperature-sensitive liposomes: An in vivo proof-of-concept study. *J Control Release*, 2011. 150(1): pp. 102–110.

11. Negussie AH, Yarmolenko PS, Partanen A, Ranjan A, Jacobs G, Woods D, et al., Formulation and characterisation of magnetic resonance imageable thermally sensitive liposomes for use with magnetic resonance-guided high intensity focused ultrasound. *Int J Hyperthermia*, 2011. 27(2): pp. 140–155.
12. Ta T, Bartolak-Suki E, Park EJ, Karrobi K, McDannold NJ, and Porter TM, Localized delivery of doxorubicin in vivo from polymer-modified thermosensitive liposomes with MR-guided focused ultrasound-mediated heating. *J Control Release*, 2014. 194: pp. 71–81.
13. Kheirolomoom A, Lai CY, Tam SM, Mahakian LM, Ingham ES, Watson KD, and Ferrara KW, Complete regression of local cancer using temperature-sensitive liposomes combined with ultrasound-mediated hyperthermia. *J Control Release*, 2013. 172(1): pp. 266–273.
14. Elias WJ, Huss D, Voss T, Loomba J, Khaled M, Zadicario E, et al., A pilot study of focused ultrasound thalamotomy for essential tremor. *N Engl J Med*, 2013. 369(7): pp. 640–648.
15. Magara A, Buhler R, Moser D, Kowalski M, Pourtehrani P, and Jeanmonod D, First experience with MR-guided focused ultrasound in the treatment of Parkinson's disease. *J Ther Ultrasound*, 2014. 2: p. 11.
16. McDannold N, Clement GT, Black P, Jolesz F, and Hynynen K, Transcranial magnetic resonance imaging-guided focused ultrasound surgery of brain tumors: Initial findings in 3 patients. *Neurosurgery*, 2010. 66(2): pp. 323–332; discussion 332.
17. Villanueva FS, Lu E, Bowry S, Kilic S, Tom E, Wang J, et al., Myocardial ischemic memory imaging with molecular echocardiography. *Circulation*, 2007. 115(3): pp. 345–352.
18. Villanueva FS and Wagner WR, Ultrasound molecular imaging of cardiovascular disease. *Nat Clin Pract Cardiovasc Med*, 2008. 5 Suppl 2: pp. S26–S32.
19. Weller GE, Wong MK, Modzelewski RA, Lu E, Klibanov AL, Wagner WR, and Villanueva FS, Ultrasonic imaging of tumor angiogenesis using contrast microbubbles targeted via the tumor-binding peptide arginine-arginine-leucine. *Cancer Res*, 2005. 65(2): pp. 533–539.
20. Kaufmann BA, Sanders JM, Davis C, Xie A, Aldred P, Sarembock IJ, and Lindner JR, Molecular imaging of inflammation in atherosclerosis with targeted ultrasound detection of vascular cell adhesion molecule-1. *Circulation*, 2007. 116(3): pp. 276–284.

21. Lindner JR, Microbubbles in medical imaging: Current applications and future directions. *Nat Rev Drug Discov*, 2004. 3(6): pp. 527–532.
22. Klibanov AL, Ligand-carrying gas-filled microbubbles: Ultrasound contrast agents for targeted molecular imaging. *Bioconjug Chem*, 2005. 16(1): pp. 9–17.
23. Wu J and Nyborg WL, Ultrasound, cavitation bubbles and their interaction with cells. *Adv. Drug Deliv. Rev.*, 2008. 60(10): pp. 1103–1116.
24. Collis J, Manasseh R, Liovic P, Tho P, Ooi A, Petkovic-Duran K, and Zhu Y, Cavitation microstreaming and stress fields created by microbubbles. *Ultrasonics*, 2010. 50(2): pp. 273–279.
25. Marmottant P and Hilgenfeldt S, Controlled vesicle deformation and lysis by single oscillating bubbles. *Nature*, 2003. 423(6936): pp. 153–156.
26. Park J, Zhang Y, Vykhodtseva N, Jolesz FA, and McDannold NJ, The kinetics of blood brain barrier permeability and targeted doxorubicin delivery into brain induced by focused ultrasound. *J Control Release*, 2012. 162(1): pp. 134–142.
27. Bekeredjian R, Kroll RD, Fein E, Tinkov S, Coester C, Winter G, et al., Ultrasound targeted microbubble destruction increases capillary permeability in hepatomas. *Ultrasound Med Biol*, 2007. 33(10): pp. 1592–1598.
28. Fan Z, Liu H, Mayer M, and Deng CX, Spatiotemporally controlled single cell sonoporation. *Proc Natl Acad Sci U S A*, 2012. 109(41): pp. 16486–16491.
29. Kooiman K, Foppen-Harteveld M, van der Steen AF, and de Jong N, Sonoporation of endothelial cells by vibrating targeted microbubbles. *J Control Release*, 2011. 154(1): pp. 35–41.
30. Hu Y, Wan JM, and Yu AC, Membrane perforation and recovery dynamics in microbubble-mediated sonoporation. *Ultrasound Med Biol*, 2013. 39(12): pp. 2393–2405.
31. Zhou Y, Kumon RE, Cui J, and Deng CX, The size of sonoporation pores on the cell membrane. *Ultrasound Med Biol*, 2009. 35(10): pp. 1756–1760.
32. Lentacker I, De Cock I, Deckers R, De Smedt SC, and Moonen CT, Understanding ultrasound induced sonoporation: Definitions and underlying mechanisms. *Adv Drug Deliv Rev*, 2014. 72: pp. 49–64.
33. Zeghimi A, Escoffre JM, and Bouakaz A, Role of endocytosis in sonoporation-mediated membrane permeabilization and uptake

of small molecules: A electron microscopy study. *Phys Biol*, 2015. 12(6): p. 066007.

34. McDannold N, Arvanitis CD, Vykhodtseva N, and Livingstone MS, Temporary disruption of the blood–brain barrier by use of ultrasound and microbubbles: Safety and efficacy evaluation in rhesus macaques. *Cancer Res*, 2012. 72(14): pp. 3652–3663.
35. Hynynen K, McDannold N, Sheikov NA, Jolesz FA, and Vykhodtseva N, Local and reversible blood-brain barrier disruption by noninvasive focused ultrasound at frequencies suitable for trans-skull sonications. *Neuroimage*, 2005. 24(1): pp. 12–20.
36. Hynynen K, McDannold N, Vykhodtseva N, Raymond S, Weissleder R, Jolesz FA, and Sheikov N, Focal disruption of the blood-brain barrier due to 260 kHz ultrasound bursts: A method for molecular imaging and targeted drug delivery. *J Neurosurg*, 2006. 105(3): pp. 445–454.
37. Aryal M, Arvanitis CD, Alexander PM, and McDannold N, Ultrasound-mediated blood-brain barrier disruption for targeted drug delivery in the central nervous system. *Adv Drug Deliv Rev*, 2014. 72: pp. 94–109.
38. Burgess A, Ayala-Grosso CA, Ganguly M, Jordao JF, Aubert I, and Hynynen K, Targeted delivery of neural stem cells to the brain using MRI-guided focused ultrasound to disrupt the blood-brain barrier. *PLoS One*, 2011. 6(11): p. e27877.
39. Alkins R, Burgess A, Ganguly M, Francia G, Kerbel R, Wels WS, and Hynynen K, Focused ultrasound delivers targeted immune cells to metastatic brain tumors. *Cancer Res*, 2013. 73(6): pp. 1892–1899.
40. Burgess A, Huang Y, Querbes W, Sah DW, and Hynynen K, Focused ultrasound for targeted delivery of siRNA and efficient knockdown of Htt expression. *J Control Release*, 2012. 163(2): pp. 125–129.
41. Baseri B, Choi JJ, Deffieux T, Samiotaki G, Tung YS, Olumolade O, et al., Activation of signaling pathways following localized delivery of systemically administered neurotrophic factors across the blood-brain barrier using focused ultrasound and microbubbles. *Phys Med Biol*, 2012. 57(7): pp. N65–N81.
42. Samiotaki G, Acosta C, Wang S, and Konofagou EE, Enhanced delivery and bioactivity of the neurturin neurotrophic factor through focused ultrasound-mediated blood–brain barrier opening in vivo. *J Cereb Blood Flow Metab*, 2015. 35(4): pp. 611–622.
43. Aryal M, Park J, Vykhodtseva N, Zhang YZ, and McDannold N, Enhancement in blood-tumor barrier permeability and delivery of

liposomal doxorubicin using focused ultrasound and microbubbles: Evaluation during tumor progression in a rat glioma model. *Phys Med Biol*, 2015. 60(6): pp. 2511–2527.

44. Aryal M, Vykhodtseva N, Zhang YZ, and McDannold N, Multiple sessions of liposomal doxorubicin delivery via focused ultrasound mediated blood-brain barrier disruption: A safety study. *J Control Release*, 2015. 204: pp. 60–69.
45. Choi JJ, Wang S, Brown TR, Small SA, Duff KE, and Konofagou EE, Noninvasive and transient blood-brain barrier opening in the hippocampus of Alzheimer's double transgenic mice using focused ultrasound. *Ultrason Imaging*, 2008. 30(3): pp. 189–200.
46. Kinoshita M, McDannold N, Jolesz FA, and Hynynen K, Noninvasive localized delivery of Herceptin to the mouse brain by MRI-guided focused ultrasound-induced blood-brain barrier disruption. *Proc Natl Acad Sci U S A*, 2006. 103(31): pp. 11719–11723.
47. Treat LH, McDannold N, Zhang Y, Vykhodtseva N, and Hynynen K, Improved anti-tumor effect of liposomal doxorubicin after targeted blood-brain barrier disruption by MRI-guided focused ultrasound in rat glioma. *Ultrasound Med Biol*, 2012. 38(10): pp. 1716–1725.
48. Park EJ, Zhang YZ, Vykhodtseva N, and McDannold N, Ultrasound-mediated blood-brain/blood-tumor barrier disruption improves outcomes with trastuzumab in a breast cancer brain metastasis model. *J Control Release*, 2012. 163(3): pp. 277–284.
49. Johnson SM, Bangham AD, Hill MW, and Korn ED, Single bilayer liposomes. *Biochim Biophys Acta*, 1971. 233(3): pp. 820–826.
50. Torchilin VP, Recent advances with liposomes as pharmaceutical carriers. *Nat Rev Drug Discov*, 2005. 4(2): pp. 145–160.
51. Barenholz Y, Doxil(R)–the first FDA-approved nano-drug: Lessons learned. *J Control Release*, 2012. 160(2): pp. 117–134.
52. Maeda H, Wu J, Sawa T, Matsumura Y, and Hori K, Tumor vascular permeability and the EPR effect in macromolecular therapeutics: A review. *J Control Release*, 2000. 65(1–2): pp. 271–284.
53. Torchilin V, Tumor delivery of macromolecular drugs based on the EPR effect. *Adv Drug Deliv Rev*, 2011. 63(3): pp. 131–135.
54. Connor J, Yatvin MB, and Huang L, pH-sensitive liposomes: Acid-induced liposome fusion. *Proc Natl Acad Sci U S A*, 1984. 81(6): pp. 1715–1718.
55. Yatvin MB, Kreutz W, Horwitz BA, and Shinitzky M, pH-sensitive liposomes: Possible clinical implications. *Science*, 1980. 210(4475): pp. 1253–1255.

56. You J, Zhang P, Hu F, Du Y, Yuan H, Zhu J, et al., Near-infrared light-sensitive liposomes for the enhanced photothermal tumor treatment by the combination with chemotherapy. *Pharm Res*, 2014. 31(3): pp. 554–565.

57. Leung SJ and Romanowski M, Light-activated content release from liposomes. *Theranostics*, 2012. 2(10): pp. 1020–1036.

58. Gerasimov OV, Boomer JA, Qualls MM, and Thompson DH, Cytosolic drug delivery using pH- and light-sensitive liposomes. *Adv Drug Deliv Rev*, 1999. 38(3): pp. 317–338.

59. Schroeder A, Avnir Y, Weisman S, Najajreh Y, Gabizon A, Talmon Y, et al., Controlling liposomal drug release with low frequency ultrasound: Mechanism and feasibility. *Langmuir*, 2007. 23(7): pp. 4019–4025.

60. Schroeder A, Honen R, Turjeman K, Gabizon A, Kost J, and Barenholz Y, Ultrasound triggered release of cisplatin from liposomes in murine tumors. *J Control Release*, 2009. 137(1): pp. 63–68.

61. Lin HY and Thomas JL, Factors affecting responsivity of unilamellar liposomes to 20 kHz ultrasound. *Langmuir*, 2004. 20(15): pp. 6100–6106.

62. Yu FT, Chen X, Wang J, Qin B, and Villanueva FS, Low intensity ultrasound mediated liposomal doxorubicin delivery using polymer microbubbles. *Mol Pharm*, 2016. 13(1): pp. 55–64.

63. Lentacker I, Geers B, Demeester J, De Smedt SC, and Sanders NN, Design and evaluation of doxorubicin-containing microbubbles for ultrasound-triggered doxorubicin delivery: Cytotoxicity and mechanisms involved. *Mol Ther*, 2010. 18(1): pp. 101–108.

64. Yan F, Li L, Deng Z, Jin Q, Chen J, Yang W, et al., Paclitaxel-liposome-microbubble complexes as ultrasound-triggered therapeutic drug delivery carriers. *J Control Release*, 2013. 166(3): pp. 246–255.

65. Wang Y, Li X, Zhou Y, Huang P, and Xu Y, Preparation of nanobubbles for ultrasound imaging and intracelluar drug delivery. *Int J Pharm*, 2010. 384(1–2): pp. 148–153.

66. Huang SL, Liposomes in ultrasonic drug and gene delivery. *Adv Drug Deliv Rev*, 2008. 60(10): pp. 1167–1176.

67. Wu HP, Rognin NG, Krupka TM, Solorio L, Yoshiara H, Guenette G, et al., Acoustic characterization and pharmacokinetic analyses of new nanobubble ultrasound contrast agents. *Ultrasound Med Biol*, 2013. 39(11): pp. 2137–2146.

68. Krupka TM, Solorio L, Wilson RE, Wu HP, Azar N, and Exner AA, Formulation and characterization of echogenic lipid-pluronic nanobubbles. *Mol Pharm*, 2010. 7(1): pp. 49–59.

69. Koebis M, Kiyatake T, Yamaura H, Nagano K, Higashihara M, Sonoo M, et al., Ultrasound-enhanced delivery of morpholino with Bubble liposomes ameliorates the myotonia of myotonic dystrophy model mice. *Sci Rep*, 2013. 3: p. 2242.

70. Sugano M, Negishi Y, Endo-Takahashi Y, Suzuki R, Maruyama K, Yamamoto M, and Aramaki Y, Gene delivery system involving Bubble liposomes and ultrasound for the efficient in vivo delivery of genes into mouse tongue tissue. *Int J Pharm*, 2012. 422(1–2): pp. 332–337.

71. Sutton JT, Raymond JL, Verleye MC, Pyne-Geithman GJ, and Holland CK, Pulsed ultrasound enhances the delivery of nitric oxide from bubble liposomes to ex vivo porcine carotid tissue. *Int J Nanomed*, 2014. 9: pp. 4671–4683.

72. Herbst SM, Klegerman ME, Kim H, Qi J, Shelat H, Wassler M, et al., Delivery of stem cells to porcine arterial wall with echogenic liposomes conjugated to antibodies against CD34 and intercellular adhesion molecule-1. *Mol Pharm*, 2010. 7(1): pp. 3–11.

73. Britton GL, Kim H, Kee PH, Aronowski J, Holland CK, McPherson DD, and Huang SL, In vivo therapeutic gas delivery for neuroprotection with echogenic liposomes. *Circulation*, 2010. 122(16): pp. 1578–1587.

74. Laing ST, Moody MR, Kim H, Smulevitz B, Huang SL, Holland CK, et al., Thrombolytic efficacy of tissue plasminogen activator-loaded echogenic liposomes in a rabbit thrombus model. *Thromb Res*, 2012. 130(4): pp. 629–635.

75. Suzuki R, Takizawa T, Negishi Y, Utoguchi N, and Maruyama K, Effective gene delivery with novel liposomal bubbles and ultrasonic destruction technology. *Int J Pharm*, 2008. 354(1–2): pp. 49–55.

76. Rapoport NY, Herron JN, Pitt WG, and Pitina L, Micellar delivery of doxorubicin and its paramagnetic analog, ruboxyl, to HL-60 cells: Effect of micelle structure and ultrasound on the intracellular drug uptake. *J Control Release*, 1999. 58(2): pp. 153–162.

77. Husseini GA, Rapoport NY, Christensen DA, Pruitt JD, and Pitt WG, Kinetics of ultrasonic release of doxorubicin from pluronic P105 micelles. *Colloids Surf B Biointerfaces*, 2002. 24(3–4): pp. 253–264.

78. Marin A, Sun H, Husseini GA, Pitt WG, Christensen DA, and Rapoport NY, Drug delivery in pluronic micelles: Effect of high-frequency

ultrasound on drug release from micelles and intracellular uptake. *J Control Release*, 2002. 84(1–2): pp. 39–47.

79. Husseini GA, de la Rosa MAD, Richardson ES, Christensen DA, and Pitt WG, The role of cavitation in acoustically activated drug delivery. *J Control Release*, 2005. 107(2): pp. 253–261.
80. Husseini GA, de la Rosa MAD, Gabuji T, Zeng Y, Christensen DA, and Pitt WG, Release of doxorubicin from unstabilized and stabilized micelles under the action of ultrasound. *J Nanosci Nanotechnol*, 2007. 7(3): pp. 1028–1033.
81. Plesset MS and Prosperetti A, Bubble dynamics and cavitation. *Ann Rev Fluid Mech*, 1977. 9: pp. 145–185.
82. Leighton TG, *The Acoustic Bubble*, 1994: Academic Press.
83. Gao Z, Fain HD, and Rapoport N, Ultrasound-enhanced tumor targeting of polymeric micellar drug carriers. *Mol Pharm*, 2004. 1(4): pp. 317–330.
84. Gao ZG, Fain HD, and Rapoport N, Controlled and targeted tumor chemotherapy by micellar-encapsulated drug and ultrasound. *J Control Release*, 2005. 102(1): pp. 203–222.
85. De Smedt SC, Demeester J, and Hennink WE, Cationic polymer based gene delivery systems. *Pharm Res*, 2000. 17(2): pp. 113–126.
86. Zhang S, Zhao B, Jiang H, Wang B, and Ma B, Cationic lipids and polymers mediated vectors for delivery of siRNA. *J Control Release*, 2007. 123(1): pp. 1–10.
87. Deshpande MC and Prausnitz MR, Synergistic effect of ultrasound and PEI on DNA transfection in vitro. *J Control Release*, 2007. 118(1): pp. 126–135.
88. Li P, Gao Y, Liu Z, Tan K, Zuo Z, Xia H, et al., DNA transfection of bone marrow stromal cells using microbubble-mediated ultrasound and polyethylenimine: An in vitro study. *Cell Biochem Biophys*, 2013. 66(3): pp. 775–786.
89. Qiu Y, Luo Y, Zhang Y, Cui W, Zhang D, Wu J, et al., The correlation between acoustic cavitation and sonoporation involved in ultrasound-mediated DNA transfection with polyethylenimine (PEI) in vitro. *J Control Release*, 2010. 145(1): pp. 40–48.
90. Qiu Y, Zhang C, Tu J, and Zhang D, Microbubble-induced sonoporation involved in ultrasound-mediated DNA transfection in vitro at low acoustic pressures. *J Biomech*, 2012. 45(8): pp. 1339–1345.
91. Zhang CB, Cao HL, Li Q, Tu J, Guo X, Liu Z, and Zhang D, Enhancement effect of ultrasound-induced microbubble cavitation on branched

polyethylenimine-mediated VEGF(165) transfection with varied N/P ratio. *Ultrasound Med Biol*, 2013. 39(1): pp. 161–171.

92. Lee JL, Lo CW, Ka SM, Chen A, and Chen WS, Prolonging the expression duration of ultrasound-mediated gene transfection using PEI nanoparticles. *J Control Release*, 2012. 160(1): pp. 64–71.
93. Mead BP, Mastorakos P, Suk JS, Klibanov AL, Hanes J, and Price RJ, Targeted gene transfer to the brain via the delivery of brain-penetrating DNA nanoparticles with focused ultrasound. *J Control Release*, 2016. 223: pp. 109–117.
94. Sirsi SR, Hernandez SL, Zielinski L, Blomback H, Koubaa A, Synder M, et al., Polyplex-microbubble hybrids for ultrasound-guided plasmid DNA delivery to solid tumors. *J Control Release*, 2012. 157(2): pp. 224–234.
95. Chen ZY, Liang K, and Qiu RX, Targeted gene delivery in tumor xenografts by the combination of ultrasound-targeted microbubble destruction and polyethylenimine to inhibit survivin gene expression and induce apoptosis. *J Exp Clin Cancer Res*, 2010. 29: p. 152.
96. Park DH, Jung BK, Lee YS, Jang JY, Kim MK, Lee JK, et al., Evaluation of in vivo antitumor effects of ANT2 shRNA delivered using PEI and ultrasound with microbubbles. *Gene Ther*, 2015. 22(4): pp. 325–332.
97. Yang F, Li Y, Chen Z, Zhang Y, Wu J, and Gu N, Superparamagnetic iron oxide nanoparticle-embedded encapsulated microbubbles as dual contrast agents of magnetic resonance and ultrasound imaging. *Biomaterials*, 2009. 30(23–24): pp. 3882–3890.
98. Cai X, Yang F, and Gu N, Applications of magnetic microbubbles for theranostics. *Theranostics*, 2012. 2: pp. 103–112.
99. Yang F, Wang Q, Gu Z, Fang K, Marriott G, and Gu N, Silver nanoparticle-embedded microbubble as a dual-mode ultrasound and optical imaging probe. *ACS Appl Mater Interfaces*, 2013. 5(18): pp. 9217–9223.
100. Mohamedi G, Azmin M, Pastoriza-Santos I, Huang V, Perez-Juste J, Liz-Marzan LM, et al., Effects of gold nanoparticles on the stability of microbubbles. *Langmuir*, 2012. 28(39): pp. 13808–13815.
101. Rosencwaig A and Gersho A, Theory of the photoacoustic effect with solids. *J Appl Phys*, 1976. 47(1): pp. 64–69.
102. Wang LV and Hu S, Photoacoustic tomography: In vivo imaging from organelles to organs. *Science*, 2012. 335(6075): pp. 1458–1462.
103. Wang LV, Multiscale photoacoustic microscopy and computed tomography. *Nat Photonics*, 2009. 3(9): pp. 503–509.

104. Yang X, Stein EW, Ashkenazi S, and Wang LV, Nanoparticles for photoacoustic imaging. *Wiley Interdiscip Rev Nanomed Nanobiotechnol*, 2009. 1(4): pp. 360–368.

105. Ntziachristos V and Razansky D, Molecular imaging by means of multispectral optoacoustic tomography (MSOT). *Chem Rev*, 2010. 110(5): pp. 2783–2794.

106. Eustis S and el-Sayed MA, Why gold nanoparticles are more precious than pretty gold: Noble metal surface plasmon resonance and its enhancement of the radiative and nonradiative properties of nanocrystals of different shapes. *Chem Soc Rev*, 2006. 35(3): pp. 209–217.

107. Hu M, Chen J, Li ZY, Au L, Hartland GV, Li X, et al., Gold nanostructures: Engineering their plasmonic properties for biomedical applications. *Chem Soc Rev*, 2006. 35(11): pp. 1084–1094.

108. Prodan E, Nordlander P, and Halas NJ, Electronic structure and optical properties of gold nanoshells. *Nano Lett*, 2003. 3(10): pp. 1411–1415.

109. Loo C, Lin A, Hirsch L, Lee MH, Barton J, Halas NJ, et al., Nanoshell-enabled photonics-based imaging and therapy of cancer. *Technol Cancer Res T*, 2004. 3(1): pp. 33–40.

110. Lee HJ, Liu Y, Zhao J, Zhou M, Bouchard RR, Mitcham T, et al., In vitro and in vivo mapping of drug release after laser ablation thermal therapy with doxorubicin-loaded hollow gold nanoshells using fluorescence and photoacoustic imaging. *J Control Release*, 2013. 172(1): pp. 152–158.

111. Rouleau L, Berti R, Ng VW, Matteau-Pelletier C, Lam T, Saboural P, et al., VCAM-1-targeting gold nanoshell probe for photoacoustic imaging of atherosclerotic plaque in mice. *Contrast Media Mol Imaging*, 2013. 8(1): pp. 27–39.

112. Wang Y, Xie X, Wang X, Ku G, Gill KL, O'Neal DP, et al., Photoacoustic tomography of a nanoshell contrast agent in the in vivo rat brain. *Nano Lett*, 2004. 4(9): pp. 1689–1692.

113. Yang X, Skrabalak SE, Li ZY, Xia Y, and Wang LV, Photoacoustic tomography of a rat cerebral cortex in vivo with au nanocages as an optical contrast agent. *Nano Lett*, 2007. 7(12): pp. 3798–3802.

114. Lu W, Huang Q, Ku G, Wen X, Zhou M, Guzatov D, et al., Photoacoustic imaging of living mouse brain vasculature using hollow gold nanospheres. *Biomaterials*, 2010. 31(9): pp. 2617–2626.

115. Eghtedari M, Oraevsky A, Copland JA, Kotov NA, Conjusteau A, and Motamedi M, High sensitivity of in vivo detection of gold nanorods

using a laser optoacoustic imaging system. *Nano Lett*, 2007. 7(7): pp. 1914–1918.

116. Agarwal A, Huang SW, O'Donnell M, Day KC, Day M, Kotov N, and Ashkenazi S, Targeted gold nanorod contrast agent for prostate cancer detection by photoacoustic imaging. *J Appl Phys*, 2007. 102(6): p. 064701.
117. Kim K, Huang SW, Ashkenazi S, O'Donnell M, Agarwal A, Kotov NA, et al., Photoacoustic imaging of early inflammatory response using gold nanorods. *Appl Phys Lett*, 2007. 90(22): p. 223901.
118. Li PC, Wang CRC, Shieh DB, Wei CW, Liao CK, Poe C, et al., In vivo photoacoustic molecular imaging with simultaneous multiple selective targeting using antibody-conjugated gold nanorods. *Opt Express*, 2008. 16(23): pp. 18605–18615.
119. Jokerst JV, Cole AJ, Van de Sompel D, and Gambhir SS, Gold nanorods for ovarian cancer detection with photoacoustic imaging and resection guidance via Raman imaging in living mice. *ACS Nano*, 2012. 6(11): pp. 10366–10377.
120. Yang HW, Liu HL, Li ML, Hsi IW, Fan CT, Huang CY, et al., Magnetic gold-nanorod/PNIPAAmMA nanoparticles for dual magnetic resonance and photoacoustic imaging and targeted photothermal therapy. *Biomaterials*, 2013. 34(22): pp. 5651–5660.
121. Skrabalak SE, Chen JY, Sun YG, Lu XM, Au L, Cobley CM, and Xia YN, Gold nanocages: Synthesis, properties, and applications. *Acc Chem Res*, 2008. 41(12): pp. 1587–1595.
122. Cho EC, Kim C, Zhou F, Cobley CM, Song KH, Chen J, et al., Measuring the optical absorption cross-sections of Au-Ag nanocages and Au nanorods by photoacoustic imaging. *J Phys Chem C Nanomater Interfaces*, 2009. 113(21): pp. 9023–9028.
123. Chen JY, Yang MX, Zhang QA, Cho EC, Cobley CM, Kim C, et al., Gold nanocages: A novel class of multifunctional nanomaterials for theranostic applications. *Adv Funct Mater*, 2010. 20(21): pp. 3684–3694.
124. Song KH, Kim C, Cobley CM, Xia Y, and Wang LV, Near-infrared gold nanocages as a new class of tracers for photoacoustic sentinel lymph node mapping on a rat model. *Nano Lett*, 2009. 9(1): pp. 183–188.
125. Kim C, Cho EC, Chen J, Song KH, Au L, Favazza C, et al., In vivo molecular photoacoustic tomography of melanomas targeted by bioconjugated gold nanocages. *ACS Nano*, 2010. 4(8): pp. 4559–4564.

126. Zhang YS, Wang Y, Wang L, Wang Y, Cai X, Zhang C, et al., Labeling human mesenchymal stem cells with gold nanocages for in vitro and in vivo tracking by two-photon microscopy and photoacoustic microscopy. *Theranostics*, 2013. 3(8): pp. 532–543.

127. Kripfgans OD, Fowlkes JB, Miller DL, Eldevik OP, and Carson PL, Acoustic droplet vaporization for therapeutic and diagnostic applications. *Ultrasound Med Biol*, 2000. 26(7): pp. 1177–1189.

128. Sheeran PS, Luois SH, Mullin LB, Matsunaga TO, and Dayton PA, Design of ultrasonically-activatable nanoparticles using low boiling point perfluorocarbons. *Biomaterials*, 2012. 33(11): pp. 3262–3269.

129. Kaneda MM, Caruthers S, Lanza GM, and Wickline SA, Perfluorocarbon nanoemulsions for quantitative molecular imaging and targeted therapeutics. *Ann Biomed Eng*, 2009. 37(10): pp. 1922–1933.

130. Rapoport N, Phase-shift, stimuli-responsive perfluorocarbon nanodroplets for drug delivery to cancer. *Wiley Interdiscip Rev Nanomed Nanobiotechnol*, 2012. 4(5): pp. 492–510.

131. Rapoport N, Gao Z, and Kennedy A, Multifunctional nanoparticles for combining ultrasonic tumor imaging and targeted chemotherapy. *J Natl Cancer Inst*, 2007. 99(14): pp. 1095–1106.

132. Rapoport N, Nam KH, Gupta R, Gao Z, Mohan P, Payne A, et al., Ultrasound-mediated tumor imaging and nanotherapy using drug loaded, block copolymer stabilized perfluorocarbon nanoemulsions. *J Control Release*, 2011. 153(1): pp. 4–15.

133. Rapoport NY, Kennedy AM, Shea JE, Scaife CL, and Nam KH, Controlled and targeted tumor chemotherapy by ultrasound-activated nanoemulsions/microbubbles. *J Control Release*, 2009. 138(3): pp. 268–276.

134. Fabiilli ML, Haworth KJ, Sebastian IE, Kripfgans OD, Carson PL, and Fowlkes JB, Delivery of chlorambucil using an acoustically-triggered perfluoropentane emulsion. *Ultrasound Med Biol*, 2010. 36(8): pp. 1364–1375.

135. Fabiilli ML, Lee JA, Kripfgans OD, Carson PL, and Fowlkes JB, Delivery of water-soluble drugs using acoustically triggered perfluorocarbon double emulsions. *Pharm Res*, 2010. 27(12): pp. 2753–2765.

136. Javadi M, Pitt WG, Belnap DM, Tsosie NH, and Hartley JM, Encapsulating nanoemulsions inside eLiposomes for ultrasonic drug delivery. *Langmuir*, 2012. 28(41): pp. 14720–14729.

137. Javadi M, Pitt WG, Tracy CM, Barrow JR, Willardson BM, Hartley JM, and Tsosie NH, Ultrasonic gene and drug delivery using eLiposomes. *J Control Release*, 2013. 167(1): pp. 92–100.

138. Lin CY, Javadi M, Belnap DM, Barrow JR, and Pitt WG, Ultrasound sensitive eLiposomes containing doxorubicin for drug targeting therapy. *Nanomedicine*, 2014. 10(1): pp. 67–76.

139. Burgess MT and Porter TM, Acoustic cavitation-mediated delivery of small interfering ribonucleic acids with phase-shift nano-emulsions. *Ultrasound Med Biol*, 2015. 41(8): pp. 2191–2201.

140. Dayton PA, Zhao S, Bloch SH, Schumann P, Penrose K, Matsunaga TO, et al., Application of ultrasound to selectively localize nanodroplets for targeted imaging and therapy. *Mol Imaging*, 2006. 5(3): pp. 160–174.

141. Chen J, Pan H, Lanza GM, and Wickline SA, Perfluorocarbon nanoparticles for physiological and molecular imaging and therapy. *Adv Chronic Kidney Dis*, 2013. 20(6): pp. 466–478.

142. Lanza GM, Winter PM, Caruthers SD, Hughes MS, Hu G, Schmieder AH, and Wickline SA, Theragnostics for tumor and plaque angiogenesis with perfluorocarbon nanoemulsions. *Angiogenesis*, 2010. 13(2): pp. 189–202.

143. Kripfgans OD, Fowlkes JB, Woydt M, Eldevik OP, and Carson PL, In vivo droplet vaporization for occlusion therapy and phase aberration correction. *IEEE Trans Ultrason Ferroelectr Freq Control*, 2002. 49(6): pp. 726–738.

144. Sheeran PS, Rojas JD, Puett C, Hjelmquist J, Arena CB, and Dayton PA, Contrast-enhanced ultrasound imaging and in vivo circulatory kinetics with low-boiling-point nanoscale phase-change perfluorocarbon agents. *Ultrasound Med Biol*, 2015. 41(3): pp. 814–831.

145. Williams R, Wright C, Cherin E, Reznik N, Lee M, Gorelikov I, et al., Characterization of submicron phase-change perfluorocarbon droplets for extravascular ultrasound imaging of cancer. *Ultrasound Med Biol*, 2013. 39(3): pp. 475–489.

146. Zhang M, Fabiilli ML, Haworth KJ, Padilla F, Swanson SD, Kripfgans OD, et al., Acoustic droplet vaporization for enhancement of thermal ablation by high intensity focused ultrasound. *Acad Radiol*, 2011. 18(9): pp. 1123–1132.

147. Kopechek JA, Park E, Mei CS, McDannold NJ, and Porter TM, Accumulation of phase-shift nanoemulsions to enhance MR-guided

ultrasound-mediated tumor ablation in vivo. *J Healthc Eng*, 2013. 4(1): pp. 109–126.

148. Kopechek JA, Park EJ, Zhang YZ, Vykhodtseva NI, McDannold NJ, and Porter TM, Cavitation-enhanced MR-guided focused ultrasound ablation of rabbit tumors in vivo using phase shift nanoemulsions. *Phys Med Biol*, 2014. 59(13): pp. 3465–3481.

149. Kopechek JA, Zhang P, Burgess MT, and Porter TM, Synthesis of phase-shift nanoemulsions with narrow size distributions for acoustic droplet vaporization and bubble-enhanced ultrasound-mediated ablation. *J Vis Exp*, 2012(67): p. e4308.

150. Moyer LC, Timbie KF, Sheeran PS, Price RJ, Miller GW, and Dayton PA, High-intensity focused ultrasound ablation enhancement in vivo via phase-shift nanodroplets compared to microbubbles. *J Ther Ultrasound*, 2015. 3: p. 7.

151. Zhou Y, Wang Z, Chen Y, Shen H, Luo Z, Li A, et al., Microbubbles from gas-generating perfluorohexane nanoemulsions for targeted temperature-sensitive ultrasonography and synergistic HIFU ablation of tumors. *Adv Mater*, 2013. 25(30): pp. 4123–4130.

Chapter 9

Light-Triggered Drug and Gene Delivery

Chengqiong Mao, Xubin Suo, and Xin Ming

Department of Cancer Biology and Comprehensive Cancer Center, Wake Forest University School of Medicine, Winston-Salem, North Carolina 27157, USA

xming@wakehealth.edu

9.1 Introduction

Light-triggered drug delivery is a main type of stimuli-responsive drug delivery due to the non-invasiveness and the possibility of remote spatiotemporal control of drug release [1]. This approach has been applied for delivery of small-molecule drugs and macromolecular therapeutics including genes. Based on the triggering mechanism, light-triggered delivery includes photothermal delivery and photochemical delivery. The first is based on hyperthermia caused by light irradiation of nanomaterials,

Stimuli-Responsive Nanomedicine
Edited by Lin Zhu

ISBN 978-981-4800-70-9 (Hardcover), 978-0-429-29529-4 (eBook)
www.jennystanford.com

and the latter is mediated with free radical generation through light irradiation of photosensitizers. Nanomaterials are the main delivery system to incorporate both light-responsive functionality and drugs, and thus become light-responsive nanomedicine [1]. Although light-responsive nanomedicine can be categorized by their nanostructure, including liposomes, micelles, nanogels, and nanoconjugates, in this chapter, we will review the recent works based on the types of the light responsiveness. We will discuss their design, delivery mechanisms and applications of both photothermal delivery and photochemical delivery.

9.2 Photothermal Drug Delivery

Photothermal therapy (PTT) is an emerging therapeutic procedure that employs electromagnetic radiation, visible and near-infrared (NIR) light, to generate thermal damages at diseased tissues [2, 3]. Unlike photodynamic therapy (PDT) in which reactive oxygen species (ROSs) are generated by excitation of a photosensitizer (PS), PTT produces both hyperthermia and photothermal ablation by converting the photoenergy to heat [4]. PTT causes biological changes ranging from protein structural changes to carbonization of the target tissues [5, 6]. As early as in 1984, it was found that PTT could induce cell necrosis when tissues were heated to 43°C for 4 h, and the same effect could be achieved when the tissue temperature reaches 54°C for 1 s [7]. During the past two decades, PTT has become an emerging therapeutic approach to ablate tumors in the body, for it can avoid inducing cancer resistance that often causes treatment failure with conventional cancer therapy such as chemotherapy.

In spite of the tremendous promise, PTT has not entered oncologic clinic as a standard therapy. The main translational barrier is low specificity of the heating produced by converting laser energy into thermal energy, and this results in insufficient treatment volume and exposure time in order to prevent potential damage to the surrounding healthy tissues [8]. To confront this barrier, the delivery system for PTT can be optimized in two directions. Firstly, the specificity towards disease sites needs to be improved to minimize the side effects of this therapy, and utilization of the enhanced permeability and retention (EPR) effect of nanomaterials towards tumors and incorporation of cancer targeting moiety are main approaches to achieve it. In addition, absorption of NIR light (650–1064 nm) is desirable for PTT because of optimal tissue penetration and low direct light toxicity [9]. Thus, the nanomaterials for PTT should have strong and tunable absorption in the NIR region. Traditionally, various gold nanostructures have been the mainstream photothermal conducting agents, including gold nanoshells, nanorods, and nanocages [10]. A major breakthrough in the application of this type of materials is that a nanoshell made of silica-cored gold has entered clinical trials with the brand name AuroLase [11]. Besides gold-based nanomaterials, other inorganic materials, including noble metal nanostructures, nanocarbons, and transition metal sulfide/oxides nanomaterials, as well as some organic nanoagents have been used as photothermal conducting agents and largely expanded the family of photothermal nanotherapeutics (PTNs) [12]. We will discuss some of the examples in detail based on the drug molecules delivered by the PTNs.

9.2.1 Light-Triggered Delivery of Small Molecule Drugs

Small molecule drugs are the main class of drug entity delivered by the PTNs. Gold, carbon, and some other inorganic nanomaterials have been used for photothermal delivery of small molecule drugs [13–16] as these nanomaterials are easily triggered by NIR light. Small molecule drugs are bound to the large surface area or encapsulated inside of these nanomaterials, and they can be released in a spatiotemporal manner in the diseased sites when hyperthermia is induced with a NIR light [9, 17]. In addition, these nano-sized drug delivery systems can achieve preferable tumor delivery through the EPR effect to increase tumor specificity and reduce side effects [18, 19].

9.2.1.1 Gold NP-based light-triggered nanomedicine

Gold nanorods (GNRs) are the primary nanomaterial for PTT because of the unique thermal conducting properties [20–23]. Their optical and photothermal properties are size- and shape-dependent due to localized surface plasmon resonance (SPR) at the NIR region, and this provides a possibility to fine-tune gold nanorods for optimal photothermal effects. For example, mesoporous silica-coated gold nanorods have been constructed to load chemotherapy agent doxorubicin (DOX) [24]. After the NIR irradiation, the resultant nanoparticles (NPs) led to significant cell death of breast cancer cells through combination of PTT and Dox-based chemotherapy in cell culture, and when tested in vivo, the light-responsive delivery of the NPs significantly suppressed tumor growth in tumor-bearing mice [24]. In another example, PEG-coated hollow gold nanoshells were used to enhance the drug loading of Dox, so that high drug loading at the level of 1.7 μg DOX/1 μg Au was achieved [25]. Treatment with these NPs followed by NIR irradiation also achieved excellent anticancer effects [25]. In a similar study, copolymer PEG-PLGA modified gold nanoshells with the SPR absorption at 790 nm were prepared to load Dox [26]. Treatment of the NPs followed by NIR laser irradiation increased the intratumoral temperature to 50–70°C and also caused rapid release of DOX [26]. The combination therapy with the PTNs led to complete tumor remission without recurrence [26].

Other materials have been incorporated into gold NPs for hybrid nanomaterials in order to improve their performance as phototherapy agents. In one example, silver/gold bimetallic NPs were prepared and were coated with a layer of polystyrene for loading a hydrophobic drug curcumin. Nonlinear PEG was used to further coat the NPs to improve dispensability, in vivo stability, and photothermal sensitivity in the physiological environment. In vitro studies indicated that 70% of the loaded curcumin was release upon irradiation with 1.5 W/cm^2 NIR light for 5 min, and thermal sensitivity of the PTNs was triggered by the conversion of photoenergy into heat in the metal core. Cytotoxicity tests revealed a 4-fold increase in the phototoxicity of the PTNs compared to that of free curcumin [27]. Another design of the PTNs involves construction of PLGA-based microspheres to encapsulate both chemotherapy agent paclitaxel and phototherapy agent hollow gold NPs [28]. NIR irradiation caused elevated local temperatures and rapid release of paclitaxel. In vitro and in vivo studies demonstrated higher anticancer activity of light-responsive delivery of these PTNs than that of single modality of PTT or chemotherapy [28]. Besides these conventional designs, a novel thermal-responsive polymer was recently used to construct PTNs. Thus, multi-stimuli responsive nanogels were prepared with a PEG shell and a cross-linked thermal-responsive poly[2-(N,N-diethylamino)ethyl methacrylate] core with NIR light-absorbing gold NPs immobilized in the core [29]. Substantial drug release was demonstrated upon NIR irradiation [29]. Light-triggered delivery of small molecule drugs via gold NPs can not only increase anticancer activity but also reduce systemic toxicity. In a study, PEG-conjugated and DOX-loaded hollow gold NPs demonstrated light-responsive and pH-dependent DOX release. This limited the potent cytotoxicity within the tumor sites, and led to greater antitumor activity and lower systemic toxicity than both free and liposomal DOX [25, 30].

9.2.1.2 Carbon nanomaterials as light-triggered nanomedicine

Carbon nanotubes (CNTs), including single-walled nanotubes (SWNTs) and multi-walled nanotubes (MWNTs), are a type of carbon nanomaterials that have been used for PTT and photothermal drug delivery. Favorable physical properties, including the strong optical absorption, high photon-to-thermal energy conversion efficiency of CNTs in the NIR region, and a high-absorption cross-section, make CNTs an excellent photothermal agent for potential in vivo application. As unmodified CNTs are often cytotoxicity, proper surface functionalization of CNTs is needed to make them biocompatible and to enable targeted delivery to diseased sites [31, 32]. Carbon nanomaterials coated with biocompatible materials have been found to be cleared from body after systemic administration without noticeable toxicity in animals [33]. Recently, transactivator of transcription peptide (TAT)-chitosan functionalized MWCNTs were constructed to deliver DOX for chemo-photothermal combination therapy of cancer [34]. Studies using both cell culture and animal tumor models indicated that photothermal delivery using this system can control release of DOX, and that excellent antitumor responses can be achieved through the synergistic function of chemotherapy and photothermal ablation [34]. Carbon materials other than

CNTs have also been used for photothermal drug delivery. Recently, mesoporous carbon nanospheres were developed for photoacoustic imaging of tumors and photothermal delivery of DOX [35]. Since the photothermal and photoacoustic activities of these materials are greater than those of other inorganic materials such as SWCNTs and gold nanorods, carbon nanospheres are capable of imaging tumors, delivering DOX in a photothermal responsive manner, and therefore achieving excellent antitumor responses [35].

9.2.2 Light-Triggered Delivery of Macromolecular Drugs

Gold NPs have been one of the main delivery systems for non-viral gene therapy over the past decade [36, 37]. One of the reasons for the research thrust is that genes and other functionalities can be easily linked to gold NPs through utilizing thiol-mental interaction [38]. Thus, by simply incorporating a thiol group in nucleic acids, gold NPs can deliver a wide variety of nucleic acids, including plasmid DNA, siRNA, antisense drugs, miRNAs, and their antagomirs, leading to various therapeutic benefits. Due to their high hydrophilicity and being highly negatively charged, nucleic acids cannot enter target cells without the assistance of a delivery vehicle. Gold NPs, including nanospheres, nanorods, and nanoshells, are particularly appealing candidates as vehicles for gene delivery because they exhibit low toxicity when tuned to appropriate sizes and coated with suitable polymers and targeting ligands [39]. Many delivery systems use non-covalent interactions, such as charge interactions, to attach nucleic acids to the NPs, while linking DNA modified with a thiol moiety to the gold NP surface directly is still a popular approach [40, 41]. For the gold NPs that are used for light-triggered delivery of genes, size tuning can generate intense absorbance in the NIR window [42]. Laser irradiation at the frequency that is unique to their nanostructures can produce photothermal effects as well as "hot" electrons, which can further trigger the release of nucleic acids that are attached to the gold NP.

Light-triggered gene delivery was achieved by inducing the release of a thiolated plasmid DNA bound to gold nanorods upon short-time laser irradiation at the longitudinal plasmon wavelength of the nanorods (Fig. 9.1) [43]. The eGFP genes were attached to gold nanorods via the thiol–gold interaction for the remote control of gene expression in living cells. Upon NIR irradiation, the gold nanorods in the DNA–NP conjugates underwent shape transformation that enabled the release of eGFP DNA. When DNA–NP conjugates were delivered to cultured HeLa cells, induction of GFP expression was observed only in the cells that were exposed to NIR irradiation.

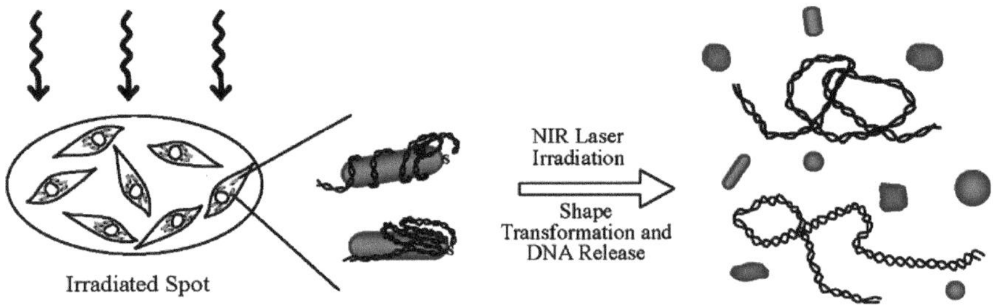

Figure 9.1 Light-triggered gene delivery. The cells containing DNA–NP conjugates within a spot are irradiated by NIR laser (left). After laser irradiation, the gold nanorods of DNA–NP conjugates undergo shape transformation that enabled the release of the DNA (right). (http://pubs.acs.org/doi/full/10.1021/ja0570180.)

Using similar strategy as the light-triggered gene delivery [43], gold NPs were used to deliver therapeutic oligonucleotides in the mammalian cells [44]. In this study, pulsed laser irradiation was used to cleave the thiol–gold bond anchoring thiolated single-stranded DNA to the particle's surface (Fig. 9.2) [44]. Selective melting of two different gold nanorods were first demonstrated in their work. Irradiated at different wavelengths, two different DNA oligonucleotides conjugated to each of the gold nanorods were released selectively with full function. Furthermore, breaking the thiol–gold bond also released free thiols, which led to combinational effects in live cells. However, this approach is limited by the stability of the NPs, since the release process depends on the release of the stabilizing surface ligands. Recent light-triggered approaches have sought to overcome this limitation by using double-stranded DNA–NP conjugates in which only one of the double strands is attached to the NP surface through thiol–gold bond. Irradiation with continuous-wave lasers triggered the photothermal effect and raised the local temperature above the melting point of the DNA duplex, and this allowed the non-thiolated strand to dissociate and be released while its other strand remained on the NP. The NP remains intact after the photothermal events, which is an advantage of this method over others. However, this approach is still limited by relatively low photo-release efficiencies, and thus a longer irradiation with pulsed lasers is needed to attain effective degrees of oligonucleotide release.

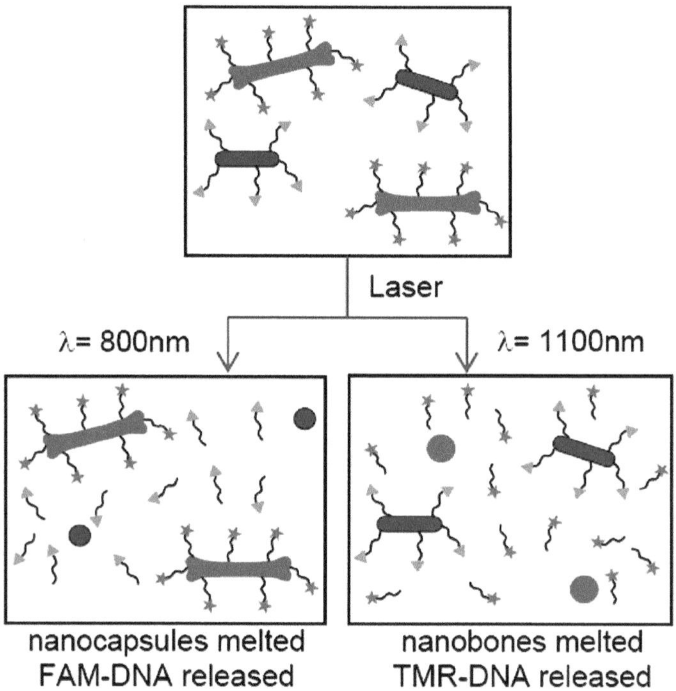

Figure 9.2 Selective and light-triggered delivery of oligonucleotides. Laser irradiation of DNA-conjugated nanocapsules (blue ovals) and nanobones (red bones) are exposed to 800 nm wavelength irradiation (left), which melts the nanocapsules and selectively releases the conjugated DNA that is labeled by a green fluorescent dye (green triangles). Exposure to 1100 nm wavelength irradiation (right) melts the nanobones, selectively releasing the conjugated DNA that is labeled by a red fluorescent dye (orange stars). (http://pubs.acs.org/doi/10.1021/nn800702n.)

Another method was introduced to not only take advantage of the efficient photothermal heating with pulsed lasers, but also avoid the disadvantage of breaking the thiol–gold bonds anchoring a double-stranded oligonucleotide to the NP (Fig. 9.3) [45]. In order to enhance the photo-release of ssDNA from DNA–NP conjugates, the conjugates' response to pulsed irradiation under varying conditions was carefully examined. The laser power and salt content were also investigated to clarify how they influence the ssDNA release through a denaturation process instead of thiol–gold bond cleavage.

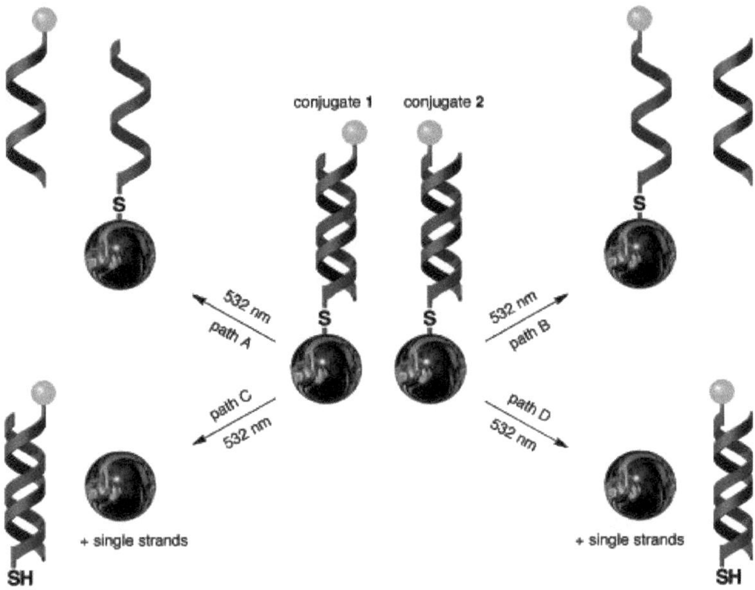

Figure 9.3 Light-activated release of DNA from the surfaces of gold NPs. The first conjugates (1) are comprised of fluorescently labeled antisense oligonucleotides that form duplex with complementary, thiolated sense strands anchored to the NP surface by a gold-thiol bond. In the other conjugates (2), the single-stranded DNA attached to the NP contains the luminescent tag, and its antisense strand is unlabeled. In both conjugates, emission from the luminescent tag should be quenched due to its close proximity to the metal NP and the existence of a non-radiative relaxation of the dye's excited state through energy and/or electron transfer. The appearance of fluorescence after light irradiation is a convenient probe for successful DNA release. Using this photothermally induced emission, the two conjugates will help identify whether denaturing of the double helix or gold-thiol bond breakup is the dominant mechanism. (http://pubs.acs.org/doi/full/10.1021/nn1016346.)

Some hybrid gold NPs were also developed for oligonucleotide delivery. Polycationic peptides or polymers coated gold nanoshells have been extensively studied for spatially and temporally controlled delivery of genes into target cell lines [46]. It was suggested that the attached dsDNA was dissociated and then the non-thiolated antisense sequence was released upon NIR irradiation at the NP's plasmon resonance wavelength. RNA delivery was also investigated using the same system. The NPs were coated with the positively charged poly-L-lysine polymers which could form the complex with therapeutic siRNA (Fig. 9.4) [46]. Upon application of an 800 nm NIR laser, the nucleic acid-NP complexes deliver the antisense ssDNA or siRNA to H1299 cells and silence target gene expression.

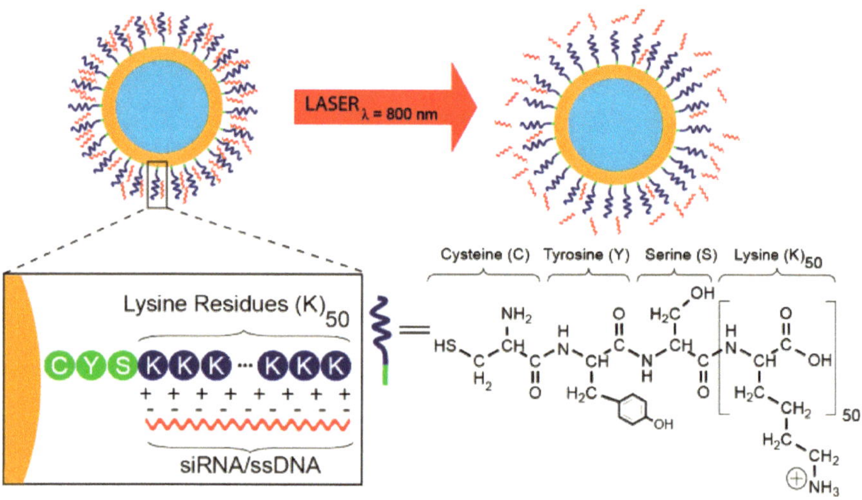

Figure 9.4 Poly-L-lysine coated gold-nanoshell for delivery of therapeutic siRNA. The negatively charged siRNA/ssDNA (red) is electrostatically attached to the cationic peptide (blue), which consists of 1 cysteine, 1 tyrosine, 1 serine, and 50 lysine amino acids. Upon laser irradiation, the siRNA/ssDNA is released. (http://pubs.acs.org/doi/full/10.1021/nn301135w.)

Although the gold NPs show great promise for clinical translation, they are non-biodegradable, and this raised concerns for their long-term application in humans. In contrast to exogenous gold, copper (Cu) is essential for human health.

In adults, the highest safe intake level of Cu is 10 mg daily, indicating that copper sulfide (CuS) NPs may be metabolized by humans. As semiconductor, the band gap of CuS varies upon its atomic composition, from 1.2 eV for Cu_2S to 2.0 eV for CuS. This allows CuS to absorb a large fraction of the solar spectrum.

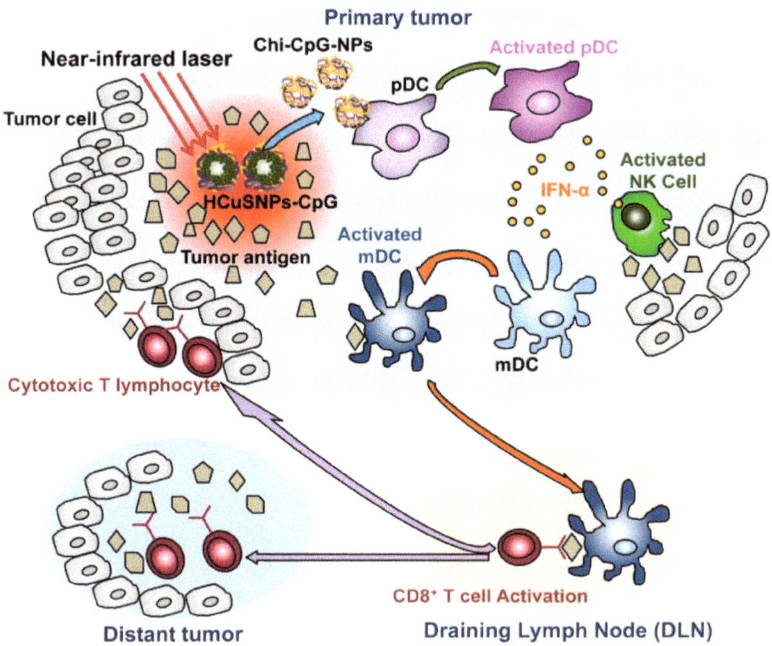

Figure 9.5 CpG-NP-mediated photothermal immunotherapy of both primary and distant tumors. Under NIR laser irradiation, the intratumorally injected CpG-NP conjugates transform to chitosan-CpG nanocomplexes (Chi-CpG-NPs), resulting in uptake into Toll-like receptor 9-rich endosomes of plasmacytoid dendritic cells (pDCs). Upon stimulation with CpG, the pDCs secrete interferon-α (IFN-α) to promote innate immunity through NK cell activation. Simultaneously, the NP-mediated photothermal ablation disrupts the cancer cells, relieving tumor burden and releasing tumor-associated antigens to myloid dendritic cells (mDCs). In the presence of IFN-α secreted by the activated pDCs, these mDCs become professional antigen-presenting cells and subsequently migrate to tumor-draining lymph nodes, where they cross-prime tumor antigen-specific T cells. The antigen-specific $CD8^+$ T cells enter the systemic circulation and are recruited to both primary tumor and untreated tumor at a distant site to trigger the "effector phase" of the adaptive immune response. (http://pubs.acs.org/doi/full/10.1021/nn5002112.)

As plasmonic material, CuS absorbs NIR light and converts it into heat via the excitation of direct transitions, indirect transitions, and plasmonic photoexcitation. These electro-optical properties allow CuS NPs to be widely used for photothermal therapy [47]. Recently, Lu et al.developed hollow CuS NPs for photothermal immunotherapy (Fig. 9.5) [48]. They coated hollow CuS NPs with CpG oligonucleotides, a type of immunoadjuvants, and used these CpG-NP conjugates for photothermal immunotherapy in a mouse breast cancer model. This approach involves photothermally triggered disintegration of NP conjugates, allowing them to reassemble and transform into chitosan-CpG nanocomplexes. Formation of chitosan-CpG nanocomplexes increases tumor retention of CpG and promotes its cellular uptake by plasmacytoid dendritic cells. The CpG-NP-mediated photothermal immunotherapy elicits more effective systemic immune responses than either immunotherapy or PTT alone, resulting in synergistic anticancer effects against primary as well as distant tumors. Strong antitumor efficacy, combined with quick action, justifies further development of this light-triggered immunotherapy approach.

9.3 Photochemical Drug Delivery

Photochemical drug delivery has been used to enhance cancer selectivity via selective release of drugs in cancer cells [49]. Photochemical internalization (PCI), a type of PDT, is used to enhance the release of endocytosed molecules into the cytosol [50]. This approach is based on the use of a PS that transports to endosomes, where PDT-generated singlet oxygen disrupts endosomes to release the drug molecules [50]. Thus, the laser is used to activate therapy only in diseased sites while unexposed normal tissues are spared. Lower light doses and oxygen levels are required to produce endosomal disruption than cell killing [51] and thus therapy in deep tissue can be achieved by endosomal release of drugs, not necessarily through photokilling [50, 52]. Encouraging anticancer response was reported in an ongoing phase I/II clinical trial evaluating PCI of bleomycin [53].

Besides small molecule drugs, photochemical drug delivery has been used to deliver macromolecules. For example, a non-caged DNA phosphorothioate antisense agent was synthesized to target the Renilla luciferase reporter gene [54]. Light-removable caging groups were linked to three or four thymidine residues in the oligonucleotide to block the hybridization of oligonucleotide to target reporter mRNA. Brief UV irradiation at 365 nm removes the caging groups, enables gene-specific binding of the oligonucleotide to the target mRNA, and thereafter, RNase H is triggered to catalyze mRNA degradation (Fig. 9.6).

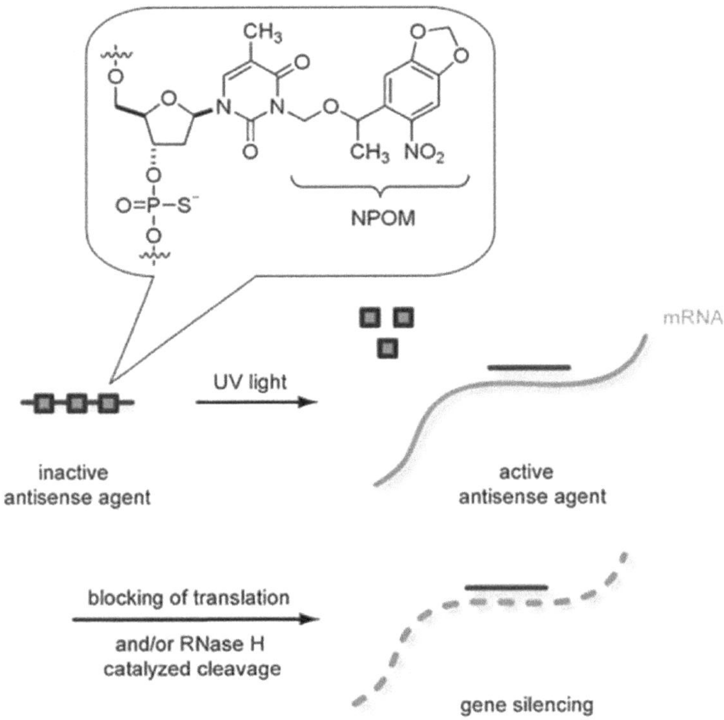

Figure 9.6 Light-regulation of antisense activity through the incorporation of light-removable groups, which block the hybridization of phosphorothioate DNA oligomers to mRNA. Brief UV irradiation at 365 nm removes the caging groups, enables sequence specific binding to the mRNA, and thus, leads to RNase H-catalyzed mRNA degradation. (http://onlinelibrary.wiley.com/doi/10.1002/cbic.200800627/full.)

Besides antisense and siRNA oligonucleotides, a light-triggered system was designed to allow spatiotemporal regulation of aptamer activity [55]. In the photo-triggerable system, the aptamer of interest is inactivated by hybridization to a photolabile complementary oligonucleotide. Upon irradiation, the complementary sequence breaks down, followed by the release of the functional aptamer AS1411, which binds with high affinity and selectivity to the tumoral receptor nucleolin (Fig. 9.7).

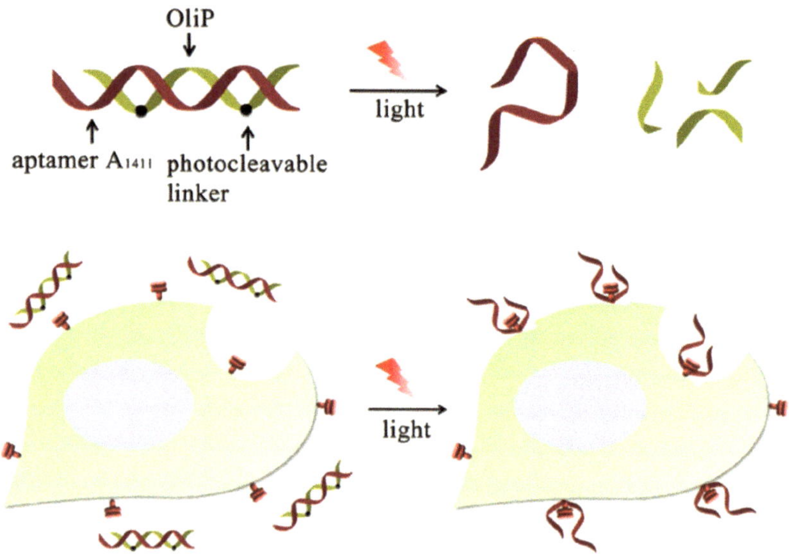

Figure 9.7 The AS1411 aptamer (A_{1411}, red DNA strand) is hybridized to a complementary oligonucleotide (OliP, green DNA strand) containing photo-cleavable bonds (black dots). The hybridized complex cannot bind cells. The A_{1411} is released by light-triggered breakage of the OliP, allowing binding to cancer cell surfaces. (http://www.pnas.org/content/111/48/17099.full.)

In addition, a DNA-camptothecin (CPT) amphiphile was designed and synthesized for light-triggered delivery of drugs and genes (Fig. 9.8) [56]. This amphiphile consists of three CPT molecules connected to DNA strands via a phenol-based self-immolative linker and a photolabile moiety. The amphiphile of the DNA-CPT conjugates can self-assemble into micellar nanostructures in an aqueous environment. The photolabile group can be cleaved by UV light (365 nm) or via a two-photon mechanism. Thereafter, the loaded DNA is released from the nanomicelles and leaves a decapped drug core, which can further undergo a rapid, spontaneous and irreversible degradation process to generate free CPT molecules.

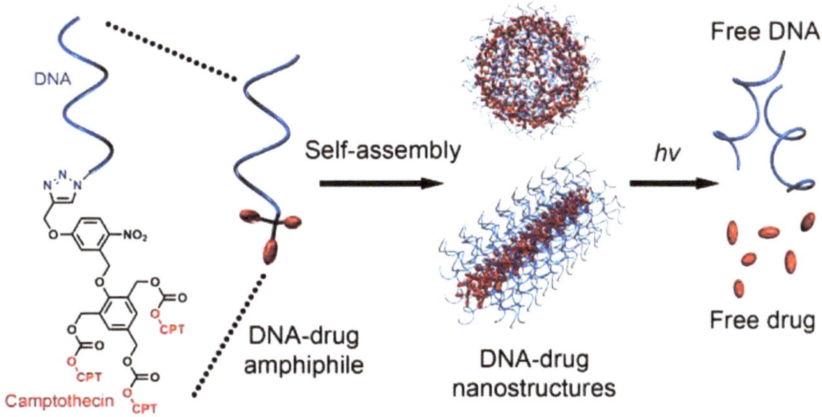

Figure 9.8 Schematics of the DNA–drug nanostructures assembled from photolabile DNA–drug amphiphiles (http://pubs.acs.org/doi/full/10.1021/jacs.5b00795).

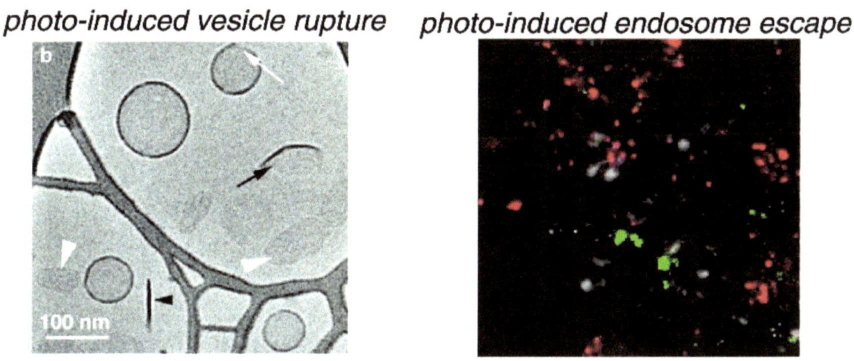

Figure 9.9 Cryo-TEM images of aqueous 31:69 (molar) azoTAB/SDS catanionic vesicles formed at 0.2 wt % overall surfactant concentration and at 38°C following 2 min of UV illumination. The white arrowheads point to flat bilayer patches imaged face on; the black arrowhead points to a bilayer patch imaged edge on; the white arrow indicates an opening in a ruptured vesicle; and the black arrow indicates a partially folded bilayer patch. Representative confocal microscopy images of fluorescein-labeled DNA transfected into HEK293T cells with azoTAB/SDBS catanionic vesicles and early endosomes labeled with Texas Red goat antimouse IgG. (http://pubs.acs.org/doi/full/10.1021/la803588d.)

Photoresponsive cationic materials have been also used to prepare "smart" gene delivery vectors, and they can enhance cellular uptake and enable light triggered gene release upon cell internalization. This type of delivery systems has been tested for DNA delivery by Liu et al. [57]. It was found that cellular entry of DNA-vesicle complexes into NIH 3T3 cells through endocytosis occurred under visible light conditions (Fig. 9.9). In contrast, subsequent UV illumination induced rupture of the vesicles and facilitated the release of free DNA from the endosome into the cytosol, where it crossed the nuclear membrane and produced enhanced gene expression.

9.4 Conclusion

Light-triggered drug delivery is one of the most dynamic research areas in drug delivery. Although promising from a conceptual point of view, clinical translation of this approach is hindered by poor tissue penetration of light and the uncertain safety and/or biodegradability of the typical nanomaterials used in light-triggered drug delivery. Photodynamic therapy, a procedure that uses a photosensitizer to enable light therapy selectively at diseased sites, has been applied to treat patients in clinic. Thus, photochemical drug/gene delivery, utilizing the same photosensitizers and light sources as photodynamic therapy, has higher translational potential because the clinical procedures can be adapted easily from the FDA-approved therapy. However, photosensitizers that are used in free form in photodynamic therapy are usually not efficient and specific. Currently, NP-based photoresponsive systems are mainstream for photochemical drug/gene delivery. Although the NP-based drug delivery has been intensively investigated, NP-based delivery systems need to be further optimized to fit the requirement of photoresponsiveness and should be the future focus of translational research of photochemical drug/gene delivery. On the other hand, inorganic NPs are the main delivery systems for achieving photothermal delivery and the majority of them haven't been tested in humans. Thus, selection or preparation of biocompatible nanomaterials is key for further development of photothermal delivery systems. In conclusion, developing biocompatible photosensitive nanomaterials and improving the efficiency and specificity of light- triggered delivery will therefore be the critical parts in the potential clinical translation of light- responsive nanomedicine.

References

1. Mura, S., Nicolas, J., and Couvreur, P. (2013) Stimuli-Responsive Nanocarriers for Drug Delivery. *Nat. Mater.*, **12**, 991–1003.
2. Zou, L., Wang, H., He, B., Zeng, L., Tan, T., Cao, H., He, X., Zhang, Z., Guo, S., and Li, Y. (2016) Current Approaches of Photothermal Therapy in Treating Cancer Metastasis with Nanotherapeutics. *Theranostics*, **6**, 762–772.
3. Jaque, D., Martinez Maestro, L., del Rosal, B., Haro-Gonzalez, P., Benayas, A., Plaza, J. L., Martin Rodriguez, E., and Garcia Sole, J. (2014) Nanoparticles for Photothermal Therapies. *Nanoscale*, **6**, 9494–9530.
4. Chatterjee, D. K., Diagaradjane, P., and Krishnan, S. (2011) Nanoparticle-Mediated Hyperthermia in Cancer Therapy. *Ther. Deliv.*, **2**, 1001–1014.
5. Huang, X., El-Sayed, I. H., Qian, W., and El-Sayed, M. A. (2006) Cancer Cell Imaging and Photothermal Therapy in the Near-Infrared Region by Using Gold Nanorods. *J. Am. Chem. Soc.*, **128**, 2115–2120.
6. Hauck, T. S., Jennings, T. L., Yatsenko, T., Kumaradas, J. C., and Chan, W. C. W. (2008) Enhancing the Toxicity of Cancer Chemotherapeutics with Gold Nanorod Hyperthermia. *Adv. Mater.*, **20**, 3832–3838.
7. Sapareto, S. A., and Dewey, W. C. (1984) Thermal Dose Determination in Cancer Therapy. *Int. J. Radiat. Oncol. Biol. Phys.*, **10**, 787–800.
8. Melancon, M. P., Zhou, M., and Li, C. (2011) Cancer Theranostics with Near-Infrared Light-Activatable Multimodal Nanoparticles. *Acc. Chem. Res.*, **44**, 947–956.
9. Melancon, M. P., Zhou, M., and Li, C. (2011) Cancer Theranostics with Near-Infrared Light-Activatable Multimodal Nanoparticles. *Acc. Chem. Res.*, **44**, 947–956.
10. Jelveh, S., and Chithrani, D. B. (2011) Gold Nanostructures as a Platform for Combinational Therapy in Future Cancer Therapeutics. *Cancers*, **3**, 1081–1110.
11. Diagaradjane, P., Shetty, A., Wang, J. C., Elliott, A. M., Schwartz, J., Shentu, S., Park, H. C., Deorukhkar, A., Stafford, R. J., Cho, S. H., et al. (2008) Modulation of in Vivo Tumor Radiation Response via Gold Nanoshell-Mediated Vascular-Focused Hyperthermia: Characterizing an Integrated Antihypoxic and Localized Vascular Disrupting Targeting Strategy. *Nano Lett.*, **8**, 1492–1500.
12. Anselmo, A. C., and Mitragotri, S. (2015) A Review of Clinical Translation of Inorganic Nanoparticles. *AAPS J.*, **17**, 1041–1054.

13. von Maltzahn, G., Park, J. H., Agrawal, A., Bandaru, N. K., Das, S. K., Sailor, M. J., and Bhatia, S. N. (2009) Computationally Guided Photothermal Tumor Therapy Using Long-Circulating Gold Nanorod Antennas. *Cancer Res.*, **69**, 3892–3900.
14. Tao, Y., Ju, E. G., Liu, Z., Dong, K., Ren, J. S., and Qu, X. G. (2014) Engineered, Self-Assembled Near-Infrared Photothermal Agents for Combined Tumor Immunotherapy and Chemo-Photothermal Therapy. *Biomaterials*, **35**, 6646–6656.
15. Hauck, T. S., Jennings, T. L., Yatsenko, T., Kumaradas, J. C., and Chan, W. C. W. (2008) Enhancing the Toxicity of Cancer Chemotherapeutics with Gold Nanorod Hyperthermia. *Adv. Mater.*, **20**, 3832–3838.
16. Weissleder, R. (2001) A Clearer Vision for in vivo Imaging. *Nat. Biotechnol.*, **19**, 316–317.
17. Sailor, M. J., and Park, J. H. (2012) Hybrid Nanoparticles for Detection and Treatment of Cancer. *Adv. Mater.*, **24**, 3779–3802.
18. Greish, K. (2007) Enhanced Permeability and Retention of Macromolecular Drugs in Solid Tumors: A Royal Gate for Targeted Anticancer Nanomedicines. *J. Drug Target.*, **15**, 457–464.
19. Matsumura, Y., Oda, T., and Maeda, H. (1987) General Mechanism of Intratumor Accumulation of Macromolecules: Advantage of Macromolecular Therapeutics. *Gan To Kagaku Ryoho*, **14**, 821–829.
20. Cobley, C. M., Chen, J., Cho, E. C., Wang, L. V., and Xia, Y. (2011) Gold Nanostructures: A Class of Multifunctional Materials for Biomedical Applications. *Chem. Soc. Rev.*, **40**, 44–56.
21. Prodan, E., Nordlander, P., and Halas, N. J. (2003) Electronic Structure and Optical Properties of Gold Nanoshells. *Nano Lett.*, **3**, 1411–1415.
22. Chen, H. J., Shao, L., Li, Q., and Wang, J. F. (2013) Gold Nanorods and their Plasmonic Properties. *Chem. Soc. Rev.*, **42**, 2679–2724.
23. Skrabalak, S. E., Chen, J. Y., Sun, Y. G., Lu, X. M., Au, L., Cobley, C. M., and Xia, Y. N. (2008) Gold Nanocages: Synthesis, Properties, and Applications. *Acc. Chem. Res.*, **41**, 1587–1595.
24. Monem, A. S., Elbialy, N., and Mohamed, N. (2014) Mesoporous Silica Coated Gold Nanorods Loaded Doxorubicin for Combined Chemo-Photothermal Therapy. *Int. J. Pharm.*, **470**, 1–7.
25. You, J., Zhang, G. D., and Li, C. (2010) Exceptionally High Payload of Doxorubicin in Hollow Gold Nanospheres for Near-Infrared Light-Triggered Drug Release. *ACS Nano*, **4**, 1033–1041.

26. Lee, S. M., Park, H., and Yoo, K. H. (2010) Synergistic Cancer Therapeutic Effects of Locally Delivered Drug and Heat Using Multifunctional Nanoparticles. *Adv. Mater.*, **22**, 4049–4053.

27. Wu, W. T., Shen, J., Banerjee, P., and Zhou, S. Q. (2011) Water-Dispersible Multifunctional Hybrid Nanogels for Combined Curcumin and Photothermal Therapy. *Biomaterials*, **32**, 598–609.

28. You, J., Shao, R. P., Wei, X., Gupta, S., and Li, C. (2010) Near-Infrared Light Triggers Release of Paclitaxel from Biodegradable Microspheres: Photothermal Effect and Enhanced Antitumor Activity. *Small*, **6**, 1022–1031.

29. Oishi, M., Nakamura, T., Jinji, Y., Matsuishi, K., and Nagasaki, Y. (2009) Multi-Stimuli-Triggered Release of Charged Dye from Smart PEGylated Nanogels Containing Gold Nanoparticles to Regulate Fluorescence Signals. *J. Mater. Chem.*, **19**, 5909–5912.

30. You, J., Zhang, R., Zhang, G. D., Zhong, M., Liu, Y., Van Pelt, C. S., Liang, D., Wei, W., Sood, A. K., and Li, C. (2012) Photothermal-Chemotherapy with Doxorubicin-Loaded Hollow Gold Nanospheres: A Platform for Near-Infrared Light-Trigged Drug Release. *J. Control. Release*, **158**, 319–328.

31. Gong, H., Peng, R., and Liu, Z. (2013) Carbon Nanotubes for Biomedical Imaging: The Recent Advances. *Adv. Drug Deliv. Rev.*, **65**, 1951–1963.

32. Yang, K., Feng, L. Z., Shi, X. Z., and Liu, Z. (2013) Nano-Graphene in Biomedicine: Theranostic Applications. *Chem. Soc. Rev.*, **42**, 530–547.

33. Liu, Y., Zhao, Y. L., Sun, B. Y., and Chen, C. Y. (2013) Understanding the Toxicity of Carbon Nanotubes. *Acc. Chem. Res.*, **46**, 702–713.

34. Dong, X., Sun, Z., Wang, X., and Leng, X. (2017) An Innovative MWCNTs/DOX/TC Nanosystem for Chemo-Photothermal Combination Therapy of Cancer. *Nanomed.: Nanotechnol. Biol. Med.*, **13**, 2271–2280.

35. Zhou, L., Jing, Y., Liu, Y., Liu, Z., Gao, D., Chen, H., Song, W., Wang, T., Fang, X., Qin, W., et al. (2018) Mesoporous Carbon Nanospheres as a Multifunctional Carrier for Cancer Theranostics. *Theranostics*, **8**, 663–675.

36. Mandal, S., Bakeine, G. J., Krol, S., Ferrari, C., Clerici, A. M., Zonta, C., Cansolino, L., Ballarini, F., Bortolussi, S., Stella, S., et al. (2011) Design, Development and Characterization of Multi-Functionalized Gold Nanoparticles for Biodetection and Targeted Boron Delivery in BNCT Applications. *Appl. Radiat. Isot.*, **69**, 1692–1697.

37. Ghosh, P., Han, G., De, M., Kim, C. K., and Rotello, V. M. (2008) Gold Nanoparticles in Delivery Applications. *Adv. Drug Deliv. Rev.*, **60**, 1307–1315.

38. Han, G., Ghosh, P., and Rotello, V. M. (2007) In Chan, W. C. W. (ed.), *Bio-Applications of Nanoparticles*. Springer New York, New York, NY, pp. 48–56.

39. Pissuwan, D., Niidome, T., and Cortie, M. B. (2011) The Forthcoming Applications of Gold Nanoparticles in Drug and Gene Delivery Systems. *J. Control. Release*, **149**, 65–71.

40. de Puig, H., Cifuentes Rius, A., Flemister, D., Baxamusa, S. H., and Hamad-Schifferli, K. (2013) Selective Light-Triggered Release of DNA from Gold Nanorods Switches Blood Clotting on and Off. *PLoS One*, **8**, e68511.

41. Gui, C., and Cui, D.-X. (2012) Functionalized Gold Nanorods for Tumor Imaging and Targeted Therapy. *Cancer Biol. Med.*, **9**, 221–233.

42. Dreaden, E. C., Austin, L. A., Mackey, M. A., and El-Sayed, M. A. (2012) Size Matters: Gold Nanoparticles in Targeted Cancer Drug Delivery. *Ther. Deliv.*, **3**, 457–478.

43. Chen, C.-C., Lin, Y.-P., Wang, C.-W., Tzeng, H.-C., Wu, C.-H., Chen, Y.-C., Chen, C.-P., Chen, L.-C., and Wu, Y.-C. (2006) DNA–Gold Nanorod Conjugates for Remote Control of Localized Gene Expression by near Infrared Irradiation. *J. Am. Chem. Soc.*, **128**, 3709–3715.

44. Wijaya, A., Schaffer, S. B., Pallares, I. G., and Hamad-Schifferli, K. (2009) Selective Release of Multiple DNA Oligonucleotides from Gold Nanorods. *ACS Nano*, **3**, 80–86.

45. Poon, L., Zandberg, W., Hsiao, D., Erno, Z., Sen, D., Gates, B. D., and Branda, N. R. (2010) Photothermal Release of Single-Stranded DNA from the Surface of Gold Nanoparticles Through Controlled Denaturating and Au–S Bond Breaking. *ACS Nano*, **4**, 6395–6403.

46. Huschka, R., Barhoumi, A., Liu, Q., Roth, J. A., Ji, L., and Halas, N. J. (2012) Gene Silencing by Gold Nanoshell-Mediated Delivery and Laser-Triggered Release of Antisense Oligonucleotide and siRNA. *ACS Nano*, **6**, 7681–7691.

47. Ortiz de Solorzano, I., Prieto, M., Mendoza, G., Alejo, T., Irusta, S., Sebastian, V., and Arruebo, M. (2016) Microfluidic Synthesis and Biological Evaluation of Photothermal Biodegradable Copper Sulfide Nanoparticles. *ACS Appl. Mater. Interfaces*, **8**, 21545–21554.

48. Guo, L., Yan, D. D., Yang, D., Li, Y., Wang, X., Zalewski, O., Yan, B., and Lu, W. (2014) Combinatorial Photothermal and Immuno Cancer

Therapy Using Chitosan-Coated Hollow Copper Sulfide Nanoparticles. *ACS Nano*, **8**, 5670–5681.

49. Fomina, N., Sankaranarayanan, J., and Almutairi, A. (2012) Photochemical Mechanisms of Light-Triggered Release from Nanocarriers. *Adv. Drug Deliv. Rev.*, **64**, 1005–1020.

50. Selbo, P. K., Weyergang, A., Hogset, A., Norum, O. J., Berstad, M. B., Vikdal, M., and Berg, K. (2010) Photochemical Internalization Provides Time- and Space-Controlled Endolysosomal Escape of Therapeutic Molecules. *J. Control. Release*, **148**, 2–12.

51. Berg, K., Weyergang, A., Prasmickaite, L., Bonsted, A., Hogset, A., Strand, M. T., Wagner, E., and Selbo, P. K. (2010) Photochemical Internalization (PCI): A Technology for Drug Delivery. *Methods Mol. Biol.*, **635**, 133–145.

52. Lu, H. L., Syu, W. J., Nishiyama, N., Kataoka, K., and Lai, P. S. (2011) Dendrimer Phthalocyanine-Encapsulated Polymeric Micelle-Mediated Photochemical Internalization Extends the Efficacy of Photodynamic Therapy and Overcomes Drug-Resistance in Vivo. *J. Control. Release*, **155**, 458–464.

53. Agostinis, P., Berg, K., Cengel, K. A., Foster, T. H., Girotti, A. W., Gollnick, S. O., Hahn, S. M., Hamblin, M. R., Juzeniene, A., Kessel, D., et al. (2011) Photodynamic Therapy of Cancer: An Update. *CA: A Cancer J. Clin.*, **61**, 250–281.

54. Young, D. D., Lusic, H., Lively, M. O., Yoder, J. A., and Deiters, A. (2008) Gene Silencing in Mammalian Cells with Light-Activated Antisense Agents. *ChemBioChem*, **9**, 2937–2940.

55. Li, L., Tong, R., Chu, H., Wang, W., Langer, R., and Kohane, D. S. (2014) Aptamer Photoregulation in Vivo. *Proc. Natl. Acad. Sci. U. S. A.*, **111**, 17099–17103.

56. Tan, X., Li, B. B., Lu, X., Jia, F., Santori, C., Menon, P., Li, H., Zhang, B., Zhao, J. J., and Zhang, K. (2015) Light-Triggered, Self-Immolative Nucleic Acid-Drug Nanostructures. *J. Am. Chem. Soc.*, **137**, 6112–6115.

57. Liu, Y.-C., Le Ny, A.-L. M., Schmidt, J., Talmon, Y., Chmelka, B. F., and Lee, C. T. (2009) Photo-Assisted Gene Delivery Using Light-Responsive Catanionic Vesicles. *Langmuir*, **25**, 5713–5724.

Chapter 10

Stimuli-Responsive Liposomes for Cancer

Jonathan P. May and Shyh-Dar Li

Faculty of Pharmaceutical Sciences, The University of British Columbia, 5519-2405 Wesbrook Mall, Vancouver, British Columbia, V6T 1Z3, Canada

shyh-dar.li@ubc.ca

The development of nanomedicine approaches for cancer offers a renewed potential for treatments with improved therapeutic window. Many forms of nanoparticles have been reported, but one of the most established are liposomes, of which there are already several examples in the marketplace. However, the first generation of liposomal nanomedicine had limitations in terms of their stability and release properties, which has led to the development of more advanced formulations that respond to internal or external stimuli to release their drug payload in a controlled and localized manner. In this chapter, the latest developments of this

Stimuli-Responsive Nanomedicine
Edited by Lin Zhu

ISBN 978-981-4800-70-9 (Hardcover), 978-0-429-29529-4 (eBook)
www.jennystanford.com

field are discussed, with reference to some of the most clinically advanced formulations, as well as discussion of innovative new systems that hold hope for new avenues to future treatments for cancer.

10.1 Treating Cancer with Nanomedicine

Cancer is one of the leading causes of mortality in the western world, which presents itself as many different forms of the disease, belonging to a multitude of different organs and tissues, with myriad phenotypes. Cancers may be solid tumors of a size visible to the human eye, in which case surgical intervention may be possible, alternatively in developed disease micro-metastases may form which are only visible with specialized equipment, techniques and staining. Furthermore, many cancers become resistant to common chemotherapeutics, making their treatment that much more of a challenge.

As the disease itself is so complex and varied, it goes to reason that the treatments involved should also be diverse and advanced. Often one of the most significant problems with conventional chemotherapy is not the fact that it does not work, but that the toxicity from the drug(s) is too high for normal healthy tissue and this lack of selectivity causes more damage than good. This is where encapsulation of chemotherapeutics in nanoparticles can make significant in-roads to improving the treatment, as this can create a less toxic formulation with greater uptake in the vicinity of the tumor, increasing its therapeutic window and leading to an improved treatment for patients.

Many early cancer nanomedicines were designed to take advantage of the so called "Enhanced Permeability and Retention (EPR) effect" [1]; an effect describing the accumulation of nanoparticles (50–200 nm) in the tumor cavity by their passage through fenestrations in the poorly formed tumor vasculature. Furthermore, the usual mechanism of clearance through the lymphatic system is thought to be less effective in tumor tissue and together these two aspects led to significant accumulation of nanoparticles in the tumor interstitial space.

One of the most well-known cancer nanomedicines works in this way; DOXIL® (or Caelyx in Europe; hydrogenated soy phosphatidylcholine (HSPC)/cholesterol (Chol)/1,2-distearoyl-sn-glycerol-3-phosphoethanolamine-N-(polyethylene glycol-2000) (DSPE-PEG_{2000}), 56:38:3 molar ratio, size: ~100 nm) is a liposomal formulation of doxorubicin (DOX) and was approved for AIDs-related Kaposi's sarcoma in 1995, and more recently it has also received FDA approval for the indications of recurrent ovarian cancer and multiple myeloma [2]. Although the formulation improves the therapeutic window of the drug, an improvement in efficacy is not observed. Instead, it is through a significant reduction in toxicity, particularly cardiotoxicity, where the formulation shows improvements and this was the major factor which led to FDA approval. In later studies, it was demonstrated that the long circulating liposome was very stable and accumulates in the tumor tissue, but release of the drug from the formulation is slow and the true potential cytotoxicity is not reached in the tumor — despite plenty of drug being present most remains stably encapsulated in the liposomes [3, 4]. While this is manageable for DOXIL as the efficacy is still about as good as free drug with much reduced toxicity (particularly the dose limiting cardiotoxicity), this was not the case for a similar liposomal formulation of cisplatin (CDDP). The formulation known as SPI-077 (HSPC/Chol/DSPE-PEG_{2000}, 51:44:5 molar %), reached phase I-II trials for the treatment of squamous cell carcinoma of the head and neck. A prolonged blood circulation and excellent tolerability were observed, but unfortunately only with a limited therapeutic activity [5]. This lack in efficacy was shown to be due to insufficient release of CDDP, leading to drug levels below the cytotoxic threshold for activity and this consequently led to the termination of any further trials [6, 7].

So while the advantages of liposomal drug delivery are considerable, leading to reduced side effects and long circulation times, a liposome that is too stable can result in a reduced bioavailability of the drug and poor efficacy. This highlights the critical role of drug release kinetics for therapeutic liposomal delivery. Hence, formulations such as DOXIL and SPI-077 have alerted the field to the fact that improvements in the release/stability properties of nanoparticle formulations are required, and one way this may be achieved is through the design of stimuli-sensitive nanoparticles.

10.2 Stimuli-Sensitive Liposomes

In a general sense for cancer treatment, these liposomes are designed to respond to a stimulus either applied from outside the body, i.e., an external stimulus such as heat, light, ultrasound or a magnetic field, or respond to an endogenous trait of the cancer cell type, i.e., an inherent stimulus such as low pH, or upregulation/overexpression of a particular ligand/antigen/enzyme. It is by no means exclusive, but a general trend is that primary solid tumors can be easily targeted with an externally applied stimulus, but for invasive, spreading, or metastatic disease, an approach utilizing the inherent stimuli localized to the tumor cells' vicinity may be preferable. In this chapter, both types of stimuli-sensitive agents will be discussed, from the well established to some innovative recent developments.

10.3 Liposomes Responding to External Stimuli

10.3.1 Temperature-Sensitive Liposomes (TSLs)

Some of the earliest stimuli-sensitive nanoparticles were liposomes, composed of specially selected lipids to undergo a phase change just above physiological temperature (42–45°C). This pioneering work of Yatvin, Weinstein, and co-workers led to a formulation capable of delivering three-times more drug (methotrexate), in a localized manner to a microwave-heated tumor, relative to control [8, 9]. This work demonstrated the potential of using heat to trigger drug release in a localized manner, and proved the concept of creating a drug delivery vehicle that could be controlled with an external trigger. However, there was still considerable room for improvement, as the liposomes were rapidly cleared from the blood and the drug release was relatively slow, which required heating at temperatures high enough to burn tissue.

In the few decades since this work, several new TSL formulations have been designed and studied with much improved physical characteristics and enhanced therapeutic effect. Noticeably, there have been two schools of thought as to how an optimal TSL should work, and this has led to two distinct variations in the types of formulation. For the purposes of this chapter, we will refer to these two types as the ultrafast releasing TSL (uTSL) and the long-circulating or slow-releasing TSL (sTSL). As the names suggest, the ultrafast releasing TSLs have been designed in such a way as to release their drug contents in an extremely fast and complete manner while the liposome is passing through the vasculature of the heated tumor area (mild-hyperthermia 39–42°C). This sudden burst-release of drug leads to high drug concentrations in the vasculature, forcing the drug across the endothelial cell layer and into the surrounding tumor tissue [10]. The long-circulating sTSL, on the other hand, is designed more like a conventional long-circulating liposome intended to accumulate in the tumor via the EPR-effect. However, the sTSL is also thermosensitive and can be triggered to release drug, following its accumulation in the chosen area, through the application of some mild heating. Sometimes the EPR-effect is further accentuated through an initial hyperthermia step before injection of the formulation to increase the permeability of fenestrations in the tumor vasculature. Both methods are currently being studied and it is not clear whether one method is superior to the other, as it could well have a lot to do with other aspects such as the heating protocol, method of heating, tumor type and drug of choice [10, 11]. Examples of each will now be given and further descriptions of the formulation, drug release and therapeutic effect will be discussed.

10.3.1.1 Lipid-modified TSLs

10.3.1.1.1 Optimized stealth TSLs

Koning and co-workers developed a sTSL for use with a 2-step heating protocol in the hope that it would improve overall DOX uptake and therapeutic effect. The theory behind this work was based upon findings that hyperthermia (HT) improved the intratumoral extravasation and penetration of nanoparticles [12], which could then be treated with a further dose of HT to release the encapsulated drug directly in the interstitial space [11]. Liposomes designed for release with such a protocol were based on some of the traditional TSL designs incorporating 1,2-dipalmitoyl-sn-glycerol-3-phosphatidylcholine (DPPC), 1,2-distearoyl-sn-glycerol-3-phosphatidylcholine (DSPC) or HSPC and DSPE-PEG$_{2000}$ [13–15].

These compositions create a more stable liposome that has good stability at 37°C, and releases drug over several minutes at 42°C (reaching 72% drug release after 1 h @ 42°C), and are defined as long circulating or slow releasing TSL (sTSL). A typical lipid composition for an sTSL is DPPC/DSPC/DSPE-PEG_{2000} (55:40:5 molar %), with a size of ~80 nm. These liposomes have been studied with a 2-step hyperthermia protocol, the initial HT dose to increase liposome extravasation and penetration, then the second dose of HT to release drug from the liposome. It was found that this 2-step protocol was particularly important when using a sTSL, and a similar amount of DOX was measured in the tumor 3 h post-treatment as for a uTSL (for more information on this uTSL formulation, see LTSL below) [11]. However, when these treatments/strategies were compared in a tumor model, a more prolonged inhibition of tumor growth was observed for the single HT treatment with uTSL, than any other combination [11]. Possible reasons for this could be that higher DOX concentrations are reached during and immediately after the uTSL/intravascular release approach, relative to the sTSL/interstitial release approach. Alternatively, it could be that a significant antitumor effect is caused by attack of the tumor vasculature (endothelial cells), as well as the tumor cells, which would be much more prevalent for the intravascular approach [11].

One area that the longer-circulating sTSL has shown promise for further development is in the case of MR guided hyperthermia treatments, where the studies have found a more stable liposomal formulation is preferable over intravascular releasing uTSLs which have a tendency for more leakiness and shorter drug half-life ($t_{1/2}$) [16–18].

Recently, a related formulation was used to create an optimized formulation of Idarubicin, an analogue of DOX [19]. The formulation contained DPPC/DSPC/DSPE-PEG_{2000} in a 60:35:5 molar ratio, used an ammonium/EDTA loading gradient, and formed liposomes of ~85 nm in size. Drug release was low at 37°C (~20%), but on heating to 42°C a rapid and complete release was observed (100% in less than 1 min). Interestingly, although a stabilized formulation was used, due to the drug being more hydrophobic and more readily able to penetrate the lipid bilayer, an intravascular drug release was used for this formulation; rapid drug release was observed and significant tumoral drug uptake was measured. The efficacy resulting from this particular formulation showed a significant improvement over free drug and non-thermosensitive DOXIL formulations, with two thirds of the animals studied showing a complete abrogation of tumor growth over the 30 days of the study.

10.3.1.1.2 Lyso-lipid temperature-sensitive liposomes (LTSL)

At the turn of the last millennium Needham and co-workers released a seminal paper on a new TSL design with extremely rapid drug release. Although it was not that different to the liposome trends at the time — using DPPC as the temperature-sensitive lipid (much like the earlier work of Yatvin/Weinstein, etc.) and a PEGylated lipid to confer prolonged circulation properties (in a similar way to the DOXIL formulation) — the lipid membrane contained one further component: monostearoyl phosphatidylcholine (MSPC). This is a lyso-lipid or single stranded lipid, which was the key ingredient for providing an ultrafast drug release rate. This formulation paved the way for an intravascular release mechanism that was to prove so effective in preclinical models for both LTSL and other uTSL formulations.

LTSL is predominantly made up of DPPC (86%), with 10% MSPC and 4% DSPE-PEG_{2000}. This formulation is most often used to deliver the drug, DOX, which is a potent chemotherapeutic and seems particularly suited to this form of intravascular, heat-triggered drug delivery. The promising initial results quickly led to LTSL-DOX being studied in several clinical trials under the trade name ThermoDox® (Celsion Inc.), which are currently at Phase III in hepatocellular carcinoma, Phase II for treatment of recurrent chest wall breast cancer [20], and a proof-of-principle study for liver cancer with high-intensity focused ultrasound (HIFU).

It is thought that through the addition of the lyso-lipid the defects formed during the heat-induced gel to liquid phase transition are stabilized, thereby increasing the permeability of the micro-pores formed in the gel-liquid phase boundaries of the lipid bilayer. It is these pores that allow the aqueous contents of the liposome core to escape — in this case, DOX. The lyso-lipid also reduces the phase transition temperature by 1°C, meaning that LTSL-DOX displays temperature dependent drug release in the 39–42°C range, perfectly suited to the preferred range for mild-hyperthermia (above physiological temperature, but below the temperatures at which tissue ablation occurs). Drug release studies have shown rapid release of greater than 80% of drug contents in just 20 s at 41.3°C, while at 37°C the formulation was shown to release just 20% over 30 min [21]. This large change in drug release rate over just a few degrees — right in the mild-hyperthermia range — is the real key to the success of this formulation.

LTSL-DOX has been studied in a number of preclinical models in vivo, including xenograft mouse models [22–24] and orthotopic murine models of breast cancer [24] leading to significant improvements in tumor regression over non-TSL (i.e., DOXIL) and DOX controls. The formulation was also studied in canine spontaneously grown solid tumors, again leading to promising tumor responses, which warranted further evaluation of the treatment in humans [25]. Thus far, all clinical studies with this formulation (ThermoDox®) have been studied in combination with radio-frequency ablation (i.e., high temperature heating >45°C) or microwave applicators predominantly for surface heating. Only recently have studies begun on the combination of this technology with state-of-the-art high-intensity focused ultrasound (HIFU) heating methods for deep penetration and localized heating. It would seem that the limitations with heating technologies, particularly with the ability to precisely control the localized heating area, and the logistics/timing of formulation dosing relative to subsequent heating have all limited success in previous trials (e.g., NCT00617981). It will be interesting to see how successful this formulation is when these challenges have been overcome in ongoing and future trials (e.g., NCT02112656).

The limitations of the LTSL formulation stem from its lipid membrane stability, and it has been shown that despite the addition of PEGylated lipid, the $t_{1/2}$ is only around 1–2 h in vivo [26]. Furthermore, studies have shown that MSPC may leach from the lipid bilayer while passing through the systemic circulation [27, 28], and the lipid membrane may interact with blood components leading to a change in composition over time [29]. However, the most common criticism of this formulation is its lack of stability at 37°C, and if the 20% drug leakage over 30 min could be reduced further, then the formulation would become considerably more attractive than it is at present.

10.3.1.1.3 $DPPG_2$-TSLs

One of the first attempts to improve on the LTSL formulation was the $DPPG_2$-TSL reported by Lindner et al. in 2004 [30]. This liposome contains DPPC, DSPC and the synthetic lipid 1,2-dipalmitoyl-sn-glycerol-3-phospho-diglycerol ($DPPG_2$) in a 50:20:30 molar ratio. $DPPG_2$ is an analogue of DPPG with an extra glycerol moiety attached to the head group, and this extra hydrophilicity allows it to replace PEGylated lipids, without seriously foregoing the long-circulation character afforded by such molecules. As the $DPPG_2$ is much smaller than DSPE-PEG_{2000}, significantly higher molar ratios can be tolerated in the liposome membrane and this has led to a corresponding increase in $t_{1/2}$ compared to many other uTSLs ($t_{1/2}$ = 9.6 h in hamsters and 5 h in rats). As may be expected, this also led to a considerable increase in stability at 37°C, and even after 1 h at 39°C little release was observed [30]. However, $DPPG_2$-TSL was found to have a T_m of 42°C, a little higher than LTSL, and this led to a slightly shifted release profile; overall release of DOX from $DPPG_2$-TSL was slower than LTSL, particularly at 40–41°C [30, 31]. Furthermore, $DPPG_2$-TSL showed a better tolerance to changes in liposome diameter (100–150 nm) and the presence of some serum components relative to LTSL [29, 31].

While most of these studies have been performed with either fluorescent dyes or DOX in vitro, a more recent report discusses the loading of this liposome with gemcitabine (GEM) and the evaluation of this formulation in vivo. A passively loaded $DPPG_2$-TSL-GEM formulation was studied in rats, giving a plasma half-life of 0.53 h or 2.59 h, depending on liposome diameter 109 or 129 nm, respectively. A delay in tumor growth was observed for this formulation relative to free drug and unheated controls [32]. This is a promising result, but studies demonstrating the therapeutic effect of $DPPG_2$-TSL head-to-head with LTSL, where both are delivering the same drug (e.g., DOX), would be useful to help demonstrate its superiority if that is the case.

10.3.1.2 Surfactant-modified TSLs

10.3.1.2.1 Heat-activated cytoToxic (HaT)

Tagami et al. introduced the HaT (Heat-activated cytoToxic) formulation in 2011 [26, 33]; a TSL composed of DPPC and just one other surfactant (Brij78). The Brij surfactant, or polyoxyethylene (20) stearyl ether, is a 20-unit PEG chain attached to a single aliphatic chain. It was hypothesized that this single chain surfactant could replace the functions played by both the lyso-lipid and PEGylated lipid in the LTSL formulation. Several different surfactants were studied in order to find the optimal combination and ratio, which was found to be based upon a critical balance between saturated aliphatic and PEGylated chains (C_{16}:PEG_{440}; C18:$PEG_{440-4400}$). The final composition chosen for the HaT liposome was DPPC/Brij78 (96:4 molar %) [33].

HaT-DOX was studied directly against LTSL-DOX with in vitro release studies, demonstrating a slightly improved release profile, particularly at 40–41°C. This was most clearly demonstrated by comparing the release rate constants which were 2-fold higher at 40°C and 1.2-fold higher at 41°C for HaT relative to LTSL, respectively. The stability of HaT and LTSL were both found to be very similar (~20% release in 30 min) and this was consistent with the clearance found for both formulations in pharmacokinetic (PK) studies with non-heated tumors (~40% injected dose remaining after 1 h). The delivery of DOX to the tumor was increased by 1.4-fold with HaT relative to LTSL, presumably due to the improved release across the hyperthermia range (39–42°C), and this led to an improved antitumor response for the HaT formulation (50% cure rate at day 7 for HaT, relative to 25% cure rate at day 14 for LTSL).

A second iteration of the HaT formulation (HaT-II) was reported in 2012, using an improved method for loading DOX using a copper(II) gradient [34]. Following active loading, DOX formed a stable copper complex, leading to significantly increased in vitro stability at 37°C over both the first generation HaT and LTSL formulations. The in vitro release properties of this formulation were optimized by post-insertion of further quantities of Brij78, leading to the HaT-II-DOX formulation which was ~2-fold faster drug release across the mild-hyperthermia range, relative to LTSL-DOX. These characteristics were reflected in vivo, with 2.5-fold decreased blood clearance and 2-fold increased DOX uptake in the heated tumor for HaT-II relative to LTSL, respectively. Finally, HaT-II-DOX showed improved tumor growth inhibition over all other formulations studied (HaT ~1.4-fold; LTSL ~2-fold; free DOX ~4-fold).

The HaT formulation was also studied following passive loading with the poorly membrane permeable drugs GEM and oxaliplatin (OXA) [35]. In vitro release profiles of these drugs showed fast release at 39–42°C, but very little release over 30 min at 37–38°C. However, when studied in vivo, clearance from the blood was not that different from the HaT-DOX formulation, with about 82% injected dose present after 1 h for GEM and 17% present after 1 h for OXA. Uptake in the heated tumor was 25-fold greater for HaT-GEM relative to the free GEM control, and this led to a complete tumor regression after 5 days; however, little improvement in drug uptake or efficacy was observed for HaT-OXA over free drug control. Consequently, the authors conclude that uTSLs can in some cases be suitable for delivering these poorly membrane permeable drugs, but drugs should be assessed on a case-by-case basis.

10.3.1.3 Polymer-modified TSLs

Several heat-sensitive polymers have been investigated as doping agents for liposomes, in order to prepare temperature-sensitive formulations. These polymers change the hydrophilic and hydrophobic properties of the TSL, reducing its lower critical solution temperature (LCST). A TSL that contains such a polymer in the liposome membrane will release its contents when the liposome is heated above the LCST. One of the first polymers to be studied was poly(N-isopropylacrylamide) or pNIPAAm, which was incorporated in relatively high amounts (~40%) and had a significant effect on the transition temperature of a DPPC TSL [36, 37]. However, as this polymer has poor biodegradability, concerns were raised over its toxicity and this led to the pursuit of biodegradable polymers instead [38]. To this end, Kono et al. have described a liposomal formulation prepared from egg-yolk phosphatidylcholine (EPC), cholesterol and DSPE-PEG$_{5000}$ (50:45:4 molar %), which has been doped with 1–2% of a block copolymer (poly[2-(2-ethoxy)ethoxyethyl-block-octadecyl vinyl ether] = p(EOEOVE)-OD4) [39]. This polymer contains temperature-sensitive and anchor sections, and by varying the length of each section the properties can be tuned [40]. A typical composition shows 80% drug release within a minute of heating at 45°C, while less than 30% release at temperatures below 40°C. The PK has been shown to be relatively long-circulating with over

10% injected dose still observed in the blood 14 h post-injection. The clearance is considerably increased if the PEGylation is removed. Most recently, a multi-modal version of this polymer-modified TSL has been prepared (called "MTPL") with fluorescent tag and MRI contrast agent incorporated with the drug, DOX. The LCST was observed to be ~41°C above which large changes to the size and PDI of the formulation were observed (initially 139 nm, PDI 0.09). Due to the relatively long circulation properties, the authors chose to wait until most of the liposome had accumulated in the tumor via the EPR-effect; at which point MTPL had been eliminated from surrounding tissues and organs. This optimal time for the heat treatment was found to be around 12 h post MTPL administration, and for the group with MTPL and heating the rate of tumor growth was observed to be slower over the following 2 weeks and more apoptosis was identified in the tumor region by TUNEL staining 2 days after treatment [41].

A recent report using a related co-polymer of pNIPAAm, namely p(NIPAAm-co-poly(acrylic acid)), has been demonstrated to prepare liposomes which possess temperature and pH-sensitive properties. A traditional TSL formulation (TTSL = DPPC/HSPC/Chol/DSPE-PEG_{2000}, 100:50:30:6) was used for comparison and formed the backbone of the polymer-modified TSL (PTSL) by post-inserting the co-polymer at a 1:1 mass ratio [42, 43]. DOX release was shown to be relatively rapid (~80% in 1 min) at 42°C, which was twice as fast as TTSL, although its stability at physiological temperature was not as good (~25% release in 3 min). An in vivo efficacy study in rats showed a considerable improvement in tumor growth inhibition, compared to both free DOX and TTSL-DOX (all 5 mg DOX/kg). However, as mentioned earlier these polymers have poor biodegradability and unfortunately no further study/discussion of this was mentioned in this report.

10.3.1.4 Peptide-modified TSLs

10.3.1.4.1 "STL" temperature-sensitive liposome

In 2013 another TSL was introduced, composed of DPPC, DSPE-PEG_{2000}, cholesterol and a fatty acid-conjugated elastin-like peptide (55:2:15:0.41 molar ratio), which was given the name STL [44]. This liposome had a larger size (160 nm), and extended $t_{1/2}$

relative to many of the uTSLs previously discussed. This was caused by the addition of cholesterol into the membrane which is a common addition for long-circulating formulations such as DOXIL. Cholesterol has been shown to interfere with temperature-dependent release properties of TSLs [14, 27, 45] and can de-sensitize the lipid bilayer to thermal shocks through the formation of lipid "rafts" [45]. At higher concentrations of cholesterol (25–30 molar %) liposomes are typically not thermosensitive as the phase transition is reduced or completely eliminated, but in this case a low concentration of cholesterol has been used to add just a small amount of stabilization. Normally, this would make the liposome slow to release its contents, but the fatty-acid/elastin-like-peptide (ELP) conjugate undergoes a change in conformation at mild-hyperthermia temperatures and destabilizes the membrane overriding any stabilization effects from other lipid components. Through balancing the quantities of cholesterol and the lipid-peptide conjugate, the formulation could be prepared for optimal release properties [46].

STL-DOX (STL loaded with doxorubicin) displays an impressive drug release profile, with excellent stability at 37°C (<20% drug release in 30 min) and fast release at 42°C (85% drug release in <20 s). This translates to extended plasma half-life for STL-DOX (2.03 h) compared to LTSL-DOX (0.92 h), a useful feature that would potentially allow more flexibility in the time spent between injection/infusion and mild-hyperthermia treatment. Finally, the tumor growth curves for STL-DOX show significantly more inhibition than those for LTSL-DOX over a 7 day timecourse, although the effects of HIFU appear to be much more significant for STL-DOX relative to LTSL-DOX, suggesting more investigation may be needed into the exact cause [44]. The authors also investigated variation of the timing of hyperthermia application and initial results suggested improvements with a longer lag time between i.v. injection and hyperthermia; the reasoning for this was that the EPR-effect could allow more accumulation of liposome before heat triggered drug release in the tumor, but the PK data is not entirely in agreement with this hypothesis as by 6 h, the formulation will have undergone 3 half-lives and there would not be a significant amount remaining in the circulation [44, 46].

10.3.1.4.2 Leucine zipper-peptide hybrid vesicles

Another recent development in the area of peptide-doped liposomes for temperature-sensitive release has been presented by Kostarelos and co-workers [47, 48]. In this work, a leucine zipper (Lp-) peptide has been incorporated at various ratios (optimal 200:1 lipid:peptide molar ratio) into a long-circulating sTSL (DPPC, DSPC, DSPE-PEG$_{2000}$; 90:10:5). The average size of the vesicles formed are ~120 nm with a PDI <0.1 and a T_m between 41–43°C. Release is controlled by the leucine peptides undergoing a conformational change from an ordered α-helix to a disordered state, resulting in excellent stability at 37°C (~10% drug release in 1 h), and a moderate release rate at 42°C (<80% in 5 min). Despite this moderate release rate, in an intravascular release xenograft tumor model, a similar efficacy was observed for both LTSL and Lp-peptide hybrid vesicles over 34 days. The authors decided to use these liposomes to study different release protocols, by changing the timing of hyperthermia application, and the Lp-peptide hybrid vesicles showed better efficacy with an interstitial release protocol than the other liposomes tested; however, this protocol was not as effective at inhibiting tumor growth as the intravascular protocol [47].

10.3.1.5 Thermoresponsive bubble-generating liposomes: ABC liposomes

An interesting take on a thermosensitive liposomal delivery system has been made through the encapsulation of ammonium bicarbonate (ABC, NH_4HCO_3) in the liposomal core. ABC has its own temperature sensitivity, and above 40°C it begins to decompose into ammonia, water and carbon dioxide. The CO_2 generated forms bubbles which create defects in the liposome bilayer allowing the contents of the liposome to be released. As it is the CO_2 bubbles and not just the membrane lipids/surfactants that are causing the drug release, the membrane was stabilized with cholesterol; DPPC/Chol/DSPE-PEG$_{2000}$, 60:40:5 molar ratio with a T_m of 52°C, size 150 nm, PDI <0.1. Release from these ABC liposomes was fast at 42°C (50% release in <1 min), with good stability at 37°C, even in whole blood (85% retention after 1 h). Studies with a confocal laser scanning microscope were used to confirm that the ammonium bicarbonate formed gas bubbles which permeated the membrane and formed transient pores. Furthermore, the bubbles generated can provide a useful diagnostic to be probed with ultrasound [49]. In a follow-up study, the authors investigated their ABC liposomes in vivo, studying biodistribution and efficacy [50]. Uptake in the tumor was significantly improved for the ABC liposome with mild-hyperthermia compared to non-heated and non-thermosensitive controls. This increased uptake also corresponded to a significant inhibition of tumor growth following 3 treatments spaced 1 week apart. The results suggest the ABC liposomes delivered DOX more effectively than all other formulations, including LTSL-DOX, and also had less toxicity as a result of the greater stability, although no free drug control was included in this report.

10.3.2 Light-Sensitive Liposomes

The use of light as a trigger for stimuli-sensitive drug delivery has been extensively investigated, but in many cases these technologies do not make it past in vitro studies. This is in part because many of the molecules used to introduce sensitivity to light, respond to light in the UV range. This wavelength of light has poor penetration of biological tissue and at high intensities can lead to tissue damage. However, more recent work has attempted to overcome this by using chromophores which respond to longer wavelengths of light in the near-IR. In the 650–900 nm range of the spectrum, there is minimal attenuation by water and hemoglobin, leading to less absorption by biological tissues. This allows light of these wavelengths to penetrate further (over 1 cm) [51], making its use as a trigger more meaningful in vivo. However, at these wavelengths the energy is less and this may be insufficient for some of the classic phototriggerable liposomes to perform. Photo-oxidation, photo-isomerization, and photo-cleavage based trigger mechanisms have all been described for shorter wavelengths of light, although most of these appear to have limited potential for in vivo applications. A recent review by Yavlovich et al. discusses each of these in more detail [51]. This also explains why current light-sensitive technologies are most successful for skin and oral applications, where tissue depth presents less of a challenge.

To get around the limitation of light-mediated technologies, metals have been used which react to these relatively long wavelengths of near-IR light and trigger release from a partner stimuli-sensitive liposome. One of the preferred metals for this is gold, due to its relative biocompatibility and the plasmonic properties of gold when prepared in the form of nanoshells or nanorods. In this way, gold particles have been shown to cause rapid release of contents from liposomal carriers [52]. This class of light mediated triggerable liposomes is suitable for use in vivo and a few examples are now discussed.

Gold nanorods are plasmonic nanomaterials which efficiently convert near-IR radiation into heat. These nanorods have been coated with PEG to make them more biocompatible, which allows them to passively accumulate in tumors via the EPR-effect (usually 24–48 h). Once a sufficient quantity has accumulated they can be used to generate a localized heat source, by excitation with a near-IR laser. This has been combined with a temperature-sensitive liposome (TSL), so that when the TSL reaches the tumor region, the heat generated by the irradiated gold nanoparticles causes the TSL to release its drug payload. A number of different reports on this technique are represented in the literature, which differ mainly by the TSLs used with each example; one with a traditional TSL similar to that used by Gaber et al. [53], and another with a unique composition of DPPC/1,2-dimyristoyl-sn-glycerol-3-phosphatidylcholine (DMPC)/Chol/DSPE-PEG_{2000} (54/3/30/3) [54]. Both of these examples are slow releasing TSLs (sTSL) that release their contents over several minutes, with significant tumor growth inhibition observed in both cases.

More recently, You et al. have reported on a related gold nanoshell TSL combination, where the gold has been attached to the lipid membrane through the use of a thiol-functionalized tethering lipid [55]. In this way, the gold could be functionalized with the lipid and combined with the other lipid components to prepare a liposome via the thin-film hydration method. Again a traditional sTSL formulation was used DPPC/HSPC/Chol/DSPE-PEG_{2000} (100:50:15:6). In this case the regression of tumors was thought to be due to the synergistic combination of a photothermal ablation effect of the gold nanoshell and the temperature-induced stimuli-sensitive release of DOX from the sTSL used [55]. This sTSL formulation has since been further modified with a targeting ligand (HER2 Antibody) on its surface [56], increasing the size (~150 nm) and negative surface charge (-60 mV). However, significant regression of tumors was observed with lower DOX dose and less intense laser irradiation, causing a marked reduction in systemic toxicity of the treatment, as measured by animal body weight.

10.3.3 Magnetic-Sensitive Liposomes

Magnetic-sensitive nanomedicine is different to other techniques as it can work via several different mechanisms. Magnetic particles can be used to guide the distribution of nanomedicine through the use of an external magnetic field, a process referred to as magnetic targeting. Alternatively, the particles may be designed to heat up through the application of an oscillating magnetic field and this kind of magnet-induced hyperthermia can be used for heat ablation of tumors, or in conjunction with thermosensitive nanomedicine to release drug in a localized fashion. An advantage of using magnetic fields is that they can penetrate further through biological tissue than light for example, and so have potential to be more versatile in vivo. However, magnetic particles are normally prepared from an oxide of iron (e.g., Fe_3O_4, magnetite), which is a ferromagnetic material, and has been observed to increase the formation of reactive oxygen species (ROS), via degradation and interaction with biological milieu [57], as well as present dangers of coagulation and agglomeration, leading to potential health risks [58]. Magnetite particles below a threshold size (~30 nm) become superparamagnetic and lose their permanent magnetism, only becoming magnetic when an external field is applied. Such particles are commonly used in biological pre-clinical studies and are called SPIONs (SuperParamagnetic Iron Oxide Nanoparticles).

Examples of using magnetic-sensitive liposomes in vivo are becoming increasingly common; there are examples of liposomes being loaded with magnetic nanoparticle conjugated drugs for magnetic targeting across the blood brain barrier (BBB) [59], and liposomes being loaded with magnetite for localization with an external static magnet [58].

In 2004, Nobuto and co-workers prepared passively loaded DOX/magnetite liposomes (EPC/Chol 1:1 molar ratio, ~85 nm) and studied their accumulation in vivo with an osteosarcoma tumor model using an externally applied 0.4 T static magnetic field. Maximum DOX concentrations in the tumor were achieved after 60 min of magnetic field exposure, with DOX magnetoliposomes (MLs) being retained in the tumor site by the applied field. Levels dropped at the 100 min timepoint which could be due to the removal of the magnetic field, or limitations of liposome stability in vivo. These magnetically localized liposomes led to complete inhibition of tumor growth for 12 days, and histological examination revealed high levels of necrosis in the DOX magnetic liposome with magnetic field treated animals, relative to the group where no magnetic field was applied (69 ± 19.5% vs. 37 ± 8.4%) [60].

MLs have been used in combination to deliver both GEM and OXA simultaneously [61]. A phosphatidylcholine (PC)/Chol/1,2-dimyristoyl-sn-glycerol-3-phosphoglycerol (DMPG) (6:4:1) liposome was chosen as the optimal composition, which was loaded with a solution of Fe_3O_4 nanoparticles (~10 nm) and either GEM or OXA leading to a final size/PDI of 227.6 nm/0.133 or 169.3 nm/0.0597, respectively. Through the use of an external magnetic field around the heads of the animals studied, a significant influence on both the PK and biodistribution was demonstrated. A small increase in clearance half-life was observed for formulations when the magnetic field was applied relative to those without, but the most significant change was the amount of drug identified 90 min after injection (magnetic field was sustained for just 30 min post-injection) in the brain tissue. For the GEM ML the levels of GEM detected were over 8-times that for the same formulation without the applied magnetic field, and similar results were observed for the OXA formulation. The effect of these formulations on tumor growth inhibition demonstrated a significant effect for all drug-ML combinations, but was most pronounced for the combination of GEM-ML with OXA-ML, where a 85% inhibition in MCF-7 (breast cancer xenograft) tumor growth was observed relative to the untreated control tumors (final weight 201 mg vs. 1335 mg). Studies were also performed to check the drug combination mechanism of action, with the combination treatment exhibiting low mRNA expression of Bcl-2 and survivin (apoptosis inhibitors), but high expression of Bax (apoptosis promoter). These results suggested GEM and OXA were efficiently delivered to the tumor site, triggered MCF-7 cell apoptosis, and demonstrated the synergistic effects from combining GEM and OXA, using an external magnetic field to localize the treatment. The same group also reported on a similar magnetoliposome formulation to deliver the protein rhIFNα_2b for treatment of a hepatocellular carcinoma model in nude mice demonstrating a significant tumor growth inhibition [62].

More recently, examples of combining the localization of the magnetic particles with another stimuli-sensitive property to improve drug release at the target site have been presented. Lin et al. have prepared TSLs of a similar composition to LTSL (DPPC/MSPC/DSPE-PEG$_{2000}$, 87:3:10 mass ratio) and encapsulated either DOX, DOX-CPP (DOX conjugated to cell penetrating peptide), Fe_3O_4 magnetic fluid (a solution of magnetite particles), or a combination of these. In this way, the temperature of the magnetic fluid could be increased from 25 to 42°C within 20 min upon remotely applying an AC magnetic field. The increased temperature could then release the encapsulated drug from the liposome core which could penetrate into the target cells. Drug release was demonstrated with both conventional heating and by applying the AC magnetic field to excite the Fe_3O_4. In both cases significant increases in drug accumulation in the MCF-7 xenograft tumor model was observed. This corresponded to increased tumor growth inhibition for both heated and magnetically activated TSLs relative to all other control groups and the magnetically activated TSL showed 1.3-fold more inhibition than the TSL with just hyperthermia [63].

10.3.4 Ultrasound-Sensitive Liposomes

Nanomedicine strategies have also been designed to be triggered by an ultrasonic triggering mechanism. This is usually through a process termed inertial cavitation, whereby an ultrasound (US) wave causes the expansion of a gas bubble followed by its sudden collapse causing shock waves that disrupt the membranes of the co-localized drug loaded liposomes. Low-frequency or high-intensity US can be used to generate gas bubbles in situ, but there are concerns over potential damage to non-targeted tissue. Alternatively, high-frequency US at lower pressures can be used with pre-existing gas bubbles. Graham et al. have recently described this approach using a commercially available gas-filled microbubble (used for ultrasound diagnostics) with a dye (luciferin) containing liposome [64]. Two different liposomal compositions were studied, one containing HSPC, much like DOXIL (HSPC/Chol/DSPE-PEG, 56:39:5), and the other substituting the phospholipid for a combination of 1,2-distearoyl-sn-glycerol-3-phosphoethanolamine (DSPE) and DSPC (DSPE/DSPC/Chol/DSPE-PEG, 65/3/25/7). The diameters of these liposomes were a little larger (~150 nm) to avoid the cut-off for the fenestrae of murine hepatocytes for longer-circulation character. The DOXIL-like formulation was found to be too stable to the cavitation technique used, but the DSPE formulation demonstrated significant release at 1.2–1.5 MPa (0.0486 kW/cm^2). The DSPE formulation was tested in vivo and although the PK and biodistribution were essentially unchanged with application of multiple cavitation events, the concentration of luciferin in the tumor was increased (16-fold increase in luminescence) compared to those that did not receive ultrasound. This is promising, although the authors note that there were limitations in their approach with regard to the variability of inertial cavitation across different regions of the tumor and the requirement for replenishing the microbubble concentration in the circulation after it is destroyed in each cavitation process. Furthermore, the authors acknowledge this technology is still some way behind that of the more developed TSL technologies where near 100% drug release is observed.

An alternative way of achieving the same cavitation approach without the need for separate microbubbles could be by encapsulating small emulsion droplets inside the liposome. Pitt and co-workers [92] have developed such an emulsion-containing liposome or eLiposome (DPPC/Chol/DSPE-PEG_{2000})/α-tocopherol, 3:1:1:0.004), loaded with both a perfluoropentane (PFC_5) fluorocarbon emulsion (by sonication) and DOX (via remote loading), with a final size of 120–500 nm. Upon application of US (20 KHz, 1 W/cm^2) the eLiposomes released 80% of the encapsulated DOX within 1 minute in vitro, which increased with increasing time and power. This work has considerable promise and has been demonstrated using micellar delivery vehicles [65], as well as liposomes. However, there are no reports on studies of this technology in vivo, so this will not be further discussed here.

Another structure which has been created for ultrasound triggering is the bubble liposome — essentially a liposome encapsulating a fluorocarbon gas [66]. These can be used to deliver drugs across cell membranes via sonoporation. In a recent example, such bubble liposomes have been prepared containing perfluoropropane gas: 1,2-distearoyl-3-trimethylammonium-propane (DSTAP)/DSPC/DSPE-PEG$_{2000}$-X (7:2:1 molar ratio), where X = NH_2 or mannose [67]. These liposomes could be mixed with pDNA (plasmid DNA) to form a lipoplex, i.e., complexes of the negatively charged DNA with the positively charged liposomes. In this way a gene delivery vector was generated, which could be targeted via mannose to the macrophage and dendritic cells, delivering the DNA directly into the cell through the transient pores formed via sonoporation of the bubble liposomes upon US application. In vitro results showed a 10-fold increase in transfection for the bubble lipoplex relative to the lipoplex with no US triggering ability. To demonstrate this system's potential for DNA vaccine therapy against cancer, an OVA expressing pDNA was used in a model with OVA-expressing cells. A high antitumor effect was observed for the E.G7-OVA cells, whereas for the EL4 cells (expressing no OVA) there was no effect. Furthermore re-transplantation of E.G7-OVA cells into animals treated by DNA vaccination with the mannose-PEG bubble lipoplex and US exposure were completely rejected up to 80 days after treatment. This method demonstrates the potential for APC (antigen presenting cell)-selective gene expression, for effective anti-tumor response in this DNA vaccine therapy.

In a similar way, nanoscale gas (C_3F_8) cored bubble liposomes (DPPC/DSPE/1,2-dipalmitoyl-sn-glycerol-3-phosphoalanine (DPPA), 18:1:1 mass ratio) have recently been shown to deliver paclitaxel and siRNA encapsulated in a separate micelle (PEG-polylysine). The electrostatic attraction between the drug containing bubble liposome (−19 mV) and the oligonucleotide containing micelle (+22 mV) was sufficient to form a "hetero-assembled nanobubble complex" (size ~400 nm) [68]. An optimal ratio of liposome to micelle was found to be 0.2, leading to a product with weak positive charge (+9 mV). When combined with low-frequency US, this nanobubble complex was shown to have improved cytotoxicity in vitro relative to treatments without US and treatments lacking the anti-apoptotic siRNA, BCL-2.

Treatments studied in vivo, showed a poor accumulation of the nanobubble complexes via the EPR effect, presumably due to their large size. However, when treated with US, two factors contributed to a much increased accumulation: (1) enhanced permeability of the endothelium and extracellular matrix in the tumor region and (2) fragmentation of the large nanobubble complex into smaller particles (~50–70 nm) with higher diffusibility, but still containing both therapeutic agents. Consequently, the physical force imposed by the US, promoted co-delivery of the drug/siRNA payload to deep sites of the tumor tissue. The efficacy was then demonstrated in a tumor model featuring repeated formulation dosing and low-frequency US exposure of the tumor area, with nanobubble complex maintaining tumor size to <80 mm^3 by day 28, whereas tumors for all other control treatments had increased to more than 500 mm^3 by this time. The nanobubble complex with US exposure also led to the best survival rate, with 85.7% surviving longer than 35 days. Hence, using US as a delivery trigger can also have other effects on the tissue physiology, in this case leading to successful co-delivery of low-dose paclitaxel and BCL-2 siRNA to overcome drug resistance.

10.4 Liposomes Responding to Internal Stimuli

10.4.1 pH-Sensitive Liposomes

The concept behind pH-sensitive liposomes (PSLs) for cancer treatment is different from the other trigger mechanisms discussed thus far, as pH is an endogenous quality of the cell type and cannot easily be controlled. In theory, it may be possible to target the tumor using pH, as a reduced pH is one of the many characteristics of the tumor microenvironment being around pH 6.4, relative to normal cells/physiological levels at pH 7.4. However, examples of pH-sensitive liposomes are most often aimed at overcoming the problems of drug release from the liposomal carrier once the nanoparticle has reached the target cell (particularly with regard to targeting the cytoplasm), and not as a cell targeting method per se. In other words, liposomes are normally accumulated in the tumor cells via the EPR-effect or through the interaction with some targeting moiety adhered to the liposome surface, whereupon the liposome is endocytosed. In the endosomal compartment pH is lower (pH 5.5–6.5) and when this matures into a lysosome, pH is reduced further still (pH 4.5–5.4), which provides a suitable change in pH for triggering drug release.

Liposomes that can be triggered by this change in pH must be stable in normal physiological conditions, but become destabilized in these mildly acidic environments. This functionality is usually achieved through the incorporation of groups which can become protonated at lower pH and therefore undergo a change in conformation, destabilizing the liposome membrane. The early pH-sensitive liposomes consisted of 1,2-dioleoyl-sn-glycerol-3-phosphoethanolamine (DOPE) and a mildly acidic amphiphile such as oleic acid or cholesterylhemisuccinate (CHEMS) to stabilize the bilayer phase. Upon mild acidification, this membrane undergoes a conformational change to a non-bilayer structure which can fuse with the endosomal membrane and release the liposome contents into the cytoplasm (Fig. 10.1).

However, in vivo these structures often had compromised stability or reduced pH-sensitivity due to interactions with serum, which reduce the membrane sensitivity to pH. PEGylated versions were prepared to improve the long-circulating character, but with increased stealth character came a reduction in pH-sensitivity. Some of the more advanced pH-sensitive liposomes therefore contain PEGylated lipids that slowly detach from the liposome, allowing improved pH-sensitivity once the nanoparticle reaches the endosome (Fig. 10.2) [69, 70]. In certain cases, good results can also be achieved through the attachment of targeting ligands (Antibody, RGD-peptide etc.) to the liposome exterior in order to recognize receptors on a particular cell type [71]. Many pH-sensitive liposomes have been studied over the last quarter of a century, which would be too much to discuss in any detail in this chapter, for this the reader is directed to recent reviews that cover many of the key findings [72, 73]. Instead, to represent the current work in this field, some of the most recent studies in this area for the delivery of chemotherapeutics will be discussed.

Barros and co-workers recently compared the delivery of a PSL (DOPE/CHEMS/DSPE-PEG$_{2000}$, 5.7:3.8:0.5 molar ratio, size: ~115 nm) relative to a DOXIL-like non-PSL, both delivering a radiolabeled complex of DOX ([^{99m}Tc]DOX). The loading efficiency of DOX was relatively low (~30%), because a passive technique was used, but a stable formulation was formed which demonstrated pH-sensitive release below pH 6.5. The PK profiles of PSL and non-PSL were similar in profile although the clearance of non-PSL was more rapid. It was thought this may hold the answer as to why the uptake of DOX in the tumor for PSL was ~4-fold higher than that for non-PSL, with tumor-to-muscle ratios significantly higher in value at 4 and 24 h for PSL relative to non-PSL [74].

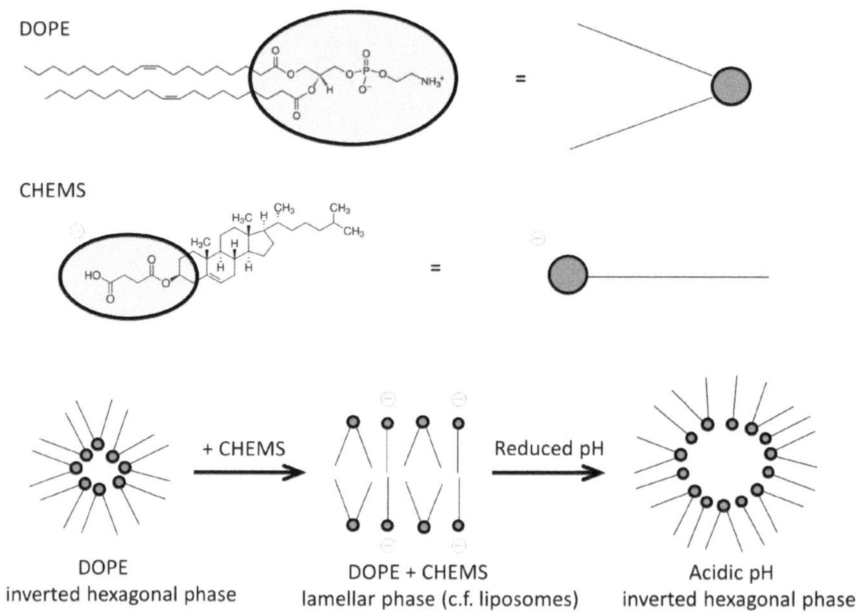

Figure 10.1 Mechanism of action for pH-sensitive liposomes. DOPE has a small headgroup and non-linear lipid tails creating a structure which prefers the formation of inverted hexagonal phase micelles. At neutral pH CHEMS has a negatively charged headgroup which is heavily solvated, effectively forming a large headgroup and linear lipid tail. Together these two lipids can combine to form a lamellar phase suitable for a liposome membrane. However, when the pH decreases, the charge on CHEMS is neutralized, disturbing the membrane and causing it to revert back to an inverted hexagonal phase. This concept forms the basis of the mechanism for drug release from many pH-sensitive liposomes.

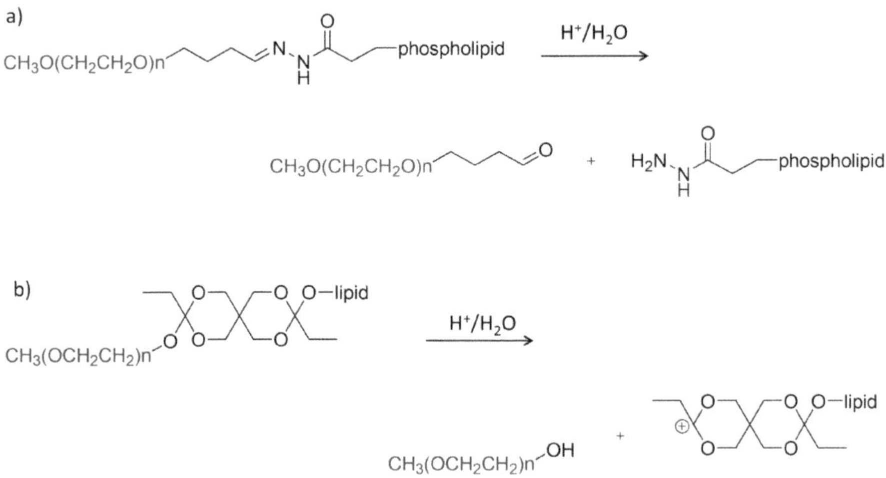

Figure 10.2 Mechanism of pH-sensitive PEG deshielding. Two examples of structures used to achieve cleavable PEGylated lipid membranes are shown; (a) A PEGylated lipid with incorporated acid labile hydrazone linker; (b) An ortho-ester linker sensitive to acid environments.

A pH-sensitive formulation of gemcitabine has also recently been prepared, again using a DOPE/CHEMS/DSPE-PEG_{2000} (6:4:0.3) liposome, with the aim of delivering the drug intracellularly and avoiding MDR (multi-drug resistant) mechanisms [75]. The PSL formulation showed pH-sensitive character, significantly greater toxicity to resistant prostate cancer cells than a non-PSL or free drug control and an extended PK profile relative to free drug. These results are promising; however, biodistribution and efficacy studies have not yet been reported.

Udofot et al. recently described the preparation of a pH-sensitive liposome prepared from CHEMS/Chol/polyethylene glycol sorbitan monolaurate (Tween20)/DSPE-PEG_{2000} (60:20:10:10, size: 164 nm), used to deliver 5-fluorouracil (5-FU), the drug of choice for colorectal cancer [76]. This is interesting because the lipid components omit the use of DOPE, so one would presume the surfactant Tween 20 plays a similar role promoting an inverted hexagonal phase which can be converted to a lamellar phase in combination with CHEMS. The drug was loaded passively and a further targeting coating of anti-EGFR was applied to the nanoparticle's surface via electrostatic interaction. The liposomal formulation was moderately more potent against HCT-11 cells in vitro and increased the half-life in vivo by 2.7-fold over free drug (following an i.p. bolus dose). Biodistribution profiles were favorable, with almost 2-fold the amount of drug reaching the tumor with the liposome, relative to the free drug. Accumulation in other organs was less than free drug, with the exception of colon. Following a dosing schedule of i.p 3 times per week, an efficacy study was performed in HCT-11 tumor bearing mice demonstrating a slight improvement in tumor growth inhibition for the pH-sensitive liposome formulation relative to the free drug.

While many reports use the same lipids to invoke pH sensitivity, there are some groups investigating new lipids for the formation of PSLs. Moku et al. recently described the synthesis of 4 cationic dipeptidyl lipids ("CDLs" containing amino acids: histidine/arginine/lysine and lipids: palmitoyl or stearoyl) [77]. Liposomes were prepared from these and other more familiar lipids (CDL/1,2-dioleoyl-sn-glycerol-3-phosphatidylcholine (DOPC)/DOPE/Chol/DSPE-PEG_{2000}, 1:1:1:1:0.05 molar ratio) and passively loaded with curcumin and paclitaxel (size: ~208 nm, PDI: 0.315). Once pH sensitivity was demonstrated, studies on in vitro cytotoxicity and drug release were performed to select the most promising new lipid: lipid 3, which was a histidine-arginine distearoyl lipid conjugate. Further in vitro studies demonstrated the best cell cycle arrest and anti-proliferation for the liposome containing lipid 3, curcumin and paclitaxel. It was also demonstrated that the expression levels of anti-apoptotic protein Bcl-2 was decreased and the apoptotic proteins Bax and Bid were increased. Biodistribution of the PSL was studied for both curcumin and paclitaxel in a B16F10 tumor mouse model and demonstrated significantly increased levels of both compounds in the tumor relative to other tissues (>7-fold) after 24 h. In an efficacy study with the same tumor model, the liposome containing lipid 3, curcumin and paclitaxel demonstrated the best therapeutic effect and longest overall survival. The liposome with just paclitaxel and liposome with just curcumin were intermediate and the liposome alone showed some effect over control. In other words, the liposome carrier itself contributes to the tumor inhibition, in contrast to the majority of other drug carriers. While some immunogenicity studies have been performed, it seems that further toxicity studies with this formulation may be the next priority.

10.4.2 Enzyme-Sensitive Liposomes

Another method of triggering release relying on an internal stimulus is through the sensitivity to locally expressed proteins or enzymes. The triggering mechanism will often involve an enzymatic cleavage of part of the liposomal carrier leading to release of the encapsulated drug payload. Several enzymatic classes have been used in this way, including proteases [78] and lipases [79] and recent reviews discuss their use for liposomes and other nanocarriers [80, 81].

Cancer-associated proteases (CAPs) have been used to trigger release from a polymer-stabilized liposome [82]. The liposome (DPPC/DOPC/Chol, 47.5:5:47.5 molar %) was designed to be relatively unstable due to a high tonic buffer encapsulated inside, which was stabilized by a polymer coating containing peptides that were substrates for CAPs. Hence, in cancerous tissue the peptides are cleaved leading to a degradation of the polymer cage, causing release of liposome payload. While extensive work on the optimization of the system has been reported, the efficacy of this formulation in vivo has not yet been studied.

The work of Andresen et al. has explored an interesting approach to liposomal drug release through the incorporation of lipid prodrugs into the liposome membrane. These prodrugs were designed to be activated by secretory phospholipase A_2 ($sPLA_2$) — known to be upregulated in the tumor environment — releasing the drug in a localized manner. In one recent example a lipid prodrug was prepared from the drug, capsaicin, which spontaneously formed small unilamellar vesicles upon hydration in buffer (size 66 nm) [83]. Although the system was stable for weeks in buffer solution, upon exposure to $sPLA_2$ ~90% capsaicin release was observed after 24 h. Further studies on the in vitro and in vivo efficacy of these formulations have proven only modest at best, although this work has led to the development of contact-facilitated drug delivery techniques as discussed in detail in a recent review [84]. Similarly, investigations on using $sPLA_2$ to promote drug delivery from liposomes by cleavage of the structural membrane lipid components has demonstrated only limited improvements, which when challenged with the $sPLA_2$ inhibitor (LY311727) displayed no change in cytotoxicity/efficacy, suggesting there are other processes at play for the observed differences in therapeutic response [85].

There are a number of examples where enzyme responsive liposomes have been designed to respond to local levels of matrix metalloprotease-2 (MMP-2). Its first use was for cleaving the stealth PEG coating of the liposome to reveal tumor targeting ligands on the surface for improved delivery [86]. More recently, this concept was built upon by Zhu and co-workers, wherein they combined an MMP-2-sensitive PEG lipid with a TAT cell penetrating peptide functionalized liposome to promote targeting and internalization of the nanocarriers in cancer cells [87]. The MMP-2 sensitivity was used to deblock the long PEG chains used to shield the liposomes and provide long circulation properties. Once the nanocarriers accumulate in the tumor microenvironment (either via EPR-effect or an optional targeting Ab) the high concentrations of MMP-2 will cause cleavage of the PEG shield, exposing the TAT peptides promoting cellular internalization, with the aim of minimizing nonspecific uptake while maximizing the penetration of cancer cells in areas of significant MMP-2 concentration. The concept was shown to work in vitro and the elucidation in vivo is pending. Again a similar concept was used, this time for an improved gene delivery system, through a target specific loss of PEG chains via an MMP mediated cleavage [88]. Anionic liposomes were loaded with adenoviral vectors and a PEG-peptide-Chol conjugate was post-inserted into the membrane to provide a stealth layer. Once in the MMP-2 rich tumor microenvironment the peptide linker was cleaved, releasing the PEG and promoting a higher gene expression than a non-cleavage PEGylated liposome. The PEG layer and anionic charge of the liposomes was necessary to avoid stimulation of the immune system and reduced systemic toxicity. While in vivo studies were performed to assess the immunostimulatory effects, the effectiveness of the system to kill cancer cells in vivo has yet to be assessed.

Liposomes sensitive to prostate specific antigen (PSA) have been prepared for treatment of prostate cancer [89]. Liposomes were designed with a cell-penetrating peptide (CPP) sequence (multiple positive charges), which was masked by a shielding peptide (negatively charged). Between these two peptides was a third region coding for a known substrate sequence of PSA. The liposomes also carried a folate targeting ligand to help localize the nanocarrier and improve transport into the prostate cancer cells via the prostate specific membrane antigen (PSMA) — a recognized property of this receptor [90]. In this way, the liposomes could be doubly targeted to the tumor where upon the high levels of PSA present cleave the shielding peptide away from the CPP, leading to uptake of the entire liposome into the cell. Once inside, the siRNA contents was released and began to act on the gene of interest (polo-like kinase 1, PLK-1, a regulator of mitotic progression). The combination of the folate ligand and the PSA-responsive peptide attached to the liposomes significantly increased the level of gene silencing and growth inhibition in vitro. Furthermore, in vivo biodistribution studies demonstrated an improved selectivity in the distribution of formulations incorporating the PSA-responsive peptide. This correlated to the best tumor growth inhibition (2-fold more efficacious than control after 1 week) and apoptosis (via TUNEL assay), with negligible toxicity represented by body weight changes.

10.5 Summary, Conclusions, and Future Perspectives

The potential of stimuli-sensitive nanomedicine approaches for cancer treatment have made it a leading avenue of research in recent years. Some of the most clinically advanced stimuli-sensitive formulations are based on liposomal nanoparticles, due partly to their well-established biocompatibility, controllable size with a narrow distribution, and potential for high drug loading with a range of chemotherapeutics. Many systems have been investigated, using both internal and external methods of stimuli, which appear to work via two distinct types of mechanism.

The external stimuli triggered nanoparticles commonly use the stimulus as a way of targeting the treatment to a localized area, often ignoring any potential EPR-effect, and relying on a localized application of heat, light, ultrasound or magnetic field. The most developed formulations of this type are the thermosensitive liposomes, usually triggered by an application of mild-hyperthermia. However, it would seem that many of the light and magnetic field triggered formulations are also heat sensitive, but in addition to the liposome being a TSL, they have been combined with gold or iron oxide nanoparticles, necessary to convert the light or magnetic stimuli into heat, causing localized hyperthermia. For these systems it then seems to be a matter of what is the most convenient, effective and economic method of stimulation; light has limitations in its penetration through tissues and magnetic fields have some biocompatibility issues, so perhaps mild-hyperthermia itself should be the obvious choice, particularly when MR-guided high-intensity focused ultrasound approaches for inducing localized and highly controllable mild-hyperthermia are available. Ultrasound as a triggering method in itself is more often of low intensity capable of bursting bubbles of gas causing cavitation and potentially breaking cell membranes. While this kind of ultrasound does not generate heat, it cannot be so finely focused and is often used at powers where damage to healthy cells could be more of an issue.

Liposomes triggered by endogenous stimuli have also undergone much research over the last 20 years. These forms of nanomedicine are often not reliant on the stimulus to localize the region of treatment, but instead rely on the EPR-effect or targeting ligands for the formulation to reach the area of interest. Once the liposome has reached this tissue, then the endogenous chemistry of that tissue triggers the release of the drug. In the case of the pH-sensitive liposomes this is most commonly used, not to target a cell, but to assist release of drug into the cytoplasm from the endosomal compartment following endocytosis. Meanwhile, for the enzyme-sensitive liposomes, the increased levels of certain enzymes in the tumor environment are used to promote drug release, although effects observed in vivo don't appear to have yet reached their true potential. For both of these systems the fundamental components and mechanism of these technologies have remained much the same since they were first introduced, with only minor tweaks to improve distribution, uptake, or drug release.

Future work in this area will likely follow recent trends, combining multiple methods of stimuli to create more complex systems with increased specificity. As an example, a recent multi-functional stimuli-sensitive liposome that incorporates both a pH- and magnetic-sensitive element has been presented by

Wang et al. [91]. In this complex system, a folate-conjugated dendrimer and magnetic nanoparticles (oleic acid conjugated Fe_3O_4 nanoparticles, 10–15 nm) were encapsulated inside a pH-sensitive long-circulating liposome (DOPE/CHEMS/Chol/DSPE-PEG_{2000} 10:4:3.6:4). In this way, the so called dendrosome (DS) could accumulate at the tumor site via both the EPR effect (size ~180 nm at pH 7.4) and in response to the external magnetic field applied, whereupon the acidic microenvironment of the tumor would cause release of the folate-dendrimer conjugate. While this study has not studied delivery of a drug, the concept of a multi-functional drug delivery platform has been demonstrated through the fluorescence-based biodistribution studies performed. It will be interesting to see how this and other complex systems like this perform with regard to tumor efficacy when a drug is used instead.

The examples discussed in this chapter represent the recent developments of the field of stimuli-sensitive liposomes that have progressed far enough to be studied in vivo. There are many other systems which may have been investigated for function in controlled in vitro assays, but these may not yet be ready for assessment in vivo. Furthermore there are a plentitude of examples of other stimuli-sensitive nanomedicine systems which are not focused on a liposomal delivery vehicle and so have not been discussed here. Bearing this in mind it shows how dynamic this area of research is at present, and signifies how great the potential is for stimuli-sensitive nanomedicine for cancer to make an impact in the years to come.

References

1. Maeda, H. (2015). Toward a full understanding of the EPR effect in primary and metastatic tumors as well as issues related to its heterogeneity, *Adv. Drug Deliv. Rev.*, **91**, pp. 3–6.
2. Barenholz, Y. (2012). Doxil® — the first FDA-approved nano-drug: Lessons learned, *J. Control. Release*, **160**(2), pp. 117–134.
3. Laginha, K. M., Verwoert, S., Charrois, G. J., and Allen, T. M. (2005). Determination of doxorubicin levels in whole tumor and tumor nuclei in murine breast cancer tumors, *Clin. Cancer Res.*, **11**(19 Pt 1), pp. 6944–6949.

4. Seynhaeve, A. L., Dicheva, B. M., Hoving, S., Koning, G. A., and ten Hagen, T. L. (2013). Intact Doxil is taken up intracellularly and released doxorubicin sequesters in the lysosome: Evaluated by in vitro/in vivo live cell imaging, *J. Control. Release*, **172**(1), pp. 330–340.
5. Harrington, K. J., Lewanski, C. R., Northcote, A. D., Whittaker, J., Wellbank, H., et al. (2001). Phase I-II study of pegylated liposomal cisplatin (SPI-077) in patients with inoperable head and neck cancer, *Ann. Oncol.*, **12**(4), pp. 493–496.
6. Meerum Terwogt, J. M., Groenewegen, G., Pluim, D., Maliepaard, M., Tibben, M. M., et al. (2002). Phase I and pharmacokinetic study of SPI-77, a liposomal encapsulated dosage form of cisplatin, *Cancer Chemother. Pharmacol.*, **49**(3), pp. 201–210.
7. White, S. C., Lorigan, P., Margison, G. P., Margison, J. M., Martin, F., et al. (2006). Phase II study of SPI-77 (sterically stabilised liposomal cisplatin) in advanced non-small-cell lung cancer, *Br. J. Cancer*, **95**(7), pp. 822–828.
8. Yatvin, M. B., Weinstein, J. N., Dennis, W. H., and Blumenthal, R. (1978). Design of liposomes for enhanced local release of drugs by hyperthermia, *Science*, **202**(4374), pp. 1290–1293.
9. Weinstein, J. N., Magin, R. L., Yatvin, M. B., and Zaharko, D. S. (1979). Liposomes and local hyperthermia: Selective delivery of methotrexate to heated tumors, *Science*, **204**(4389), pp. 188–191.
10. Manzoor, A. A., Lindner, L. H., Landon, C. D., Park, J. Y., Simnick, A. J., et al. (2012). Overcoming limitations in nanoparticle drug delivery: Triggered, intravascular release to improve drug penetration into tumors, *Cancer Res.*, **72**(21), pp. 5566–5575.
11. Li, L., ten Hagen, T. L., Haeri, A., Soullie, T., Scholten, C., et al. (2014). A novel two-step mild hyperthermia for advanced liposomal chemotherapy, *J. Control. Release*, **174**, pp. 202–208.
12. Li, L., ten Hagen, T. L., Bolkestein, M., Gasselhuber, A., Yatvin, J., et al. (2013). Improved intratumoral nanoparticle extravasation and penetration by mild hyperthermia, *J. Control. Release*, **167**(2), pp. 130–137.
13. Li, L., ten Hagen, T. L., Schipper, D., Wijnberg, T. M., van Rhoon, G. C., et al. (2010). Triggered content release from optimized stealth thermosensitive liposomes using mild hyperthermia, *J. Control. Release*, **143**(2), pp. 274–279.
14. Gaber, M. H., Hong, K., Huang, S. K., and Papahadjopoulos, D. (1995). Thermosensitive sterically stabilized liposomes: Formulation and

in vitro studies on mechanism of doxorubicin release by bovine serum and human plasma, *Pharm. Res.*, **12**(10), pp. 1407–1416.

15. Li, L., ten Hagen, T. L., Hossann, M., Suss, R., van Rhoon, G. C., et al. (2013). Mild hyperthermia triggered doxorubicin release from optimized stealth thermosensitive liposomes improves intratumoral drug delivery and efficacy, *J. Control. Release*, **168**(2), pp. 142–150.
16. de Smet, M., Langereis, S., van den Bosch, S., and Grull, H. (2010). Temperature-sensitive liposomes for doxorubicin delivery under MRI guidance, *J. Control. Release*, **143**(1), pp. 120–127.
17. de Smet, M., Heijman, E., Langereis, S., Hijnen, N. M., and Grull, H. (2011). Magnetic resonance imaging of high intensity focused ultrasound mediated drug delivery from temperature-sensitive liposomes: An in vivo proof-of-concept study, *J. Control. Release*, **150**(1), pp. 102–110.
18. de Smet, M., Langereis, S., van den Bosch, S., Bitter, K., Hijnen, N. M., et al. (2013). SPECT/CT imaging of temperature-sensitive liposomes for MR-image guided drug delivery with high intensity focused ultrasound, *J. Control. Release*, **169**(1–2), pp. 82–90.
19. Lu, T., Lokerse, W. J., Seynhaeve, A. L., Koning, G. A., and ten Hagen, T. L. (2015). Formulation and optimization of idarubicin thermosensitive liposomes provides ultrafast triggered release at mild hyperthermia and improves tumor response, *J. Control. Release*, **220**(Pt A), pp. 425–437.
20. Zagar, T. M., Vujaskovic, Z., Formenti, S., Rugo, H., Muggia, F., et al. (2014). Two phase I dose-escalation/pharmacokinetics studies of low temperature liposomal doxorubicin (LTLD) and mild local hyperthermia in heavily pretreated patients with local regionally recurrent breast cancer, *Int. J. Hyperthermia*, **30**(5), pp. 285–294.
21. Landon, C., Park, C. J., Needham, D., and Dewhirst, M. W. (2011). Nanoscale drug delivery and hyperthermia: The materials design and preclinical and clinical testing of low temperature-sensitive liposomes used in combination with mild hyperthermia in the treatment of local cancer, *Open Nanomed. J.*, **3**, pp. 38–64.
22. Needham, D., and Dewhirst, M. W. (2001). The development and testing of a new temperature-sensitive drug delivery system for the treatment of solid tumors, *Adv. Drug Deliv. Rev.*, **53**(3), pp. 285–305.
23. Yarmolenko, P. S., Zhao, Y., Landon, C., Spasojevic, I., Yuan, F., et al. (2010). Comparative effects of thermosensitive doxorubicin-containing liposomes and hyperthermia in human and murine tumours, *Int. J. Hyperthermia*, **26**(5), pp. 485–498.

24. Chen, Q., Krol, A., Wright, A., Needham, D., Dewhirst, M. W., et al. (2008). Tumor microvascular permeability is a key determinant for antivascular effects of doxorubicin encapsulated in a temperature sensitive liposome, *Int. J. Hyperthermia*, **24**(6), pp. 475–482.

25. Hauck, M. L., LaRue, S. M., Petros, W. P., Poulson, J. M., Yu, D., et al. (2006). Phase I trial of doxorubicin-containing low temperature sensitive liposomes in spontaneous canine tumors, *Clin. Cancer Res.*, **12**(13), pp. 4004–4010.

26. Tagami, T., Ernsting, M. J., and Li, S. D. (2011). Efficient tumor regression by a single and low dose treatment with a novel and enhanced formulation of thermosensitive liposomal doxorubicin, *J. Control. Release*, **152**(2), pp. 303–309.

27. Sandstrom, M. C., Ickenstein, L. M., Mayer, L. D., and Edwards, K. (2005). Effects of lipid segregation and lysolipid dissociation on drug release from thermosensitive liposomes, *J. Control. Release*, **107**(1), pp. 131–142.

28. Banno, B., Ickenstein, L. M., Chiu, G. N., Bally, M. B., Thewalt, J., et al. (2010). The functional roles of poly(ethylene glycol)-lipid and lysolipid in the drug retention and release from lysolipid-containing thermosensitive liposomes in vitro and in vivo, *J. Pharm. Sci.*, **99**(5), pp. 2295–2308.

29. Hossann, M., Syunyaeva, Z., Schmidt, R., Zengerle, A., Eibl, H., et al. (2012). Proteins and cholesterol lipid vesicles are mediators of drug release from thermosensitive liposomes, *J. Control. Release*, **162**(2), pp. 400–406.

30. Lindner, L. H., Eichhorn, M. E., Eibl, H., Teichert, N., Schmitt-Sody, M., et al. (2004). Novel temperature-sensitive liposomes with prolonged circulation time, *Clin. Cancer Res.*, **10**(6), pp. 2168–2178.

31. Hossann, M., Wang, T., Wiggenhorn, M., Schmidt, R., Zengerle, A., et al. (2010). Size of thermosensitive liposomes influences content release, *J. Control. Release*, **147**(3), pp. 436–443.

32. Xu, H., Paxton, J., Lim, J., Li, Y., Zhang, W., et al. (2014). Development of high-content gemcitabine PEGylated liposomes and their cytotoxicity on drug-resistant pancreatic tumour cells, *Pharm. Res.*, **31**(10), pp. 2583–2592.

33. Tagami, T., Ernsting, M. J., and Li, S. D. (2011). Optimization of a novel and improved thermosensitive liposome formulated with DPPC and a Brij surfactant using a robust in vitro system, *J. Control. Release*, **154**(3), pp. 290–297.

34. Tagami, T., May, J. P., Ernsting, M. J., and Li, S. D. (2012). A thermosensitive liposome prepared with a Cu^{2+} gradient demonstrates improved pharmacokinetics, drug delivery and antitumor efficacy, *J. Control. Release*, **161**(1), pp. 142–149.

35. May, J. P., Ernsting, M. J., Undzys, E., and Li, S. D. (2013). Thermosensitive liposomes for the delivery of gemcitabine and oxaliplatin to tumors, *Mol. Pharm.*, **10**(12), pp. 4499–4508.

36. Kono, K. (2001). Thermosensitive polymer-modified liposomes, *Adv. Drug Deliv. Rev.*, **53**(3), pp. 307–319.

37. Han, H. D., Choi, M. S., Hwang, T., Song, C. K., Seong, H., et al. (2006). Hyperthermia-induced antitumor activity of thermosensitive polymer modified temperature-sensitive liposomes, *J. Pharm. Sci.*, **95**(9), pp. 1909–1917.

38. Paasonen, L., Romberg, B., Storm, G., Yliperttula, M., Urtti, A., et al. (2007). Temperature-sensitive poly(N-(2-hydroxypropyl)methacrylamide mono/dilactate)-coated liposomes for triggered contents release, *Bioconjug. Chem.*, **18**(6), pp. 2131–2136.

39. Kono, K., Ozawa, T., Yoshida, T., Ozaki, F., Ishizaka, Y., et al. (2010). Highly temperature-sensitive liposomes based on a thermosensitive block copolymer for tumor-specific chemotherapy, *Biomaterials*, **31**(27), pp. 7096–7105.

40. Kono, K., Murakami, T., Yoshida, T., Haba, Y., Kanaoka, S., et al. (2005). Temperature sensitization of liposomes by use of thermosensitive block copolymers synthesized by living cationic polymerization: Effect of copolymer chain length, *Bioconjug. Chem.*, **16**(6), pp. 1367–1374.

41. Kokuryo, D., Nakashima, S., Ozaki, F., Yuba, E., Chuang, K. H., et al. (2015). Evaluation of thermo-triggered drug release in intramuscular-transplanted tumors using thermosensitive polymer-modified liposomes and MRI, *Nanomedicine*, **11**(1), pp. 229–238.

42. Ta, T., Bartolak-Suki, E., Park, E. J., Karrobi, K., McDannold, N. J., et al. (2014). Localized delivery of doxorubicin in vivo from polymer-modified thermosensitive liposomes with MR-guided focused ultrasound-mediated heating, *J. Control. Release*, **194**, pp. 71–81.

43. Ta, T., Convertine, A. J., Reyes, C. R., Stayton, P. S., and Porter, T. M. (2010). Thermosensitive liposomes modified with poly(N-isopropylacrylamide-co-propylacrylic acid) copolymers for triggered release of doxorubicin, *Biomacromolecules*, **11**(8), pp. 1915–1920.

44. Park, S. M., Kim, M. S., Park, S. J., Park, E. S., Choi, K. S., et al. (2013). Novel temperature-triggered liposome with high stability: Formulation, in vitro evaluation, and in vivo study combined with high-intensity focused ultrasound (HIFU), *J. Control. Release*, **170**(3), pp. 373–379.

45. Beck, J. G., Mathieu, D., Loudet, C., Buchoux, S., and Dufourc, E. J. (2007). Plant sterols in "rafts": A better way to regulate membrane thermal shocks, *FASEB J.*, **21**(8), pp. 1714–1723.

46. Park, S. M., Cha, J. M., Nam, J., Kim, M. S., Park, S. J., et al. (2014). Formulation optimization and in vivo proof-of-concept study of thermosensitive liposomes balanced by phospholipid, elastin-like polypeptide, and cholesterol, *PLoS One*, **9**(7), e103116.

47. Al-Ahmady, Z. S., Scudamore, C. L., and Kostarelos, K. (2015). Triggered doxorubicin release in solid tumors from thermosensitive liposome-peptide hybrids: Critical parameters and therapeutic efficacy, *Int. J. Cancer*, **137**(3), pp. 731–743.

48. Al-Ahmady, Z. S., Al-Jamal, W. T., Bossche, J. V., Bui, T. T., Drake, A. F., et al. (2012). Lipid-peptide vesicle nanoscale hybrids for triggered drug release by mild hyperthermia in vitro and in vivo, *ACS Nano*, **6**(10), pp. 9335–9346.

49. Chen, K. J., Liang, H. F., Chen, H. L., Wang, Y., Cheng, P. Y., et al. (2013). A thermoresponsive bubble-generating liposomal system for triggering localized extracellular drug delivery, *ACS Nano*, **7**(1), pp. 438–446.

50. Chen, K. J., Chaung, E. Y., Wey, S. P., Lin, K. J., Cheng, F., et al. (2014). Hyperthermia-mediated local drug delivery by a bubble-generating liposomal system for tumor-specific chemotherapy, *ACS Nano*, **8**(5), pp. 5105–5115.

51. Yavlovich, A., Smith, B., Gupta, K., Blumenthal, R., and Puri, A. (2010). Light-sensitive lipid-based nanoparticles for drug delivery: Design principles and future considerations for biological applications, *Mol. Membr. Biol.*, **27**(7), pp. 364–381.

52. Wu, G., Mikhailovsky, A., Khant, H. A., Fu, C., Chiu, W., et al. (2008). Remotely triggered liposome release by near-infrared light absorption via hollow gold nanoshells, *J. Am. Chem. Soc.*, **130**(26), pp. 8175–8177.

53. Park, J. H., von Maltzahn, G., Ong, L. L., Centrone, A., Hatton, T. A., et al. (2010). Cooperative nanoparticles for tumor detection and photothermally triggered drug delivery, *Adv. Mater.*, **22**(8), pp. 880–885.

54. Agarwal, A., Mackey, M. A., El-Sayed, M. A., and Bellamkonda, R. V. (2011). Remote triggered release of doxorubicin in tumors by synergistic application of thermosensitive liposomes and gold nanorods, *ACS Nano*, **5**(6), pp. 4919–4926.

55. You, J., Zhang, P., Hu, F., Du, Y., Yuan, H., et al. (2014). Near-infrared light-sensitive liposomes for the enhanced photothermal tumor treatment by the combination with chemotherapy, *Pharm. Res.*, **31**(3), pp. 554–565.

56. Li, Q., Tang, Q., Zhang, P., Wang, Z., Zhao, T., et al. (2015). Human epidermal growth factor receptor-2 antibodies enhance the specificity and anticancer activity of light-sensitive doxorubicin-labeled liposomes, *Biomaterials*, **57**, pp. 1–11.

57. Chignell, C. F., and Sik, R. H. (1998). Effect of magnetite particles on photoinduced and nonphotoinduced free radical processes in human erythrocytes, *Photochem. Photobiol.*, **68**(4), pp. 598–601.

58. Zhu, L., Huo, Z., Wang, L., Tong, X., Xiao, Y., et al. (2009). Targeted delivery of methotrexate to skeletal muscular tissue by thermo-sensitive magnetoliposomes, *Int. J. Pharm.*, **370**(1–2), pp. 136–143.

59. Saiyed, Z. M., Gandhi, N. H., and Nair, M. P. (2010). Magnetic nanoformulation of azidothymidine 5'-triphosphate for targeted delivery across the blood-brain barrier, *Int. J. Nanomed.*, **5**, pp. 157–166.

60. Nobuto, H., Sugita, T., Kubo, T., Shimose, S., Yasunaga, Y., et al. (2004). Evaluation of systemic chemotherapy with magnetic liposomal doxorubicin and a dipole external electromagnet, *Int. J. Cancer*, **109**(4), pp. 627–635.

61. Ye, H., Tong, J., Liu, J., Lin, W., Zhang, C., et al. (2016). Combination of gemcitabine-containing magnetoliposome and oxaliplatin-containing magnetoliposome in breast cancer treatment: A possible mechanism with potential for clinical application, *Oncotarget*, **7**(28), pp. 43762–43778.

62. Ye, H., Tong, J., Wu, J., Xu, X., Wu, S., et al. (2014). Preclinical evaluation of recombinant human IFNalpha2b-containing magnetoliposomes for treating hepatocellular carcinoma, *Int. J. Nanomed.*, **9**, pp. 4533–4550.

63. Lin, W., Xie, X., Yang, Y., Fu, X., Liu, H., et al. (2016). Thermosensitive magnetic liposomes with doxorubicin cell-penetrating peptides conjugate for enhanced and targeted cancer therapy, *Drug Deliv.*, 23(9), pp. 3436–3443.

64. Graham, S. M., Carlisle, R., Choi, J. J., Stevenson, M., Shah, A. R., et al. (2014). Inertial cavitation to non-invasively trigger and monitor intratumoral release of drug from intravenously delivered liposomes, *J. Control. Release*, **178**, pp. 101–107.

65. Rapoport, N., Gao, Z., and Kennedy, A. (2007). Multifunctional nanoparticles for combining ultrasonic tumor imaging and targeted chemotherapy, *J. Natl. Cancer Inst.*, **99**(14), pp. 1095–1106.

66. Suzuki, R., Takizawa, T., Negishi, Y., Utoguchi, N., Sawamura, K., et al. (2008). Tumor specific ultrasound enhanced gene transfer in vivo with novel liposomal bubbles, *J. Control. Release*, **125**(2), pp. 137–144.

67. Un, K., Kawakami, S., Suzuki, R., Maruyama, K., Yamashita, F., et al. (2010). Development of an ultrasound-responsive and mannose-modified gene carrier for DNA vaccine therapy, *Biomaterials*, **31**(30), pp. 7813–7826.

68. Yin, T., Wang, P., Li, J., Wang, Y., Zheng, B., et al. (2014). Tumor-penetrating codelivery of siRNA and paclitaxel with ultrasound-responsive nanobubbles hetero-assembled from polymeric micelles and liposomes, *Biomaterials*, **35**(22), pp. 5932–5943.

69. Guo, X., and Szoka, F. C. (2001). Steric stabilization of fusogenic liposomes by a low-pH sensitive PEG-diortho ester-lipid conjugate, *Bioconjug. Chem.*, **12**(2), pp. 291–300.

70. Sawant, R. M., Hurley, J. P., Salmaso, S., Kale, A., Tolcheva, E., et al. (2006). "SMART" drug delivery systems: Double-targeted pH-responsive pharmaceutical nanocarriers, *Bioconjug. Chem.*, **17**(4), pp. 943–949.

71. Sakurai, Y., Hatakeyama, H., Sato, Y., Hyodo, M., Akita, H., et al. (2014). RNAi-mediated gene knockdown and anti-angiogenic therapy of RCCs using a cyclic RGD-modified liposomal-siRNA system, *J. Control. Release*, **173**, pp. 110–118.

72. Paliwal, S. R., Paliwal, R., and Vyas, S. P. (2015). A review of mechanistic insight and application of pH-sensitive liposomes in drug delivery, *Drug Deliv.*, **22**(3), pp. 231–242.

73. Movahedi, F., Hu, R. G., Becker, D. L., and Xu, C. (2015). Stimuli-responsive liposomes for the delivery of nucleic acid therapeutics, *Nanomedicine*, **11**(6), pp. 1575–1584.

74. Udofot, O., Affram, K., Smith, T., Tshabe, B., Krishnan, S., et al. (2016). Pharmacokinetic, biodistribution and therapeutic efficacy of

5-fluorouracil-loaded pH-sensitive PEGylated liposomal nanoparticles in HCT-116 tumor bearing mouse, *J. Nat. Sci.*, **2**(1), pii: e171.

75. Xu, H., Paxton, J. W., and Wu, Z. (2016). Development of long-circulating pH-sensitive liposomes to circumvent gemcitabine resistance in pancreatic cancer cells, *Pharm. Res.*, **33**(7), pp. 1628–1637.

76. Silva, J. O., Fernandes, R. S., Lopes, S. C., Cardoso, V. N., Leite, E. A., et al. (2016). pH-sensitive, long-circulating liposomes as an alternative tool to deliver doxorubicin into tumors: A feasibility animal study, *Mol. Imaging Biol.*, **18**(6), 898–904.

77. Moku, G., Gulla, S. K., Nimmu, N. V., Khalid, S., and Chaudhuri, A. (2016). Delivering anti-cancer drugs with endosomal pH-sensitive anti-cancer liposomes, *Biomater. Sci.*, **4**(4), pp. 627–638.

78. Pak, C. C., Ali, S., Janoff, A. S., and Meers, P. (1998). Triggerable liposomal fusion by enzyme cleavage of a novel peptide-lipid conjugate, *Biochim. Biophys. Acta*, **1372**(1), pp. 13–27.

79. Andresen, T. L., Davidsen, J., Begtrup, M., Mouritsen, O. G., and Jorgensen, K. (2004). Enzymatic release of antitumor ether lipids by specific phospholipase A2 activation of liposome-forming prodrugs, *J. Med. Chem.*, **47**(7), pp. 1694–1703.

80. de la Rica, R., Aili, D., and Stevens, M. M. (2012). Enzyme-responsive nanoparticles for drug release and diagnostics, *Adv. Drug Deliv. Rev.*, **64**(11), pp. 967–978.

81. Andresen, T. L., Thompson, D. H., and Kaasgaard, T. (2010). Enzyme-triggered nanomedicine: Drug release strategies in cancer therapy (Invited Review), *Mol. Membr. Biol.*, **27**(7), pp. 353–363.

82. Basel, M. T., Shrestha, T. B., Troyer, D. L., and Bossmann, S. H. (2011). Protease-sensitive, polymer-caged liposomes: A method for making highly targeted liposomes using triggered release, *ACS Nano*, **5**(3), pp. 2162–2175.

83. Linderoth, L., Peters, G. H., Madsen, R., and Andresen, T. L. (2009). Drug delivery by an enzyme-mediated cyclization of a lipid prodrug with unique bilayer-formation properties, *Angew. Chem. Int. Ed. Engl.*, **48**(10), pp. 1823–1826.

84. Pan, D., Pham, C. T., Weilbaecher, K. N., Tomasson, M. H., Wickline, S. A., et al. (2016). Contact-facilitated drug delivery with Sn2 lipase labile prodrugs optimize targeted lipid nanoparticle drug delivery, *Wiley Interdiscip. Rev. Nanomed. Nanobiotechnol.*, **8**(1), pp. 85–106.

85. Mock, J. N., Costyn, L. J., Wilding, S. L., Arnold, R. D., and Cummings, B. S. (2013). Evidence for distinct mechanisms of uptake and

antitumor activity of secretory phospholipase A2 responsive liposome in prostate cancer, *Integr. Biol. (Camb.)*, **5**(1), pp. 172–182.

86. Terada, T., Iwai, M., Kawakami, S., Yamashita, F., and Hashida, M. (2006). Novel PEG-matrix metalloproteinase-2 cleavable peptide-lipid containing galactosylated liposomes for hepatocellular carcinoma-selective targeting, *J. Control. Release*, **111**(3), pp. 333–342.
87. Zhu, L., Kate, P., and Torchilin, V. P. (2012). Matrix metalloprotease 2-responsive multifunctional liposomal nanocarrier for enhanced tumor targeting, *ACS Nano*, **6**(4), pp. 3491–3498.
88. Wan, Y., Han, J., Fan, G., Zhang, Z., Gong, T., et al. (2013). Enzyme-responsive liposomes modified adenoviral vectors for enhanced tumor cell transduction and reduced immunogenicity, *Biomaterials*, **34**(12), pp. 3020–3030.
89. Xiang, B., Dong, D. W., Shi, N. Q., Gao, W., Yang, Z. Z., et al. (2013). PSA-responsive and PSMA-mediated multifunctional liposomes for targeted therapy of prostate cancer, *Biomaterials*, **34**(28), pp. 6976–6991.
90. Yao, V., Berkman, C. E., Choi, J. K., O'Keefe, D. S., and Bacich, D. J. (2010). Expression of prostate-specific membrane antigen (PSMA), increases cell folate uptake and proliferation and suggests a novel role for PSMA in the uptake of the non-polyglutamated folate, folic acid, *Prostate*, **70**(3), pp. 305–316.
91. Wang, M., Li, J., Li, X., Mu, H., Zhang, X., et al. (2016). Magnetically and pH dual responsive dendrosomes for tumor accumulation enhanced folate-targeted hybrid drug delivery, *J. Control. Release*, **232**, pp. 161–174.
92. Javadi, M., Pitt, W. G., Tracy, C. M., Barrow, J. R., Willardson, B. M., Hartley, J. M., Tsosie, N. H. (2015). Ultrasonic gene and drug delivery using eLiposomes, *J. Control. Release*, **167**(1), pp. 92–100.

Chapter 11

Stimuli-Responsive Nanomedicine for Treating Non-Cancer Diseases

Himanshu Bhatt,* Balaram Ghosh,* and Swati Biswas

Department of Pharmacy, Birla Institute of Technology and Science, Hyderabad Campus, Hyderabad, Telangana 500078, India

swati.biswas@hyderabad.bits-pilani.ac.in

Nanomedicines have emerged with great potential to treat various diseases, including cancer, cardiovascular diseases, metabolic disorders, disorders of central nervous systems, and infections. Development of stimuli-responsive nanomedicines is a smart approach towards reaching the treatment goal of site-specific drug delivery. Even though stimuli-sensitive nanocarrier-mediated drug delivery has found wide application in cancer owing to the unique tumor microenvironment, the potential of this treatment strategy is being expanded to other disease areas for controlled drug delivery and to reduce off-target drug effect. This chapter

*These authors contributed equally.

Stimuli-Responsive Nanomedicine
Edited by Lin Zhu

ISBN 978-981-4800-70-9 (Hardcover), 978-0-429-29529-4 (eBook)
www.jennystanford.com

reviews the current trend and the pre-clinical status in the development of various stimuli-sensitive nanomedicines for treating non-cancerous diseases.

11.1 Introduction

The field of nanotechnology brings the science of small device closer to reality, and it is a relatively new technological advance for the improvement of human health. In the past decade, promising developments in nanotechnology have had profound impact on disease diagnosis, prevention, targeted drug delivery, molecular imaging, and treatment and have been intensively investigated. There are some nanotechnology-based products under clinical trials and some are in clinics (e.g., AmBisome®, DepoDur®, Emend®, and Feraheme™). Presently, 95% of drugs possess poor pharmacokinetics as well as biopharmaceutical properties [1]. Hence, considerable effort has been made for the therapeutically active molecules to be effective at the site of action, without altering the normal physiology of healthy organs by means of lowering doses or increasing the therapeutic index and safety profiles.

Targeted drug delivery has presented attractive results in preclinical studies. Nevertheless, the results of unpredictable behavior of ligand recognition and ligand–receptor interactions may create difficulties in controlling the drug release from targeted nanocarriers to certain extent into the clinic and make some targeted delivery strategies questionable. Apart from that, Fickian diffusion (responsible to govern the escape of the drug from nanocarriers) is not dependent on the cell and the tissue. Hence, more effective delivery is needed to deliver drugs at the action site. One approach is to develop stimuli-responsive nanocarriers, which are also named as smart/intelligent/switch on or off nanocarriers. These nanocarriers work on the principle that they recognize microenvironment and give tailored release profiles spatially and temporally, mimicking the responsiveness of living organisms.

Stimuli-responsive nanocarriers control biodistribution of drugs in response to specific physical stimuli (e.g., light, ultrasound, electric, magnetic field, temperature) or endogenous stimuli (e.g., redox potential, enzyme, changes in pH) [3]. Various stimuli-responsive nanocarriers have been reported, e.g., dendrimers, liposomes, nanoparticles, and micelles in Fig. 11.1 [2].

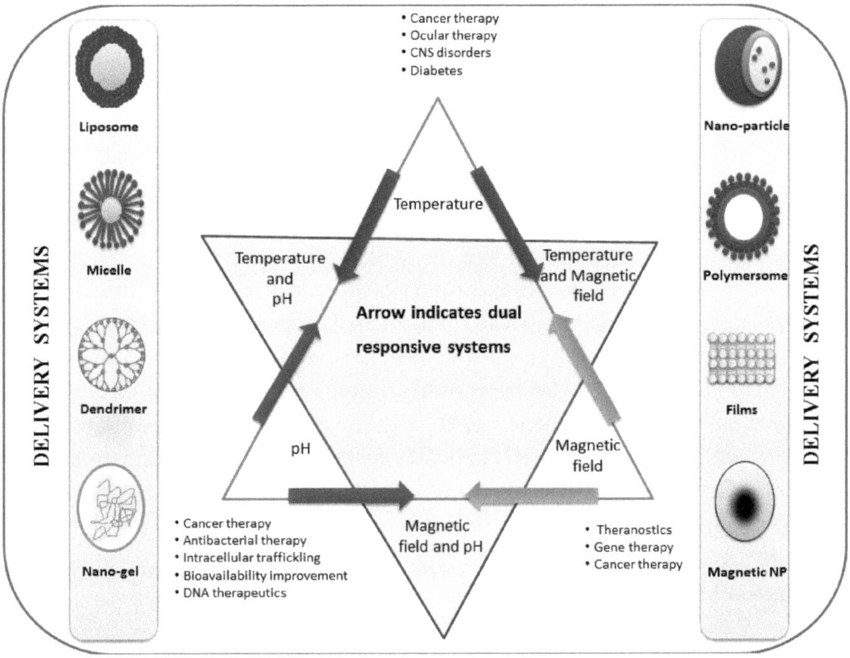

Figure 11.1 Stimuli-responsive nanocarriers. Reproduced with permission of ref. [2].

11.2 Stimuli-Responsive Nanomedicines for Antimicrobial Therapy

In 2012, the World Health Organization (WHO) reported that infectious diseases such as tuberculosis, HIV/AIDS, malaria, cholera, dengue, hepatitis, influenza, typhoid, and leishmaniasis are the major cause of morbidity of more than 8 million people in 2008 worldwide in developing countries.

Among various pathological diseases, very low pH is observed in bacterial and viral infections because some bacteria and viruses produce organic acids, e.g., lactic and acetic acids due to anaerobic fermentation [4]. Furthermore, the infection-induced inflammation increases acidity because of the increased production of lactic acids during phagocytosis [5, 6]. Inflamed tissues and wounds are characterized by a pH of 5.4–7.2, whereas non-healing wounds can achieve relevant alkaline pH values, up to 8.9 [7]. Other diseases with lowered pH include pneumonia, cystic fibrosis [8, 9], abscesses [10], and *Helicobacter pylori* infections which are responsible for peptic ulcers [11], where the activity of some drugs may be lost due to the pH change.

So far, newer drug delivery systems (DDS) have been established using pH as stimuli, e.g., micelles, liposomes, cross-linked gels, and silica nanoparticles. They are involved in the release of antibacterial agents which are microbially triggered [12]. Lectin conjugated nanoparticles could attached directly to the cell wall of *H. Pylori* because it contains carbohydrate receptors [13]. By using stimuli-responsive nanocarriers, a surface-charge switching at the sites of localized acidic environment allows nanoparticles to interact with the negatively charged bacterial cell wall, and increased cellular uptake in both Gram-positive and Gram-negative microbes [14]. To release penicillin in lysosomes, pH as stimuli was used in the treatment of resistant infection intracellularly because bacteria are present in acidic intracellular region [15].

Variables such as pH, pO_2, and temperature as well as increased levels of toxins and lipases secreted from bacteria work as unique characteristics for stimuli-responsive delivery of antimicrobial drugs to the infection microenvironment [16–19]. This also allows for the monitoring of the bacteria growth to the antimicrobial therapy [20]. Researchers prepared nanogels which contain a polyphosphoester core and a mannosyl ligand-modified PEGylated shell, efficiently taken via the mannose receptors by macrophages. This sensitive core was degraded by bacteria and released phosphatase or phospholipase enzymes [21]. In addition, after endocytosis, the vacuoles were merged and could release antimicrobial agents triggered by enzymes present in microbial and endolysosomal environments. Nanogels loaded with vancomycin were very responsive to *Staphylococcus aureus* bacteria and the bacterial growth was suppressed by the released drug efficiently [19].

Nanoparticles can be also modified for the selective attachment to bacteria. Bacterial affinity ligands are attached to graphene oxide nanoparticles tagged with Raman scattering. When nanoparticles attach to bacterial surface, this tagging acts as optical nanoprobes which change the Raman signal. This signal indicates whether the bacteria are alive or dead [22]. In the field of biomedical imaging, these tagged nanoparticles absorb near-infrared radiation using laser, which results in an increase in temperature in the bacterial microenvironment.

Biofilm targeting is another approach, both passively and actively, using nanoparticles for overcoming resistance to antimicrobial therapy [23]. For example, *Streptococcus mutans* is responsible for the development of dental caries because they are highly acidic in nature, and they have ability to ferment sucrose. They also facilitate the production of glucan polymers, which is responsible for adhesion tooth adhesion and formation of biofilm. Pluronics copolymers have been used for the development of tooth-binding micelles. These polymers were conjugated to such molecules that possess high affinity with the hydroxyapatites present in tooth enamel like alendronate [24, 25], peptides like diphosphoserine [26], and pyrophosphate (which acts as targeting agents) [26]. These copolymers self-assemble due to their favorable concentration and temperature surrounding which lead to formation of micelles that bind to the surface of tooth, and thus preventing biofilm development. Without attachment of binding moieties, Pluronic micelles can enhance the antimicrobial effect of farnesol (Quorum Sensing molecule produced by *Candida albicans*) [27]. A decrease in the pH during the growth of bacteria is observed because of the production of metabolites (CO_2, acetic acid and lactic acid) [20]. Farnesol-loaded polymeric micelles for site-specific buccal delivery were developed which bind to hydroxyapatites leads to pH-dependent release [28]. Polymeric micelles composed of copolymers like 2-(dimethylamino) ethyl methacrylate, butyl methacrylate, and 2-propylacrylic acid showed positive charge. It assisted the binding of micelles to the surfaces of teeth as well as biofilm having negative charge. The polymeric micelles destabilized in the cariogenic environment (pH 4.5) and sustained release of farnesol was observed for several hours.

Surface charge-switching is relatively new concept for treating bacterial infections. The nanoparticles have been prepared with poly(D,L-lactic-co-glycolic acid) terminated with poly(L-histidine) and poly (ethylene glycol). At pH 7.4, these nanoparticles have no targeting ability. They strongly bind to bacterial surface at acidic pH because imidazole groups of histidine are protonated and deliver the drug [14]. In another interesting investigation, these block copolymers were modified with the antibody against *Candida albicans* to deliver amphotericin B for the fungal infection [29]. At pH 6.8, the negatively charged nanoparticles were bound to fungi walls and increased internalization of nanoparticles which increased the antifungal activity of amphotericin B.

There are a number of challenges to antiretroviral therapy although the infection by human immunodeficiency virus is reduced with the highly active anti-retroviral drugs. To overcome these barriers, stimuli-sensitive nanoparticles have been developed to deliver anti-retroviral drugs efficiently. For antiretroviral therapy, the relevant stimuli, including enzyme, pH, temperature, magnetic field, ultrasound, and light, have been investigated.

A lot of changes were observed in vaginal fluids and semen after sexual intercourse, such as pH changes. The pH changes due to contact with semen have been exploited as semen-triggered antiretroviral drug delivery. The pH-responsive nanoparticles have been developed to deliver anti-retroviral agents and show the controlled drug release in a site-specific and timely manner. These nanoparticles have high stability at acidic vaginal pH (less than 4.5) and are disrupted when they come in contact with alkaline semen pH (7.4–8.4), leading to the release of the loaded drugs. Therefore, semen-triggered drug delivery is a promising approach to prevent the HIV infection. Tenofovir-loaded nanoparticles composed of copolymers of PLGA and methacrylic acid were prepared which were biocompatible to the physiological environment of vagina and released the loaded drug on contact with semen [30]. Tenofovir disoproxil fumarate-loaded electrospun fibers made up of cellulose acetate phthalate (CAP) were also developed for pH-responsive drug delivery [31]. There are many enzymes present in the human semen, e.g., hyaluronidase and prostate specific antigen. These enzymes can be used as stimuli to deliver the anti-HIV drugs [32, 33]. Recently, smart polymers have been developed as carriers for gene delivery which can release the drug when they come in contact with specific enzymes present in viruses. Asai and co-workers developed cationic polymer-peptide conjugates in which the cleavable sequence of HIV protease enzyme was incorporated [34]. The cleavage of the polymer takes place when the protease is active and the drug (e.g., DNA) is released thereafter [34]. This has great potential to specifically deliver anti-viral drugs and genes to HIV infected cells.

Thermosensitive gels have been developed which are liquid below their low critical solution temperature (LCST) but become adhesive when temperature is above LCST. To use this phase transition for drug delivery, the LCST can be set to close to 37°C. Therefore, these gels remain as liquid during storage condition and become gelation at body temperature after administration. This feature makes it feasible to develop intelligent depot systems [35]. Date and co-workers have developed a thermosensitive vaginal gel which contains raltegravir and efavirenz loaded into nanoparticles to treat HIV infection [36]. In another study, the thermoreversible PLGA nanoparticles loaded with tenofovir were developed for vaginal drug delivery. The researchers have used different combinations of Poloxamers to prepare gel matrix. These polymers remained in solution state at 2–8°C and remained in gel form in temperature range of 28–30°C, at which the sustained release of tenofovir was observed [37].

It is known that cytotoxic T-lymphocytes kill HIV-infected cells at higher temperature more efficiently. Williams et al. developed super paramagnetic iron oxide nanoparticles (SPIONs) employing magnetic fields to generate hyperthermia which aids in treatment of HIV/AIDS [38]. SPIONs were easily taken up by T-lymphocytes and were able to achieve temperatures with the help of magnetic field necessary for cell death.

For treating infectious liver diseases, Brown et al. developed lactose modified low density lipoprotein (LDL) nanocarrier for targeting LDL receptors on Kupffer cells in infectious liver diseases [39]. Similarly, Vyas et al. developed Apolipoprotein E anchored chylomicron of primaquine phosphate for infectious liver diseases [40]. In another study, Bhadra et al. developed the polypro-pyleneimine dendrimers coated peripherally with galactose for loading of primaquine phosphate and targeting asialoglycoprotein (ASGP) receptors overexpressed on liver cells [41].

11.3 Stimuli-Responsive Nanomedicines for Metabolic Disorders

There are many diseases which affect the liver metabolism and their death rates keep increasing worldwide [42]. One of the major diseases is diabetes which involves dysregulation of glucose metabolism [43]. The report of International Diabetes Federation indicates that the number of children and adults who are suffering from diabetes will reach approximately 592 million by 2035. Insulin injection, as one of the major treatment options for diabetes, is inconvenient and painful. The development of noninvasive alternatives for insulin injection is desired [44].

In recent decades, glucose-responsive drug delivery has attracted much attention. Some interesting glucose-responsive insulin delivery systems have been established [45], including: (a) drug delivery systems which incorporates glucose oxidase (GOx) enzyme, (b) drug delivery systems with glucose regulating proteins, and (c) drug delivery systems containing boronic acids which act as glucose-sensitive moieties.

PLGA-PEG-PLGA block copolymer-based thermoreversible hydrogel incorporated with exenatide was developed to cure diabetes. The hydrogel was able to inhibit the degradation of the peptide exenatide and gave sustained drug release [46]. The hydrogel systems can be further functionalized via the combination of pH-responsive hydrogel matrix and GOx. In the presence of oxygen, GOx converts glucose into gluconic acid, leading to the decrease in pH. This conversion triggers the swelling of the hydrogel which results in the release of insulin in a controlled manner. Guiseppi-Elie and co-workers entrapped GOx inside amphiphilic and pH-sensitive matrix composed of hydroxyethyl methacrylate, dimethylaminoethylmethacrylate, and 3 trimethoysilypropylmethacrylate, which impart the glucose-responsiveness to the insulin delivery system [47]. Gu et al. utilized the surface-modified dextran as an acid-degradable linkage for the self-regulated insulin delivery [48]. The nanoparticles were coated with chitosan having positive charge and alginate having negative charge. The cohesive gel-like nano-matrix was formed by electrostatic interaction in between GOx and insulin-loaded dextran nanoparticles which are oppositely charged. In another findings, Gu et al. also prepared a swellable microgel based on pH-sensitive chitosan, enzyme nanocapsule and insulin [49]. In the hyperglycemic environment, glucose is converted to gluconic acid with the help of GOx encapsulated in nanoparticles. This caused the swelling of the gel because the amines present in the chitosan are protonated. Tanna and co-workers covalently coupled carbomers which contained Concanavalin A and specific polysaccharides and developed a glucose-sensitive gel for the controlled delivery of insulin [50]. Nie et al. developed glucose-triggered self-regulated insulin delivery system based on both methacrylate derivatives and Concanavalin A, in which insulin release was dependent on the glucose level in blood [51]. Kim and co-workers developed self-assembled polymersomes made up of poly (styreneboroxole) (PBOx) and PEG chains [52]. These polymersomes could entrap water-soluble macromolecules such as insulin [53].

Lin and co-workers designed a delivery system to retard the release of cyclic adenosinemonophosphate (cAMP). This delivery system was based on mesoporous silica nanoparticle functionalized with boronic acid to control the drug release. cAMP were able to activate Ca^{2+} channels in beta cells of pancreas and hence secreted insulin based on blood glucose level [54]. Jin and co-workers developed insulin-loaded ultrasound-sensitive injectable nano-network with the help of chitosan and alginate polymers [55]. To deliver the insulin orally, Gao et al. developed insulin-loaded pH-sensitive hydrogels made up of polyacrylic acid derivatives [56]. Lin et al. developed pH-sensitive nanoparticles composed of chitosan and poly-glutamic acid (PGA) mixed with tripolyphosphate [57]. At higher pH, amines were deprotonated in the chitosan and the nanoparticles released the loaded insulin.

11.4 Stimuli-Responsive Nanomedicine for Ocular Diseases

Drug release can be tuned by using stimuli-responsive polymers. These polymers impart changes in the structure, solubility and charge of nanoparticles when changes occur physically or chemically in the environment [58]. Recently available eye implants deliver drug constantly but multiple dosing of drug is mandatory at various time points according to the patient requirements and the condition of ocular diseases [59]. The stimuli responsiveness of polymers has gain particular interest to make the drug deliverable to the eye's posterior segment. Light-activated nanodrug delivery system has been developed because of the transparency of the cornea and lens [60].

Poly (N-isopropylacrylamide) (pNIPAM) polymer has gain considerable interest to develop thermosenstive drug delivery systems. This polymer shows in situ gel-forming property at 32°C, which is close to eye temperature. Researchers developed timolol maleate-loaded thermosensitive gel made up of combination of chitosan with pNIPAM (pNIPAM–CS) for ocular application. Pharmacokinetic study showed high $C_{\max}$ and AUC in eye of the rabbit than those of conventional eye drops. Additionally, this polymer demonstrated less cytotoxicity in the eye and reduced the intra-ocular pressure compared to conventional timolol maleate eye drops [61].

Light-activated polymeric micelles have been prepared by loading light-sensitive materials in them. These materials absorb a specific wavelength of light or convert light into heat, e.g., gold nanoparticles. Ultraviolet (UV) and near infra-red (NIR) sensitive micelles have been developed via chemical incorporation of chromophores with the lipophilic matrix of the polymer. These micelles become ionizable or cleavable under the UV/NIR which creates an osmotic pressure in the polymeric system. These resulted in influx of solvent molecules and caused drug release [62, 63]. Peng and co-workers incorporated green fluorescent protein (GFP) in the hydrogel matrix consisting of biodegradable dextran and azobenzene to study photo-triggered protein release [64]. When it was exposed to electromagnetic radiation (EMR) stimulus, 70% of the incorporated GFP was released within 90 min, but in the absence of light only 5% release was observed.

On Demand Therapeutics (ODTx) has developed implants comprising drug reservoirs which are light activatable. The metal shells of the implants absorb electromagnetic radiation and convert it into thermal energy which causes the cleavage in the implant [65]. This led to the formation of small orifice in the implantable system and drug release on exposure of light. Although promising, this type of implantable system is in pre-clinical studies and its possibility to treat posterior eye segment diseases remains unexplored.

In addition, biodegradable conducting/conductive polymers (CP), e.g., poly (ethylenedixoythiophene) (PEDOT), N-methyl pyrrole and polypyrrole (PPy) offer good opportunities to develop triggerable implantable drug delivery systems to treat posterior segment diseases. These polymers are non-toxic and provide good control on drug release under electrochemical stimuli [66]. In the presence of electrochemical stimuli, these polymers alter the redox state and change their charge, permeability and lipophilicity. Sharma and co-workers developed films based on porous PPy using polymethylmethacrylate (PMMA) as a template which loads the drug more than non-template PPy films [67]. Upon exposure of an electrical stimulus, the drug was released to the target site. Ge and co-workers developed CP-coated gold microchip-based drug delivery systems and found that upon redox switching of CP under electroconductive stimuli, the volume of the system changes. By this way, the drug release can be accurately controlled [68]. As hydrogels offer swelling/deswelling properties and CPs possess electrochemical properties, combinations of both can provide increased drug loading and desirable drug release. Therefore, the systems made up of hydrogels and CPs have been also developed [69]. This technology can also be used to prepare eye implants placed into the subconjunctival space where the drug can be directly released into the vitreous based on electrical stimuli.

Photodynamic therapy (PDT) is relatively new to deliver the drugs to posterior segment of eye. Nishiyama and co-workers developed porphyrin-loaded PEG-block-poly(L-lysine) micelles for treating choroidal neovascularization [70, 71]. In another findings, pDNA/polycation polyplex was enveloped in dendritic phthalocyanine which was functionalized using a nuclear localization signal [72]. The phthalocyanine disrupts the endosomes and releases pDNA into the cytoplasm. As a result, a considerable gene expression was observed in the subconjunctival tissue. Liaw and co-workers delivered plasmid DNA with lacZ gene using micelles composed of triblock copolymers of poly (ethyleneoxide) and poly (propyleneoxide) [73]. After topical instillation, gene expression was observed in the subconjunctival tissue of rabbit eyes. In addition, Wilson et al. found that light signals generated by the nanoparticles are useful measures for imaging eye diseases [74].

11.5 Stimuli-Responsive Nanomedicine for Central Nervous System Disorders

Nowadays central nervous system (CNS) diseases, e.g., Alzheimer's disease, depression, schizophrenia, insomnia, Parkinson's disease, amyotrophic lateral sclerosis and stroke, are becoming one of major threats to human health [75, 76]. In the last decade, the therapy for CNS diseases grew tremendously. But the effective drug treatments which can deliver the drugs efficiently to brain are rare. The major obstacle is the blood–brain barrier (BBB) which only allows for the permeation of essential molecules but prevents more than 98% of drugs from reaching brain targets. Many endogenous transport proteins and receptors are involved in selective uptake of nutrients like glucose, amino acids or nucleosides, and transport of large molecules, e.g., insulin, transferrin, etc., through the BBB [77]. Furthermore, various efflux transporters like p-glycoprotein (MDR1), breast cancer–resistant protein (BCRP), multidrug-resistant protein (MRP1) residing in the BBB pump out the substrates, including many drugs, from cells back into the bloodstream.

Figure 11.2 gives information on the mechanisms of CNS drug transport. One of the most attractive strategies for CNS drug delivery is to develop prodrugs which are susceptible to CNS/BBB–specific enzymes [79]. Another promising strategy is to design stimuli-responsive nanoparticle-based delivery systems for CNS drug delivery. There are many pathological changes which can be used as endogenous stimuli for nanomedicines, e.g., lowered pH [80, 81], increased levels of glutathione [82, 83], overexpressed enzymes [84, 85], and inflammation-associated elevated temperatures encountered in the brain lesion region. In addition, the physical stimuli including light, magnetic fields, or electric fields, can be applied externally. In response to these internal and external stimuli, the stimuli-responsive nanomedicines can effectively deliver and release therapeutic agents through BBB in CNS.

Hsiao et al. developed injectable thermosensitive nanogels made up of amphiphilic polymers to deliver drugs to the brain [86]. The superparamagnetic iron oxide nanoparticles (SPIONs) have been widely used for drug delivery, bioseparation, brain magnetic resonance imaging (MRI), and magnetic hyperthermia [87]. Qiao et al. developed PEGylated magnetic nanoparticles which were modified with lactoferrin to help the transport of magnetic nanoparticles across BBB through endocytosis [88].

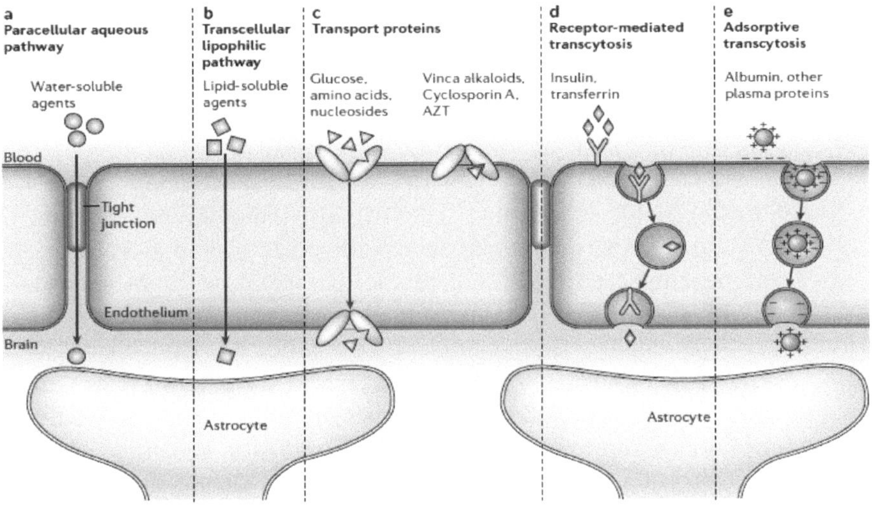

Figure 11.2 Pathways across the blood–brain barrier (reproduced with permission of ref. [78]).

Due to the cavitation-induced pore formation, ultrasound is effective to temporarily and reversibly open the BBB. Patrick and co-workers studied ultrasound-responsive nanocarriers for the delivery of drugs to the CNS by using high intensity focused ultrasound [89], while Hynynen et al. induced opening of BBB using low-power focused ultrasound applied after intravenous injection of MRI contrast agent [90]. Hwang and co-workers prepared ultrasound sensitive perfluorocarbon nanobubbles (PNs) incorporated with apomorphine (dopamine receptor agonist) to target the brain. This ultrasound-mediated delivery altered the drug release as per demand and less than 10% hemolysis was observed of PNs [91]. Kinoshita and co-workers found that even large molecules, such as dopamine receptor antibody, could be delivered to the brain to bind with its antigens after ultrasound-induced BBB disruption [92].

Stimuli-sensitive polymeric micelles have been made in response to many stimuli, e.g., pH, light, temperature, ultrasound, which trigger the drug release in the CNS [93]. Shao and co-workers developed intelligent polymeric micelles (DHA-$PLys_{(s\text{-}s)}$P) using dehydroascorbic acid as a ligand for glucose transport protein 1 (GLUT1) on the BBB. This smart "-S-S-" bond could keep micelles integrity in blood and prevented drug leakage [94]. Their research showed that the smart nanodevice could help the drug accumulation continuously in the brain and cured anti-intracranial infection in the brain.

Light-responsive nanoparticles have been also developed for CNS targeting. Yoo and co-workers prepared biocompatible gold nanorods (GNR) containing cetyltrimethylammonium bromide (CTAB), and then conjugated them with polyethylene glycol [95]. These near-infrared (NIR)-activatable nanoparticles were able to bind with the negatively charged plasma membrane. Interestingly, under NIR light, the nanoparticles inhibited the electrical activity of the neurons in the epilepsy, while in the absence of NIR light, the electrical activity was restored. The light-responsive nanoparticles represent a promising therapeutic tool to control the disordered neuronal activity in the CNS disorders.

Coating of conductive polymers on neural implants is another interesting approach for stimuli-responsive drug delivery to CNS [96]. Wadhwa and co-workers coated gold nano-electrodes with PPy based polymers to develop electrically controlled local delivery of dexamethasone [97]. This smart electrode is responsive to electrical stimuli because redox reactions cause the charging/discharging of this polymer, resulting in the release of ions, drugs and growth factors from the polymer [97]. The CP coatings on nano-electrodes increased biocompatibility and led to the increase in drug loading and faster release of the loaded drugs [98]. This feature can also improve the surface area between the electrodes and brain tissue and improve contact with live neurons [97].

In addition, the stimuli-responsive nanomaterials have been used to prepare biosensors for diagnosis of CNS disorders. Yue and colleagues developed zinc oxide (ZnO) nanowire (100 nm in diameter) upon a 3D foam made of graphene. The nanoparticle was able to selectively detect dopamine with urate and ascorbate. These nanowires generated high electrical conductance of biosensors and able to detect even 1 nM dopamine and uric acid (potential biomarkers of Parkinson's disease) in the patients' serum [99].

11.6 Stimuli-Responsive Nanomedicine for Heart Disorders

In the United States, heart disorders are the leading cause of mortality [100]. Among various types of heart disorders, myocardial infarction, stroke and ischemia are major reasons responsible for human death. Higher shear stress is observed in narrowed arteries compared to healthy arteries which can be used as mechanical/physical stimuli to develop smart drug delivery systems [101]. Hence, heart diseases such as stenosis and atherosclerosis would be great targets for developing mechanically triggered stimuli-responsive systems. Holme et al. developed liposomes that were stable in normal conditions, but they became unstable in the high shear stress condition of atherosclerosis which triggered the release of drugs in the high shear stress region of arteries [102]. Table 11.1 indicates some interesting examples of stimuli-triggered nanomedicines for atherosclerosis. Korin and co-workers developed microscale particles for the treatment of thrombosis that were cleaved into nanoscale particles under the condition of elevated shear stress in arteries [103]. After breakdown of particles in vivo, rapid clot dissolution and restoration of normal activity were observed with reduced side-effects. To inhibit the LDL uptake, multifunctional micelles were developed by Peters and co-workers which contain a targeting pentapeptide, fluorophore, and bivalirudin. These micelles bound with clotted plasma proteins in atherosclerotic lesions and released bivalirudin in the shear stress region [104]. With the same concept, Broz and co-workers developed multifunctional polymeric micelles loaded with pravastatin which reduced the phagocytosis by macrophages [105].

Mulder et al. developed paramagnetic micelles for diagnostic purposes to visualize atherosclerotic plaques in the aortas of atherosclerotic mice and they targeted the plaques using antibodies specific to scavenger receptors of macrophages [106]. Guo et al. imaged organ rejection in heart transplantation in rats using multifunctional superparamagnetic iron oxide nanoparticles loaded with plasmid DNA [107]. These nanoparticles acted as contrast agent in MRI. Meanwhile, the immune response was suppressed by loaded plasmid DNA. Heid and co-workers developed iron oxide nanoparticles conjugated with recombinant tissue plasminogen activator (tPA, sensitive to activated factor XIII) for thrombolytic therapy [108]. These nanoparticles improved thrombolytic activity.

Table 11.1 Nanomedical formulations with atherosclerosis-targeting properties

Ligand/stimuli	Nanoparticle	Purpose as described in literature	References
Peptide (cLABL)	TAT-peptide-DNA complex	Gene delivery	[125]
Cyclic peptide VHSPNKK	Cross-linked iron oxide nanoparticle	MRI imaging	[126]
Peptide VHPKQHR	Monocrystalline magnetic nanoparticle	MRI imaging	[127]
Anti-PECAM antibody	Polymer nanocarriers	Enzyme delivery	[128]
Peptidomimetic vitronectin antagonist	Paramagnetic nanoparticles	Imaging, drug delivery	[129, 130]

Note: cLABL, cyclo-(1,12)-PenITDGEATDSGC; PECAM, platelet endothelial cell adhesion molecule; VHPKQHR, valine-histidine-proline-lysine-glutamine-histidine-arginine.

Matrix metalloproteinases (MMPs) are attractive targets in the myocardial infarction because they are upregulated in heart tissues during myocardial infarction (MI). Nguyen and co-workers developed enzyme-responsive micellar nanoparticles with peptide-polymer amphiphiles as the enzyme-responsive material for acute MI in rats. These nanoparticles were responsive to MMP-2 and MMP-9 and undergo a morphological change to network-like assemblies [109]. Lin and co-workers developed nanoSTATs based on thrombin activity which released heparin in response to thrombin as well as other proteases and inhibited thrombin. These nanoSTATs also prevented fibrin deposition in pulmonary embolism and maintain normal bleeding time [110]. Bertram and co-workers developed PLGA-PLL nanoparticles in which multiple PEG chains were terminated with RGD peptides [111]. Specifically, when clotting is initiated naturally, these RGD terminated nanoparticles acted as multivalent cross-linkers and cut the bleeding time in the injury of the femoral artery. In another interesting findings by Chan et al., an engineered hemostatic polymer was developed in such a way that they mimic clotting factor XIIIa, enhanced clotting rates, reduced blood volume loss, and maintained long clearance half-life in the injury of femoral artery in mice [112].

Generally, ischemic heart diseases involve overproduction of hydrogen peroxide which is the main cause of the heart tissue damage. Lee et al. developed a novel strategy based on hydroxybenzyl alcohol (HBA) incorporated with copolyoxalate (HPOX) which was triggered by hydrogen peroxide in ischemic injury. In the presence of hydrogen peroxide, these nanoparticles were disrupted into three known safe compounds, cyclohexane-dimethanol, HBA and CO_2. Copolyoxalate scavenges hydrogen peroxide in dose-dependent manner and HBA is released. The nanoparticles show antioxidant and anti-inflammatory activities in the hind limb ischemic injury. Also, fluorophore-loaded HPOX nanoparticles are used for imaging of hydrogen peroxide in hind limb ischemic injury. It demonstrates scope for bioimaging of hydrogen peroxide-related heart diseases [113].

In responsive to reactive oxygen species (ROS) produced by certain diseases, Bae et al. chemically engineered nanoparticles composed of HBA, HPOX, and vanillyl alcohol (VA) (PVAX) which contain peroxalate ester linkages. They were activated by hydrogen peroxide at the ROS generation site and prevented the ischemic injuries, as evidenced by their antioxidant, anti-inflammatory, as well as antiapoptotic effects [114]. Bae and co-workers also developed novel hydrogen peroxide triggered polymeric nanoparticles composed of PVAX and HPOX. These nanoparticles were able to target the ROS overproduction site and reduced the oxidative stress in heart myocardial injury [114].

Ultrasound-mediated drug delivery also possesses great potential in treating heart diseases. Kawata et al. and co-workers developed smart drug delivery systems for treating coronary thrombolysis. They developed gelatin and zinc ion-based nanoparticles incorporated with tPA. These nanoparticles attenuated the activity of tPA with the help of ultrasound and improved perfusion rate in occluded coronary artery in the myocardial injury [115]. Tiukinhoy-Laing et al. entrapped tPA (as tPA has many binding sites for fibrin) into echogenic liposomes which were strongly bound to porcine clots without affecting the natural thrombin activity. The fibrin binding affinity of tPA-containing echogenic liposomes was twice as compared to that of free tPA, which was determined using echogenic analysis of porcine clots. This study demonstrated the potential for fibrin-targeted and ultrasound-triggered delivery for thrombolytic agent [116]. Herbst and co-workers developed the bifunctional echogenic liposomes for the delivery of stem cells to porcine arterial wall. To increase the targetability, the liposomes were modified with two antibodies, one against CD34 of stem cells and the other against intercellular adhesion molecule-1 (ICAM-1) antigens of the inflammatory endotheliaum [117]. Laing and co-workers also developed actin-targeted echogenic liposomes labeled with calcein. After administration, exposing porcine carotid arteries to continuous ultrasound resulted in high calcein release in the tunica intima and adventitia [118].

In addition, the echocardiography, such as ultrasound targeted microbubbles destruction (UTMD) therapy, has drawn attention in drug and gene delivery [119]. For instance, researchers improved the loading capacity for gene-based drugs in microbubbles with the help of cationic polymers (e.g., polyethyleneimine) [120, 121]. Genetic materials can be fused into nanoparticles and they are bound to microbubbles whose surfaces were modified with cationic polymers. For imaging purpose, MRI contrast agent can be loaded into the nanoparticles that are able to bind with the microbubbles [122]. Klibanov and co-workers developed liposomes that were able to bind with microbubbles for UTMD [123]. Burke et al. injected PLGA nanoparticles and improved drug delivery to skeletal muscles by the activation of microbubbles using ultrasound [124].

11.7 Conclusion and Future Perspective

In the area of nanotechnology, nanoparticulate drug delivery systems have advantages to overcome poor pharmacokinetics as well as pharmacodynamics of active agents which suffer from poor bioavailability. With understanding of disease pathology, material design and surface chemistry of nanoparticles, there are good opportunities to develop tunable nanocarriers which respond to the local stimuli. This book chapter has examined the endogenous and exogenous stimuli, e.g., pH, magnetic field, temperature, and enzyme, and has revealed that stimuli-responsive nanocarriers can provide great potential to overcome problems associated with conventional drug delivery systems, particularly in treating non-cancer diseases.

Some proofs of concepts have reported dramatic improvement in therapeutic efficacy via stimuli-responsive drug delivery; however, their development from bench to bedside is still a huge challenge and only a few (Fe_2O_3 nanoparticles and thermosensitive liposomes) have reached clinical applications. Most of the stimuli-responsive nano-systems are facing difficulties in clinical translation due to complex design and preparation. Moreover, the safety of these nanomaterials depends on many factors, e.g., biodegradability, physicochemical properties, biocompatibility, composition, and dose and route of administration. For specific medical application, the benefit-to-risk ratio has to be balanced. Making drug delivery systems to respond to the subtle alteration in stimuli is not easy to achieve and novel nanomaterials with high sensitivity are required. For exogenous stimuli, the issues related to the penetration depth of these stimuli need to be optimized to achieve the optimal therapeutic effects.

Apart from these issues, a few major hurdles or concerns are highlighted here:

(a) **Protein corona effect**. Blood proteins have ability to cover the surface of smart nanosystems when the nanoparticles circulate in the bloodstream. This phenomenon is mainly due to the non-specific interaction between blood proteins and smart nanoparticles and may result in non-specific cellular uptake, which has not been studied intensively. Some properties of nanoparticles like shape, particle size, surface charge and other parameters can affect the binding between smart nanocarriers and blood proteins. To minimize the adsorption by serum proteins which are negatively charged, there is a need to develop smart nanocarriers with neutral or electronegative surfaces. PEGylation is a common approach to minimize the protein corona effect by reducing non-specific protein adsorption on the nanoparticle surface and prolonging the circulation time.

(b) **Low accuracy and sufficiency of stimuli responsiveness**. After smart nanocarriers enter in the human body, their drug release may get disturbed by the complexity of the human body and disease microenvironments, resulting in low drug targetability and efficacy. However, we expect that these nanocarriers should give accurate response in a narrow window to a stimulus signal. To address this problem, a novel silica nanoparticle containing an analyte-responsive gate has been developed by Zink et al [131]. The smart nanocarriers first analyzed all stimuli signals. When two specific stimuli existed simultaneously, the drug would be released. Working in a "Boolean AND logic" manner, the developed AND gate has two true specific values and this logic gate gives more precisely controlled drug release from smart nanocarriers.

(c) **Biodegradation issues**. On the one hand, ideal biodegradable smart nanomaterials should remain stable or functional to control drug release; on the other hand, the nanomaterials should meet safety standards and can be easily removed from the biological system through metabolism and excretion to minimize the immune response. Therefore, the successful design of stimuli-responsive nanomaterials is a complicated process and it requires multi-disciplinary knowledge of pharmaceutical science, biological science, chemistry, material science, etc.

Current studies indicated that the combination of existing nanocarriers with appropriate stimuli (heat, light, ultrasound, magnetic, electric fields, enzymes, redox, and other biomolecules) can acquire on-demand and triggerable drug release. In the future, new breakthroughs in controlled drug release may be achieved by designing of nanosystems in such a way that the discharge of cargo is responsive to concurrent multiple stimuli of different microenvironments as well as that each cargo of cargo depot in nanocarriers is triggered to release by each specific stimulus. Except above-mentioned stimuli, well understanding of physiology and pathology of diseases may help to identify other new powerful stimuli for developing stimuli-responsive drug delivery systems. Overall, although immense progress has been made in the materials chemistry as well as drug delivery over the past decades, focus should be shifted to obtain more sensitive, biocompatible, and safe drug delivery systems with potential for facile clinical translation.

References

1. Brayden, D. J., Controlled release technologies for drug delivery. *Drug Discovery Today*, 2003. **8**(21): pp. 976–978.
2. Muntimadugu, E., A. Jain, and W. Khan, Stimuli responsive carriers: Magnetically, thermally and pH assisted drug delivery, in *Targeted Drug Delivery: Concepts and Design*. 2015, Springer. pp. 341–365.
3. Ganta, S., et al., A review of stimuli-responsive nanocarriers for drug and gene delivery. *Journal of Controlled Release*, 2008. **126**(3): pp. 187–204.
4. Barrett, K. E., et al., *Ganong's Review of Medical Physiology, 23rd Edition*, 2009: McGraw-Hill.
5. Trevani, A. S., et al., Extracellular acidification induces human neutrophil activation. *The Journal of Immunology*, 1999. **162**(8): pp. 4849–4857.
6. Dubos, R., The micro-environment of inflammation or Metchnikoff revisited. *The Lancet*, 1955. **266**(6879): p. xxiv-5.
7. Gethin, G., The significance of surface pH in chronic wounds. *Wounds UK*, 2007. **3**(3): p. 52.
8. Tate, S., et al., Airways in cystic fibrosis are acidified: Detection by exhaled breath condensate. *Thorax*, 2002. **57**(11): pp. 926–929.
9. Poschet, J., E. Perkett, and V. Deretic, Hyperacidification in cystic fibrosis: links with lung disease and new prospects for treatment. *Trends in Molecular Medicine*, 2002. **8**(11): pp. 512–519.
10. Simmen, H.-P., and J. Blaser, Analysis of pH and pO_2 in abscesses, peritoneal fluid, and drainage fluid in the presence or absence of bacterial infection during and after abdominal surgery. *The American Journal of Surgery*, 1993. **166**(1): pp. 24–27.

11. Tomb, J.-F., et al., The complete genome sequence of the gastric pathogen Helicobacter pylori. *Nature*, 1997. **388**(6642): pp. 539–547.

12. Aimetti, A. A., A. J. Machen, and K. S. Anseth, Poly (ethylene glycol) hydrogels formed by thiol-ene photopolymerization for enzyme-responsive protein delivery. *Biomaterials*, 2009. **30**(30): pp. 6048–6054.

13. Umamaheshwari, R., and N. Jain, Receptor mediated targeting of lectin conjugated gliadin nanoparticles in the treatment of Helicobacter pylori. *Journal of Drug Targeting*, 2003. **11**(7): pp. 415–424.

14. Radovic-Moreno, A. F., et al., Surface charge-switching polymeric nanoparticles for bacterial cell wall-targeted delivery of antibiotics. *ACS Nano*, 2012. **6**(5): pp. 4279–4287.

15. Sémiramoth, N., et al., Self-assembled squalenoylated penicillin bioconjugates: An original approach for the treatment of intracellular infections. *ACS Nano*, 2012. **6**(5): pp. 3820–3831.

16. Sawyer, R. G., et al., The peritoneal environment during infection. The effect of monomicrobial and polymicrobial bacteria on pO_2 and pH. *Annals of Surgery*, 1991. **213**(3): p. 253.

17. Pornpattananangkul, D., et al., Bacterial toxin-triggered drug release from gold nanoparticle-stabilized liposomes for the treatment of bacterial infection. *Journal of the American Chemical Society*, 2011. **133**(11): pp. 4132–4139.

18. Xiong, M.-H., et al., Lipase-sensitive polymeric triple-layered nanogel for "on-demand" drug delivery. *Journal of the American Chemical Society*, 2012. **134**(9): pp. 4355–4362.

19. Xiong, M. H., et al., Bacteria-responsive multifunctional nanogel for targeted antibiotic delivery. *Advanced Materials*, 2012. **24**(46): pp. 6175–6180.

20. Si, Y., et al., Rapid and accurate detection of Escherichia coli growth by fluorescent pH-sensitive organic nanoparticles for high-throughput screening applications. *Biosensors and Bioelectronics*, 2016. **75**: pp. 320–327.

21. Pinto-Alphandary, H., A. Andremont, and P. Couvreur, Targeted delivery of antibiotics using liposomes and nanoparticles: Research and applications. *International Journal of Antimicrobial Agents*, 2000. **13**(3): pp. 155–168.

22. Lin, D., et al., Graphene oxide wrapped SERS tags: Multifunctional platforms toward optical labeling, photothermal ablation of bacteria,

and the monitoring of killing effect. *ACS Applied Materials & Interfaces*, 2014. **6**(2): pp. 1320–1329.

23. Forier, K., et al., Lipid and polymer nanoparticles for drug delivery to bacterial biofilms. *Journal of Controlled Release*, 2014. **190**: pp. 607–623.
24. Chen, F., et al., Tooth-binding micelles for dental caries prevention. *Antimicrobial Agents and Chemotherapy*, 2009. **53**(11): pp. 4898–4902.
25. Chen, F., et al., Triclosan-loaded tooth-binding micelles for prevention and treatment of dental biofilm. *Pharmaceutical Research*, 2010. **27**(11): pp. 2356–2364.
26. Chen, F., et al., The development of dentotropic micelles with biodegradable tooth-binding moieties. *Pharmaceutical Research*, 2013. **30**(11): pp. 2808–2817.
27. Mogen, A. B., et al., Pluronics-formulated farnesol promotes efficient killing and demonstrates novel interactions with Streptococcus mutans biofilms. *PloS One*, 2015. **10**(7): p. e0133886.
28. Horev, B., et al., pH-activated nanoparticles for controlled topical delivery of farnesol to disrupt oral biofilm virulence. *ACS Nano*, 2015. **9**(3): pp. 2390–2404.
29. Tang, X., et al., Enhanced antifungal activity by Ab-modified amphotericin B-loaded nanoparticles using a pH-responsive block copolymer. *Nanoscale Research Letters*, 2015. **10**(1): p. 256.
30. Zhang, T., T. F. Sturgis, and B.-B. C. Youan, pH-responsive nanoparticles releasing tenofovir intended for the prevention of HIV transmission. *European Journal of Pharmaceutics and Biopharmaceutics*, 2011. **79**(3): pp. 526–536.
31. Huang, C., et al., Electrospun cellulose acetate phthalate fibers for semen induced anti-HIV vaginal drug delivery. *Biomaterials*, 2012. **33**(3): pp. 962–969.
32. Clark, M. R., et al., Enzymatic triggered release of an HIV-1 entry inhibitor from prostate specific antigen degradable microparticles. *International Journal of Pharmaceutics*, 2011. **413**(1): pp. 10–18.
33. Agrahari, V., et al., Hyaluronidase-sensitive nanoparticle templates for triggered release of HIV/AIDS microbicide in vitro. *The AAPS Journal*, 2014. **16**(2): pp. 181–193.
34. Asai, D., et al., Specific transgene expression in HIV-infected cells using protease-cleavable transcription regulator. *Journal of Controlled Release*, 2010. **141**(1): pp. 52–61.

35. Ward, M. A., and T. K. Georgiou, Thermoresponsive polymers for biomedical applications. *Polymers*, 2011. **3**(3): pp. 1215–1242.

36. Date, A. A., et al., Development and evaluation of a thermosensitive vaginal gel containing raltegravir+ efavirenz loaded nanoparticles for HIV prophylaxis. *Antiviral Research*, 2012. **96**(3): pp. 430–436.

37. Karla, P., Thermo reversible buffered tenofovir nano gel for vaginal delivery. *Journal of Pharmacology & Clinical Toxicology*, 2013. **1**(2): p. 1010.

38. Williams, J. P., et al., Application of magnetic field hyperthermia and superparamagnetic iron oxide nanoparticles to HIV-1-specific T-cell cytotoxicity. *International Journal of Nanomedicine*, 2013. **8**: p. 2543.

39. Brown, M. S., and J. L. Goldstein, A receptor-mediated pathway for cholesterol homeostasis. *Science*, 1986. **232**(4746): pp. 34–47.

40. Vyas, S., and V. Sihorkar, Endogenous carriers and ligands in non-immunogenic site-specific drug delivery. *Advanced Drug Delivery Reviews*, 2000. **43**(2): pp. 101–164.

41. Bhadra, D., et al., Glycodendrimeric nanoparticulate carriers of primaquine phosphate for liver targeting. *International Journal of Pharmaceutics*, 2005. **295**(1): pp. 221–233.

42. Levchenko, T. S., et al., ATP-loaded liposomes for targeted treatment in models of myocardial ischemia. In *Liposomes: Methods and Protocols, Volume 1: Pharmaceutical Nanocarriers*, 2010: pp. 361–375.

43. Wu, Q., et al., Organization of glucose-responsive systems and their properties. *Chemical Reviews*, 2011. **111**(12): pp. 7855–7875.

44. Mo, R., et al., Emerging micro-and nanotechnology based synthetic approaches for insulin delivery. *Chemical Society Reviews*, 2014. **43**(10): pp. 3595–3629.

45. Wu, W., and S. Zhou, Responsive materials for self-regulated insulin delivery. *Macromolecular Bioscience*, 2013. **13**(11): pp. 1464–1477.

46. Li, K., et al., A long-acting formulation of a polypeptide drug exenatide in treatment of diabetes using an injectable block copolymer hydrogel. *Biomaterials*, 2013. **34**(11): pp. 2834–2842.

47. Guiseppi-Elie, A., S. I. Brahim, and D. Narinesingh, A chemically synthesized artificial pancreas: Release of insulin from glucose-responsive hydrogels. *Advanced Materials*, 2002. **14**(10): pp. 743–746.

48. Gu, Z., et al., Injectable nano-network for glucose-mediated insulin delivery. *ACS Nano*, 2013. **7**(5): pp. 4194–4201.

49. Gu, Z., et al., Glucose-responsive microgels integrated with enzyme nanocapsules for closed-loop insulin delivery. *ACS Nano*, 2013. **7**(8): pp. 6758–6766.

50. Tanna, S., et al., Covalent coupling of concanavalin A to a Carbopol 934P and 941P carrier in glucose-sensitive gels for delivery of insulin. *Journal of Pharmacy and Pharmacology*, 2002. **54**(11): pp. 1461–1469.

51. Yin, R., et al., Glucose-responsive microhydrogels based on methacrylate modified dextran/concanavalin A for insulin delivery. *Journal of Controlled Release*, 2011. **152**: pp. e163–e165.

52. Kim, H., et al., Monosaccharide-responsive release of insulin from polymersomes of polyboroxole block copolymers at neutral pH. *Journal of the American Chemical Society*, 2012. **134**(9): pp. 4030–4033.

53. Tanner, P., et al., Polymeric vesicles: From drug carriers to nanoreactors and artificial organelles. *Accounts of Chemical Research*, 2011. **44**(10): pp. 1039–1049.

54. Zhao, Y., et al., Mesoporous silica nanoparticle-based double drug delivery system for glucose-responsive controlled release of insulin and cyclic AMP. *Journal of the American Chemical Society*, 2009. **131**(24): pp. 8398–8400.

55. Di, J., et al., Ultrasound-triggered regulation of blood glucose levels using injectable nano-network. *Advanced Healthcare Materials*, 2014. **3**(6): pp. 811–816.

56. Gao, X., et al., Biodegradable pH-responsive polyacrylic acid derivative hydrogels with tunable swelling behavior for oral delivery of insulin. *Polymer*, 2013. **54**(7): pp. 1786–1793.

57. Lin, Y.-H., et al., Multi-ion-crosslinked nanoparticles with pH-responsive characteristics for oral delivery of protein drugs. *Journal of Controlled Release*, 2008. **132**(2): pp. 141–149.

58. Kost, J., and R. Langer, Responsive polymeric delivery systems. *Advanced Drug Delivery Reviews*, 2012. **64**: pp. 327–341.

59. Ong, F. S., et al., Personalized medicine in ophthalmology: From pharmacogenetic biomarkers to therapeutic and dosage optimization. *Journal of Personalized Medicine*, 2013. **3**(1): pp. 40–69.

60. Christie, J. G., and U. B. Kompella, Ophthalmic light sensitive nanocarrier systems. *Drug Discovery Today*, 2008. **13**(3): pp. 124–134.

61. Cao, Y., et al., Poly(N-isopropylacrylamide)-chitosan as thermo-sensitive in situ gel-forming system for ocular drug delivery. *Journal of Controlled Release*, 2007. **120**(3): pp. 186–194.
62. Mamada, A., et al., Photoinduced phase transition of gels. *Macromolecules*, 1990. **23**(5): pp. 1517–1519.
63. Jiang, J., X. Tong, and Y. Zhao, A new design for light-breakable polymer micelles. *Journal of the American Chemical Society*, 2005. **127**(23): pp. 8290–8291.
64. Peng, K., et al., Dextran based photodegradable hydrogels formed via a Michael addition. *Soft Matter*, 2011. **7**(10): pp. 4881–4887.
65. Hirsch, L. R., et al., Nanoshell-mediated near-infrared thermal therapy of tumors under magnetic resonance guidance. *Proceedings of the National Academy of Sciences*, 2003. **100**(23): pp. 13549–13554.
66. Svirskis, D., et al., Electrochemically controlled drug delivery based on intrinsically conducting polymers. *Journal of Controlled Release*, 2010. **146**(1): pp. 6–15.
67. Sharma, M., et al., High surface area polypyrrole scaffolds for tunable drug delivery. *International Journal of Pharmaceutics*, 2013. **443**(1): pp. 163–168.
68. Ge, D., et al., A polypyrrole-based microchip for controlled drug release. *Electrochimica Acta*, 2009. **55**(1): pp. 271–275.
69. Valdés-Ramírez, G., et al., Multiplexed and switchable release of distinct fluids from microneedle platforms via conducting polymer nanoactuators for potential drug delivery. *Sensors and Actuators B: Chemical*, 2012. **161**(1): pp. 1018–1024.
70. Nishiyama, N., and K. Kataoka, Current state, achievements, and future prospects of polymeric micelles as nanocarriers for drug and gene delivery. *Pharmacology & Therapeutics*, 2006. **112**(3): pp. 630–648.
71. Ideta, R., et al., Nanotechnology-based photodynamic therapy for neovascular disease using a supramolecular nanocarrier loaded with a dendritic photosensitizer. *Nano Letters*, 2005. **5**(12): pp. 2426–2431.
72. Nishiyama, N., et al., Light-induced gene transfer from packaged DNA enveloped in a dendrimeric photosensitizer. *Nature Materials*, 2005. **4**(12).
73. Liaw, J., S. Chang, and F. Hsiao, In vivo gene delivery into ocular tissues by eye drops of poly(ethylene oxide)-poly(propylene oxide)-

poly(ethylene oxide)(PEO-PPO-PEO) polymeric micelles. *Gene Therapy*, 2001. **8**(13): p. 999.

74. Wilson, D. F., et al., Oxygen distribution and vascular injury in the mouse eye measured by phosphorescence-lifetime imaging. *Applied Optics*, 2005. **44**(25): pp. 5239–5248.
75. Pardridge, W. M., Why is the global CNS pharmaceutical market so under-penetrated? *Drug Discovery Today*, 2002. **7**(1): pp. 5–7.
76. Palmer, A. M., and F. A. Stephenson, CNS drug discovery: Challenges and solutions. *Drug News Perspect*, 2005. **18**(1): pp. 51–57.
77. Pardridge, W. M., Re-engineering biopharmaceuticals for delivery to brain with molecular Trojan horses. *Bioconjugate Chemistry*, 2008. **19**(7): pp. 1327–1338.
78. Abbott, N. J., L. Rönnbäck, and E. Hansson, Astrocyte–endothelial interactions at the blood–brain barrier. *Nature Reviews Neuroscience*, 2006. **7**(1): pp. 41–53.
79. Pavan, B., et al., Progress in drug delivery to the central nervous system by the prodrug approach. *Molecules*, 2008. **13**(5): pp. 1035–1065.
80. Hu, W., et al., Redox and pH-responsive poly (amidoamine) dendrimer–poly (ethylene glycol) conjugates with disulfide linkages for efficient intracellular drug release. *Colloids and Surfaces B: Biointerfaces*, 2014. **123**: pp. 254–263.
81. An, K., et al., A pH and redox dual responsive 4-arm poly(ethylene glycol)-block-poly (disulfide histamine) copolymer for non-viral gene transfection in vitro and in vivo. *International Journal of Molecular Sciences*, 2014. **15**(5): pp. 9067–9081.
82. Saito, G., J. A. Swanson, and K.-D. Lee, Drug delivery strategy utilizing conjugation via reversible disulfide linkages: Role and site of cellular reducing activities. *Advanced Drug Delivery Reviews*, 2003. **55**(2): pp. 199–215.
83. Muhammad, F., et al., Intracellular antioxidants dissolve man-made antioxidant nanoparticles: Using redox vulnerability of nanoceria to develop a responsive drug delivery system. *ACS Applied Materials & Interfaces*, 2014. **6**(21): pp. 19424–19433.
84. Kim, S. W., et al., Hyaluronated nanoparticles with pH-and enzyme-responsive drug release properties. *Colloids and Surfaces B: Biointerfaces*, 2014. **116**: pp. 359–364.
85. De La Rica, R., D. Aili, and M. M. Stevens, Enzyme-responsive nanoparticles for drug release and diagnostics. *Advanced Drug Delivery Reviews*, 2012. **64**(11): pp. 967–978.

86. Hsiao, M.-H., et al., Design and characterization of a novel amphiphilic chitosan nanocapsule-based thermo-gelling biogel with sustained in vivo release of the hydrophilic anti-epilepsy drug ethosuximide. *Journal of Controlled Release*, 2012. **161**(3): pp. 942–948.

87. C Gonçalves, M., et al., Silica/Ormosil SPIONs for biomedical applications. *Current Nanoscience*, 2013. **9**(5): pp. 599–608.

88. Qiao, R., et al., Receptor-mediated delivery of magnetic nanoparticles across the blood–brain barrier. *ACS Nano*, 2012. **6**(4): pp. 3304–3310.

89. Patrick, J. T., et al., Ultrasound and the blood-brain barrier, in *Consensus on Hyperthermia for the 1990s*. 1990, Springer. pp. 369–381.

90. Hynynen, K., et al., Noninvasive MR imaging–guided focal opening of the blood-brain barrier in rabbits 1. *Radiology*, 2001. **220**(3): pp. 640–646.

91. Hwang, T. L., et al., Development and evaluation of perfluorocarbon nanobubbles for apomorphine delivery. *Journal of Pharmaceutical Sciences*, 2009. **98**(10): pp. 3735–3747.

92. Kinoshita, M., et al., Targeted delivery of antibodies through the blood–brain barrier by MRI-guided focused ultrasound. *Biochemical and Biophysical Research Communications*, 2006. **340**(4): pp. 1085–1090.

93. Kabanov, A. V., et al., A new class of drug carriers: Micelles of poly (oxyethylene)-poly (oxypropylene) block copolymers as microcontainers for drug targeting from blood in brain. *Journal of Controlled Release*, 1992. **22**(2): pp. 141–157.

94. Shao, K., et al., Functionalized nanoscale micelles with brain targeting ability and intercellular microenvironment biosensitivity for anti-intracranial infection applications. *Advanced Healthcare Materials*, 2015. **4**(2): pp. 291–300.

95. Yoo, S., et al., Photothermal inhibition of neural activity with near-infrared-sensitive nanotransducers. *ACS Nano*, 2014. **8**(8): pp. 8040–8049.

96. Aregueta-Robles, U. A., et al., Organic electrode coatings for next-generation neural interfaces. *Frontiers in Neuroengineering*, 2014. **7**: doi: 10.3389/fneng.2014.00015.

97. Wadhwa, R., C. F. Lagenaur, and X. T. Cui, Electrochemically controlled release of dexamethasone from conducting polymer polypyrrole coated electrode. *Journal of Controlled Release*, 2006. **110**(3): pp. 531–541.

98. Zhao, Y., et al., 3D nanostructured conductive polymer hydrogels for high-performance electrochemical devices. *Energy & Environmental Science*, 2013. **6**(10): pp. 2856–2870.

99. Yue, H. Y., et al., ZnO nanowire arrays on 3D hierarchical graphene foam: Biomarker detection of Parkinson's disease. *ACS Nano*, 2014. **8**(2): pp. 1639–1646.

100. Roger, V. L., et al., Heart disease and stroke statistics—2012 update a report from the American heart association. *Circulation*, 2012. **125**(1): pp. e2–e220.

101. Saxer, T., A. Zumbuehl, and B. Müller, The use of shear stress for targeted drug delivery. *Cardiovascular Research*, 2013. **99**(2): pp. 328–333.

102. Holme, M. N., et al., Shear-stress sensitive lenticular vesicles for targeted drug delivery. *Nature Nanotechnology*, 2012. **7**(8): pp. 536–543.

103. Korin, N., et al., Shear-activated nanotherapeutics for drug targeting to obstructed blood vessels. *Science*, 2012. **337**(6095): pp. 738–742.

104. Peters, D., et al., Targeting atherosclerosis by using modular, multifunctional micelles. *Proceedings of the National Academy of Sciences*, 2009. **106**(24): pp. 9815–9819.

105. Broz, P., et al., Inhibition of macrophage phagocytotic activity by a receptor-targeted polymer vesicle-based drug delivery formulation of pravastatin. *Journal of Cardiovascular Pharmacology*, 2008. **51**(3): pp. 246–252.

106. Mulder, W. J., et al., Molecular imaging of macrophages in atherosclerotic plaques using bimodal PEG-micelles. *Magnetic Resonance in Medicine*, 2007. **58**(6): pp. 1164–1170.

107. Guo, Y., et al., Simultaneous diagnosis and gene therapy of immuno-rejection in rat allogeneic heart transplantation model using a T-cell-targeted theranostic nanosystem. *ACS Nano*, 2012. **6**(12): pp. 10646–10657.

108. Heid, S., et al., Synthesis and characterization of tissue plasminogen activator—functionalized superparamagnetic iron oxide nanoparticles for targeted fibrin clot dissolution. *International Journal of Molecular Sciences*, 2017. **18**(9): p. 1837.

109. Nguyen, M. M., et al., Enzyme-responsive nanoparticles for targeted accumulation and prolonged retention in heart tissue after myocardial infarction. *Advanced Materials*, 2015. **27**(37): pp. 5547–5552.

110. Lin, K. Y., et al., Self-titrating anticoagulant nanocomplexes that restore homeostatic regulation of the coagulation cascade. *ACS Nano*, 2014. **8**(9): pp. 8776–8785.

111. Bertram, J. P., et al., Intravenous hemostat: Nanotechnology to halt bleeding. *Science Translational Medicine*, 2009. **1**(11): pp. 11ra22–11ra22.

112. Chan, L. W., et al., A synthetic fibrin cross-linking polymer for modulating clot properties and inducing hemostasis. *Science Translational Medicine*, 2015. **7**(277): pp. 277ra29–277ra29.

113. Lee, D., et al., Hydrogen peroxide-responsive copolyoxalate nanoparticles for detection and therapy of ischemia–reperfusion injury. *Journal of Controlled Release*, 2013. **172**(3): pp. 1102–1110.

114. Bae, S., et al., Hydrogen peroxide-responsive nanoparticle reduces myocardial ischemia/reperfusion injury. *Journal of the American Heart Association*, 2016. **5**(11): p. e003697.

115. Kawata, H., et al., A new drug delivery system for intravenous coronary thrombolysis with thrombus targeting and stealth activity recoverable by ultrasound. *Journal of the American College of Cardiology*, 2012. **60**(24): pp. 2550–2557.

116. Tiukinhoy-Laing, S. D., et al., Fibrin targeting of tissue plasminogen activator-loaded echogenic liposomes. *Journal of Drug Targeting*, 2007. **15**(2): pp. 109–114.

117. Herbst, S. M., et al., Delivery of stem cells to porcine arterial wall with echogenic liposomes conjugated to antibodies against CD34 and intercellular adhesion molecule-1. *Molecular Pharmaceutics*, 2009. **7**(1): pp. 3–11.

118. Huang, S.-L., Liposomes in ultrasonic drug and gene delivery. *Advanced Drug Delivery Reviews*, 2008. **60**(10): pp. 1167–1176.

119. Main, M. L., et al., Acute mortality in critically ill patients undergoing echocardiography with or without an ultrasound contrast agent. *JACC: Cardiovascular Imaging*, 2014. **7**(1): pp. 40–48.

120. Sirsi, S. R., et al., Polyplex-microbubble hybrids for ultrasound-guided plasmid DNA delivery to solid tumors. *Journal of Controlled Release*, 2012. **157**(2): pp. 224–234.

121. Jin, Q., et al., A novel cationic microbubble coated with stearic acid-modified polyethylenimine to enhance DNA loading and gene delivery by ultrasound. *PloS One*, 2013. **8**(9): p. e76544.

122. Park, J. I., et al., Microbubbles loaded with nanoparticles: A route to multiple imaging modalities. *ACS Nano*, 2010. **4**(11): pp. 6579–6586.

123. Klibanov, A. L., et al., Ultrasound-triggered release of materials entrapped in microbubble–liposome constructs: A tool for targeted drug delivery. *Journal of Controlled Release*, 2010. **148**(1): pp. 13–17.

124. Burke, C. W., et al., Covalently linking poly (lactic-co-glycolic acid) nanoparticles to microbubbles before intravenous injection improves their ultrasound-targeted delivery to skeletal muscle. *Small*, 2011. **7**(9): pp. 1227–1235.

125. Khondee, S., et al., Calcium condensed LABL-TAT complexes effectively target gene delivery to ICAM-1 expressing cells. *Molecular Pharmaceutics*, 2011. **8**(3): pp. 788–798.

126. Kelly, K. A., et al., Detection of vascular adhesion molecule-1 expression using a novel multimodal nanoparticle. *Circulation Research*, 2005. **96**(3): pp. 327–336.

127. Nahrendorf, M., et al., Noninvasive vascular cell adhesion molecule-1 imaging identifies inflammatory activation of cells in atherosclerosis. *Circulation*, 2006. **114**(14): pp. 1504–1511.

128. Dziubla, T. D., et al., Endothelial targeting of semi-permeable polymer nanocarriers for enzyme therapies. *Biomaterials*, 2008. **29**(2): pp. 215–227.

129. Winter, P. M., et al., Endothelial αvβ3 integrin–targeted fumagillin nanoparticles inhibit angiogenesis in atherosclerosis. *Arteriosclerosis, Thrombosis, and Vascular Biology*, 2006. **26**(9): pp. 2103–2109.

130. Cyrus, T., et al., Intramural delivery of rapamycin with αvβ3-targeted paramagnetic nanoparticles inhibits stenosis after balloon injury. *Arteriosclerosis, Thrombosis, and Vascular Biology*, 2008. **28**(5): pp. 820–826.

131. Guardado-Alvarez, T. M., et al., Analyte-responsive gated hollow mesoporous silica nanoparticles exhibiting inverse functionality and an AND logic response. *Nanoscale*, 2016. **8**(43): pp. 18296–18300.

Index